SEISMOLOGICAL ATTENUATION WITHOUT Q

Igor B. Morozov

University of Saskatchewan
Saskatoon, Saskatchewan
Canada

Order this book online at www.trafford.com
or email orders@trafford.com

Most Trafford titles are also available at major online book retailers.

Printed in the United States of America.

ISBN: 978-1-4269-4525-0 (sc)
ISBN: 978-1-4269-4526-7 (hc)

Library of Congress Control Number: 2010915228

*Our mission is to efficiently provide the world's finest, most comprehensive book publishing
service, enabling every author to experience success. To find out how to publish your book,
your way, and have it available worldwide, visit us online at www.trafford.com*

Trafford rev. 10/12/2010

 www.trafford.com

North America & international
toll-free: 1 888 232 4444 (USA & Canada)
phone: 250 383 6864 ♦ fax: 812 355 4082

To my parents

Contents

7. RESULTS   269

Symbols and Abbreviations

List of abbreviations

Acronym	Description	Page
ABM	Absortption Band Model for the Earth (Anderson and Given, 1982)	238
DSS	Deep Seismic Sounding (seismic program in the Soviet Union)	264
EURS	Eastern U.S. model by Der *et al.* (1986b)	209
IASP91	Kennett and Engdahl (1991)	115
MM8	Mantle Q model used in Anderson *et al.* (1965)	
PNE	Peaceful Nuclear Explosion (seismic source used in DSS program)	89
PREM	Preliminary Reference Earth Model (Dziewonski and Anderson, 1981)	280
QL6	Mantle Q model by Durek and Ekström (1996)	280
QM1	Mantle Q model by Widmer *et al.* (1991)	280
SIRT	Simultaneous Iterative Reconstruction Technique	261

Mathematical Symbols

Symbol	Description
A	Seismic amplitude
B_0	Seismic albedo
D	Dissipation function
E_p, E_{el}, E_k	Energy: potential, elastic energy, kinetic
f	Frequency; force
f_c	Cross-over frequency
G	Green's function
H	Wavefront curvature
J	Visco-elastic compliance
k	Wavenumber; spring constant
L	Lagrangian function, density
ℓ	Mean free path in scattering

m	Mass
M, M_R, M_U	Visco-elastic modulus (also relaxed, unrelaxed)
P	Attenuation path factor
Q	Quality factor (conventional)
Q_e, Q_i, Q_L, Q_P, Q_s, Q_S	Quality factor: effective, anelastic (intrinsic), Love wave, P wave, scattering-attenuation, and S wave
Q_i^D	Generalized dissipative force
r	Source-receiver distance; reflectivity
$\mathbf{S}$	Scattering matrix
S	Hamiltonian action
t	Time
t^*	Cumulative attenuation factor
u_i, $\mathbf{u}$	Displacement in an elastic field
Z, Z_A	Impedance (also acoustic)
α, α_i	Spatial attenuation coefficient: apparent, intrinsic
χ^*	Cumulative temporal attenuation coefficient
χ, χ_i	Temporal attenuation coefficient: apparent, intrinsic
ε_{ij}	Strain tensor
ϕ	Creep function
γ, γ_i	Geometrical-attenuation coefficient (apparent, intrinsic)
κ, κ_i	Dissipation-attenuation coefficient (apparent, intrinsic)
λ	Scattering intensity; first Lamé modulus
μ	Lamé shear modulus
$\theta(...)$	Heavyside step function
ρ	Mass density
σ_{ij}	Stress tensor
τ_ε, τ_σ	Visco-elastic relaxation times for ε and σ
ω	Angular frequency
ξ	Energy dissipation constant for linear oscillator
ζ	Constant of viscous friction

Preface

This book is a monograph about the methodology of measuring, interpreting, and modeling seismic attenuation. In the last couple of years, these topics have resulted in some debate, sparked by my *Opinion* piece in *Seismological Research Letters* [1]. It appears that, after over half a century of active development, numerous publications, and impressive results, the subject of seismic attenuation may still be far from clear. Presentations of attenuation results contain striking amounts of stereotypes, imprecise terminology, speculative assumptions, and even theoretical prejudices. The purpose of this book is to discuss and clarify some of these assumptions, and to revisit the subject of attenuation from fundamental physical principles. A broad range of topics is covered from a uniform point of view, ranging from the fundamentals of attenuation theory and measurement to alternate interpretations of several well-known results. The book is based on several recent publications combined with material that would be difficult to publish in short research papers.

The term "seismological" in the title emphasizes the principal audience to which the book is addressed. In earthquake seismology, the use of attenuation properties is significantly more extensive and important for interpretation than the relatively simple Q compensation performed in exploration seismic data processing. Observational seismologists also work with significantly lower attenuation levels and more diverse and difficult datasets, but often strive to reveal such subtle properties of attenuation as its frequency dependence. All discussions and data examples in this book use this "seismological" perspective on seismic wave attenuation.

Perhaps the most important misconception which complicates the understanding of seismic attenuation lies in its association with the quality factor,

---

[1] Morozov (2009a); also see Xie and Fehler (2009), Xie (2010), Mitchell (2010), and Morozov (2009b, 2009c, 2009d; 2010a, 2010d).

or Q. Over several decades of presence in seismology textbooks, this association became instilled in minds and is practically never questioned. Whenever seismologists of any rank are asked about attenuation, they almost invariably respond that it is a "Q." However, as shown in this book, the Q is actually not a viable property to describe attenuation in the Earth's medium. The concept of "medium Q" neither arises from the physics of wave attenuation nor can it be ambiguously measured. The concept of Q-factor is also not required in order to describe most observations, except maybe those using resonant devices in the lab. This concept was introduced by analogy with the well-known quality factors of vibrating mechanical and electrical systems. However, if not understood clearly, this analogy may become misleading and cause significant problems in interpretation. One such problem appears to be the pervasive observation of the frequency-dependent Q within the Earth. As demonstrated in this book, this frequency dependence is not as pervasive as commonly thought, and it may even not be the case at all.

Another important stereotype of the conventional view challenged here is the association of attenuation with the visco-elastic theory. Once again, most seismologists believe that visco-elasticity is the only way to describe seismic attenuation. This theory provides a formal theoretical basis for Q acting as a fundamental property of the Earth's medium, and also for the popular view of attenuation as being caused by "imperfect elasticity." It also justifies the theoretical representation of attenuation by a complex-valued velocity of the medium, known as the "correspondence principle." As shown in this book, however, both the visco-elastic theory and the correspondence principle might be strongly limited when applied to real seismic-wave problems. Although visco-elastic models can be constructed for most attenuated waves, their extrapolation to the level of fundamental mechanics may be erroneous.

Seismologists rarely study theoretical mechanics and are often unfamiliar with the Lagrangian approaches that form the core of such fields as the analytical mechanics and classical field theory. This is unfortunate, because variational approaches provide the most solid foundations for describing the dynamics of complex systems, such as the elastic dissipative medium. The language of functional forms and variational principles is indispensable for discussing, for example, whether it is the kinetic or elastic energy that dissipates in the process of wave propagation, and how this dissipation can depend on the frequency. For these reasons, significant portions of this book are devoted to introducing these methods and illustrating their application to the problem of seismic attenuation.

If the validity of visco-elasticity is doubted, then the field of attenuation may require a significant revision, and this book is a first attempt of it. The problem of seismic attenuation is by no means closed, but rather has been re-opened by this study. The purpose of this book is to achieve a simple and uniform view, which is based on the true (as opposed to equivalent) mechanics of the medium. This view allows us to solve the described problems in a simple and consistent manner and also make data observations that may have not been noticed before.

The book covers a broad range of aspects, from theory to modeling, and from observations using multiple wave types to temporal variations of attenuation properties. The only subject left virtually untouched here is inversion, although we discuss its forward kernels for several key cases. Many topics covered in the book deserve further development.

This book is intended for an expert reader seeking to learn the subject of seismic wave attenuation. This is not a course text, and it contains more unsolved problems than exercises. Familiarity with seismological observation techniques and basic physics is expected. Because the presented viewpoint is quite novel, a significant portion of the discussion is devoted to explaining and debating the fundamentals of energy dissipation and presenting examples of alternate interpretations of several key datasets. Parts of this book can therefore be used in graduate courses in seismology; however, one needs to be aware of the "unorthodox" character of many of its conclusions. The book could constitute a course in "critical thinking about attenuation," yet its students should also realize that critical views may not be easily rewarded by publications and research funding. Nevertheless, in the long run, good understanding of first principles of seismic attenuation and judicious use of the data should definitely pay off.

As the reader will see almost immediately, the "critical thinking" approach leads us to questioning numerous models, observations, and conclusions about seismic attenuation dating as far back as the 1960s to 1970s. This applies, for example, to the methods of data presentation, use of subjective models, frequency-dependence of Q, separation of scattering and intrinsic attenuation, correlations with geology and tectonics, visco-elastic theory, numerical waveform modeling, and inversion for Q. Although revising so much material is a difficult task, it appears that, at some point, the problems still persistent in these areas and should be clarified and resolved. In most cases, I offer alternate solutions to these problems. On the other hand, illustrations using a very broad scope of well-known, published datasets should help the reader to appreciate the generality of the ideas proposed in this book.

A few words about the structure of this book. **Chapter 1** contains an introduction, in which the concepts of seismic attenuation are explained and the general questions are posed. The attenuation-coefficient approach, which is developed throughout the book, is described in relation to the problem of Q. At the end of this chapter, the key statements arising from this approach are summarized. Each of the following chapters also starts with a preamble summarizing its key points.

Because of the critical attitude taken with respect to mainstream visco-elastic theory and the very concept of Q, we need to carefully discuss the foundations of these concepts first. These discussions are carried out in **Chapters 2 to 4**. Readers not concerned about this theoretical background and only interested in the application of the attenuation-coefficient approach may continue to Chapter 5 without significant loss of continuity.

In **Chapter 2**, the fundamental principles of solid-state and fluid mechanics are discussed, and it is shown that some of them are insufficiently appreciated in the conventional attenuation methodology. We consider several approaches to elastic-continuum mechanics and particularly emphasize the Hamilton-Lagrange method. This method helps pose and answer two of the most fundamental questions, which are: 1) whether the attenuation should principally be associated with the elastic- or kinetic-energy dissipation and 2) whether it can be described by a single parameter, like Q. We also present the visco-elastic approach, seek its connections to the Lagrangian mechanics, and give a detailed critique. At the end of this chapter, a linear form of the frequency-dependent attenuation coefficient, $\chi(f)$, is introduced. This form is most commonly found in the models and data discussed in the rest of this book.

Chapter 3 contains a review of the current Q paradigm, with specific emphasis on the various lines of argument on which the different types of Q are based. This chapter also discusses the procedures of measuring Q, and suggests some general causes of its frequency dependence. In particular, the existence of Q and its inferred properties are related to several assumptions of the visco-elastic theory and also to the uncertainties of geometrical spreading.

In **Chapter 4**, we discuss the concept of geometrical spreading. Compared to attenuation, geometrical spreading represents a more significant effect on seismic amplitudes, and accurate corrections for it are critical for most attenuation measurements. Traditionally, the treatment of geometrical spreading is based on rather crude models and *ad hoc* conventions. This chapter presents a more realistic numerical model and shows how the geometrical-spreading effects can be incorporated in attenuation measurements.

In **Chapter 5**, the concept of attenuation coefficient is systematically developed. Starting from the "apparent" attenuation coefficient describing the observations (denoted χ), its "intrinsic" counterpart, χ_i, is introduced. It is demonstrated that in modeling and inversion, χ_i can often be treated as the conventional Q^{-1}, although χ_i also incorporates the complete effects of geometrical spreading and attenuation. Also in this chapter, several theoretical models are developed, showing that geometrical-spreading variations, short-scale reflectivity, or coda-type averaging indeed produce the expected intrinsic χ effects. Exact relations of χ_i to wavefront curvatures and to the acoustic-impedance variations are derived.

Chapter 6 represents a broad collection of case studies. Several seismological datasets are analyzed by using the attenuation-coefficient approach. For several wave types within frequency bands from ~500 s to ~100 Hz, linear $\chi(f)$ dependences hypothesized in Chapter 2 are indeed found. In many cases, the new χ view seriously changes the traditional interpretations and reveals data features that had been hitherto unnoticed.

Chapter 7 presents summaries of the most important results of this book, such as correlations of the attenuation-coefficient parameters with geology, frequency dependence of the apparent Q, and explanations of the apparent absorption band of the Earth.

In the concluding **Chapter 8**, we outline some implications of the attenuation-coefficient methodology and describe some problems that remain still unsolved. The most significant of these potential implications could be: 1) reconsideration of the need for frequency-dependent rheological Q, 2) reconsideration of pervasive scattering often interpreted within the lithosphere; 3) highlighting the need for numerical modeling approaches not based on the visco-elastic model, and 4) reconsideration of the parameterization (Q^{-1}) and Fréchet kernels used in inversions for seismic attenuation of the Earth.

Finally, **Appendices A1** to **A3** give some mathematical detail about the Kramers-Krönig causality relations, a sketch of a Lagrangian model of creep, and a model of the acoustic impedance in the presence of attenuation. The **Subject and Author Index** at the end of the book is not exhaustive and given only for quick guidance on the key terminology and the most prominent authors who contributed to this field.

Discussions with several colleagues have inspired many directions of this study and most of the data examples in this book. In particular, I thank Bob Nowack for his interest and numerous discussions of intricate theoretical subjects. At various stages of this work, discussions with Anton Dainty, Larry Lines, Cinna Lomnitz, Jim Merriam, Paul Richards, Michael Pasyanos, Scott Phillips, Sven Treitel, and Bill Walter were most stimulating. I am indebted to several reviewers of my papers, who remained anonymous, but provided valuable critique and suggested several research areas which are reflected in this book. I am particularly grateful to Brian Mitchell, who is one of the best authorities in seismic attenuation and the Editor-in-Chief of *Pure and Applied Geophysics*, for calling a special forum dedicated to discussing the present views in this journal (October–December 2010). Jack Xie made a particularly strong impact on this work — both as a most respected colleague and reviewer, but also as a leading opponent in the ongoing debate in the literature. His conscientious critique led me to better realizing the arguments of the conventional attenuation model and looking for the best ways for introducing the new concepts. Generic Mapping Tools (Wessel and Smith, 1995) and Octave[2] software were used in preparation of illustrations. I also thank Carol Brown for help with editing of the manuscript.

For epigraphs to each chapter, I selected quotes from a 19th century Russian classic Kozma Prutkov. This is a collective pen name of four well-known writers famous for their aphorisms, fables, and satiric verse. Although somewhat

---

[2] http://www.gnu.org/software/octave/, accessed 5 March 5, 2008.

lost in translation, their ironic pronouncements help setting the philosophical tone for the discussions. Perhaps the best quote to start with the book is this:

Look in the root!

Kozma Prutkov, Fruits of Reflection (1853-54)[3]

[3]Сочинения Козьмы Пруткова/Сост. и послесл. Д. А. Жукова; Примеч. А.К. Бабореко; М.: Сов. Россия, 1981

Chapter 1

Introduction

If on an elephant's cage you see a sign reading "buffalo," do not believe your eyes.

Kozma Prutkov, Fruits of Reflection (1853-54)

Attenuation of seismic waves in the Earth's medium is typically observed by indirect methods, such as using temporal amplitude decays of standing waves, widths of spectral peaks, or spatial decays of amplitudes in propagating waves at different frequencies. In about 50 years of intensive studies of these properties, many analysis methods were proposed and numerous spectacular results obtained. Among the most general results, the dependence of attenuation on wave frequency is perhaps the most fundamental and important. At the same time, with typical seismic-amplitude data paucity and scatter, differences between wave types and datasets, and the ever increasing complexity of inversion techniques, attenuation models have also become prone to significant uncertainties. In many seismological publications, the concepts of attenuation and the "quality factor," denoted Q, are not clearly differentiated, resulting in numerous misconceptions and potentially erroneous interpretations.

The concept of Q as a descriptor of the ability of the medium to dissipate elastic-wave energy has been deeply imbued in minds of more than one generation of seismologists, and it is now accepted as a matter of fact. This concept has a major influence even on the character of seismological measurements, whose results are almost always presented in the form of Q values, even though the procedures for deriving such values are well known for their ambiguities.

Somewhat surprisingly, a critical review shows that the basis of this concept is actually not that well established from both theoretical and practical points of view. In this book, a more general description using the concept of the attenuation coefficient is given, which resolves most problems arising from the reliance on Q. Revisiting several published datasets from the attenuation-coefficient viewpoint also allows making several important observations which were not available before.

1.1 Seismic attenuation: logarithmic decrement or Q?

The most general property describing the amplitude decay of a seismic wave traveling within a medium is the attenuation coefficient. In most general terms, the attenuation coefficient is the logarithmic decrement of seismic amplitudes corrected for the effects of the source, receiver, and also for estimated "background" geometrical spreading. It can be rendered in two forms, depending on whether we measure the propagation of a seismic wave in space or the variation of seismic amplitude with time at a given receiver.

Most of the difficulty of interpreting medium attenuation comes from the limited knowledge of the structure in which the wave propagation takes place, and therefore the effect of geometrical spreading should be taken into account most carefully. For example, let us denote $P_G(r,f)$ the path factor, which is the total seismic amplitude from which the source and receiver effects are removed (r is the source-receiver distance). Denoting the geometrical spreading estimated for some modeled propagating structure by $G_0(r,f)$, we can express the total path factor P_G as:

$$P_G(r,f) = G_0(r,f) P(r,f), \tag{1.1}$$

where $P(r,f)$ is the attenuation path factor. Assuming that $G_0(r,f)$ represents a close approximation to reality, path factor $P(r,f)$ should equal 1 for $r = 0$, and therefore it can be written as:

$$P(r,f) = \exp\left[-\alpha^*(r,f)\right], \tag{1.2}$$

where $\alpha^*(0,f) = 0$. Similarly, in temporal form, the path factor is:

$$P(t,f) = \exp\left[-\chi^*(t,f)\right], \tag{1.3}$$

where t is the observation time, and $\chi^*(0,f) = 0$. In respect to the variations of seismic amplitude, eqs. (1.2) and (1.3) represent the perturbation-theory (more specifically, scattering-theory) approximation, which only specifies that the variations of P with distance or time are small and proportional to P itself; for example,

$$\delta P(t, f) = -P(t, f)\delta\chi^*(t, f).$$ (1.4)

Quantities α^* and χ^* represent the cumulative effects of attenuation accumulated from distance (time) 0 to r (or t, respectively). For typical attenuation measurements involving relatively short time lags, these quantities can be further approximated through the corresponding differential quantities, which we denote α and χ:

$$\alpha = \frac{\partial\alpha^*}{\partial r}, \text{ and}$$ (1.5)

$$\chi = \frac{\partial\chi^*}{\partial t}.$$ (1.6)

For a traveling wave, these parameters are related by $\chi = \alpha V$, where V is the group velocity. As shown throughout this book, these two quantities provide the most general and consistent descriptions of the process of energy dissipation within a seismic wave.

Integrating eq. (1.6) in time, we can also express the cumulative temporal path factor,

$$\chi^* = \int_0^t \chi(\tau, f)d\tau,$$ (1.7)

leading to,

$$P(t, f) = G_0(t, f)\exp\left(-\int_0^t \chi d\tau\right).$$ (1.8)

This expression emphasizes the characteristic exponential decay of seismic amplitudes with time and will be useful later.

As one can see from eqs. (1.5) and (1.6), α and χ simply represent the logarithmic decrements, $i.e.$, the relative amplitude decay rates in space and time at a given point in the medium. Among all other attenuation parameters, these quantities are the most directly measurable, and they also represent the principal tools of the scattering theory. However, in common practice originating from about Bland (1960) and Knopoff (1964), another property is usually derived from α or χ and called the "quality factor," Q. Its origins can be traced to analogies with mechanical, electrical, and optical systems, which will be closely followed in

Chapter 3. Q represents the relative amplitude drop taken *cumulatively over one wavelength* λ, but still compared to the peak wave energy density. In terms of χ and α, this factor is:

$$Q = \frac{\pi f}{\chi} = \frac{\pi}{\alpha\lambda} \cdot \qquad (1.9)$$

Conversely, the temporal attenuation coefficient can be derived from Q:

$$\chi = \frac{\pi f}{Q} \cdot \qquad (1.10)$$

This expression already demonstrates an important implied connotation, which continues through all subsequent discussions and creates most difficulties with using the Q concept. Equation (1.10) represents χ as proportional to f, and consequently the path factor (1.8) is written as $\exp(-\pi ft/Q)$, which suggests the popular interpretation of attenuation as related to the number of wave cycles, 'ft'. This is a misconception based on assuming factor Q in eq. (1.10) to be a constant. In reality, Q typically increases with frequency, canceling most of the frequency dependence in eq. (1.10) and leaving the dependence of χ on f completely arbitrary. In other words, Q merely remains yet another representation for χ by means of relation (1.9), and its physical meaning still needs to be established.

In a similar manner, *assuming* the proportionality of χ to f, the cumulative attenuation factor χ^* can also be transformed into the corresponding parameter denoted t^*,

$$t^* = \frac{\chi^*}{\pi f} = \int_0^t \frac{d\tau}{Q} \cdot \qquad (1.11)$$

This parameter is measured in the units of time and represents the cumulative effect of Q^{-1} over the ray path. Because of its cumulative-Q^{-1} character, t^* is often used in body-wave attenuation studies (Der and Lees, 1985). Similarly to Q, t^* also turns out to be strongly frequency dependent, and its physical meaning also does not extend beyond an alternate representation for χ^*.

Two important differences distinguish parameter Q^{-1} in eq. (1.9) from α and χ, and these differences are responsible for most problems of the Q-factor model discussed below. First, the definition of Q is non-local (Figure 1.1), because it contains characteristics of the incident wave (wavelength λ or frequency f). Therefore, Q^{-1} cannot be automatically considered as a true property of the medium. Note that this situation is similar to mechanics, in which Q is a property of the entire oscillator and not of the components of which it is built. Second, definition (1.10) excludes the first-order possibility of χ and α being non-zero for

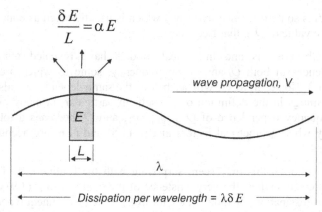

$$\frac{\delta E}{L} = \alpha E$$

wave propagation, V

E

L

λ

Dissipation per wavelength = $\lambda \delta E$

FIGURE 1.1
Energy dissipation problem (intrinsic or scattering) in eqs. (1.5) to (1.9). For typical sizes of dissipating volumes $L << \lambda$, the attenuation coefficient $\alpha = \delta E / E$ should depend on the frequency, but be independent of λ.

$f \rightarrow 0$, and the same problem applies to t^* in eq. (1.11). There is certainly no reason to assume that the zero-frequency limit of the attenuation coefficient is automatically zero and, as will be see in Chapters 5 and 6, values of $\chi|_{f \rightarrow 0}$ typically provide the most important information for interpretation. Thus, it appears that α and χ, and not Q and t^* should be given preference when describing the properties of the Earth's medium.

Although the above observations are very simple and may even appear obvious, the Q-based approach is nevertheless hugely dominant in today's attenuation studies. Q is believed to be closely associated with the rheological properties (such as the composition, physical state, temperature, fracturing, and presence of fluids) of crustal and mantle materials, and most attenuation models are constructed in terms of Q. Further, Q is subdivided into two types: one related to anelastic, or intrinsic energy dissipation, Q_i, and the second to elastic scattering, Q_s:

$$Q^{-1} = Q_i^{-1} + Q_s^{-1}. \qquad (1.12)$$

Both of these factors, and particularly Q_s often show significant frequency dependences. These dependencies are typically approximated by the power-law form:

$$Q(f) = Q_0 \left(f / f_0 \right)^{\eta}, \qquad (1.13)$$

where f_0 is some reference frequency which is usually taken as equal to 1 Hz, and Q_0 is the value of Q at that frequency.

Observations and theoretical models have revealed numerous strong dependences of both Q_i and Q_s on frequency, some of which are discussed in Chapter 3. However, as we will see below, the same dependences also indicate the shortcomings in the definition of Q itself. In particular, in many practical cases, the frequency dependence of Q disappears, and Q_s becomes a totally fictitious quantity when geometrical factors in eqs. (1.5) and (1.6) are accounted for with sufficient accuracy.

Although this may seem surprising with the routine use of Q in today's seismological studies, the very existence of the quality factor (1.9) describing the energy dissipation within the Earth is neither unequivocally obvious observationally nor follows from solid-state or fluid mechanics. On the contrary, using parameters α and χ leads to a simple and unambiguous picture well-connected to the fundamental physics.

1.2 Peculiarity of the subject of attenuation

An important peculiarity of the subject of attenuation is in its combining fairly limited data constraints with a very rich spectrum of deep, but at the same time permissive, mathematical models. Note that in seismological applications, we are interested in relatively weak attenuation effects ($Q^{-1} \approx 10^{-3}$), which may be comparable to various data and model errors. All this creates a situation in which many forward models can be constructed, but unambiguous inversion and verification is difficult to perform, sometimes even for the most basic quantities.

1.2.1 Weak data constraints

Q is never measured directly, but is derived from the observed quantities by using mathematical models of experiments. Generally, three unknowns need to be resolved in any inversion for seismic attenuation: 1) the residual geometrical spreading remaining after its correction in eq. (1.6), 2) anelastic attenuation, and 3) scattering. However, the data usually enter the analysis in some form of the attenuation coefficient, such as (1.6). Within typical data errors, the frequency dependence of α or χ can be described by two parameters at most [e.g., constant γ and κ in eq. (1.14) below], giving only two constraints for determining the three (or more) unknowns. This shows that the attenuation measurement problem is inherently *underconstrained* and additional considerations are required in order to resolve this ambiguity.

When using model parameterizations that contain more degrees of freedom than constrained by the data, the results are mostly controlled by the underlying "a

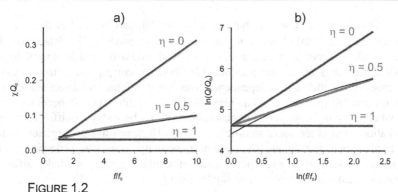

FIGURE 1.2

Attenuation data fit in: a) linear attenuation-coefficient form and b) $Q(f)$ form plotted in logarithmic scales. Gray lines are the $Q(f)$-type dependences of the type (1.10) with $\eta = 0$, 0.5, and 1. Black lines show linear $\chi(f)$ functions (1.14) with constant γ and κ and approximating the same amplitude-decay data.

priori constraints," and a danger of biased or "preferred" solutions arises. The approach taken in most published studies targeting the frequency-dependent Q is to *assume* that the geometrical spreading is corrected accurately, and consequently to attribute the entire amplitude decay to the anelastic and scattering attenuation. The solution with a strongly frequency-dependent Q thus serves as just such a "preferred" solution; however, it still does not have to be close to reality. In Chapters 4, 6, and 7, we show that in reality, the situation is the opposite — both data and modeling indicate that the geometrical spreading is variable, virtually intractable for modeling, but readily measurable. Once the geometrical spreading is measured, only one variable remains to be constrained by the data, and consequently the observed "Q" becomes frequency independent.

Note that in most cases currently found in the literature, the quality of frequency-dependent Q solutions is only judged by its ability to fit the seismic-amplitude data. This criterion is insufficient and may be misleading, because, in under-constrained inversion, multiple models can fit the data. Generally, the more redundant parameters are included in under-constrained inversion, the better data fit can be achieved (*e.g.*, Menke, 1989). In terms of the amplitude-decay data fit, both the Q- (eq. 1.10) and attenuation-coefficient based, two-parameter approximations give very close amplitude predictions for practical frequency-band widths (Figure 1.2). Differences between the predicted data values are small in both linear (f, χ) coordinates used in this book (Figure 1.2a) and logarithmic ($\ln f$, $\ln Q$) coordinates used in most conventional studies (Figure 1.2b). The differences between these curves are usually significantly smaller than the typical data errors. In Chapter 6, this point is illustrated on several real-data examples.

Thus, data fit is not the type of criterion that can be used to differentiate between the traditional Q- and our new α- or χ-type models. The data constraint is too weak to serve as a discriminator. Instead, we will need to appeal to the physical meanings of parameters, their likelihood to correspond to true physical properties of the medium, independence from assumptions, and direct association with observable quantities. From reviewing today's literature, it becomes clear that when pursuing a universal Q description, it can be tricky to differentiate the attenuation from other, more significant factors affecting wave propagation within the Earth. Such factors are the structure, velocity gradients, reflectivity, and variations of scattering amplitudes, each of which having potentially stronger effects on the observed seismic amplitudes than Q.

1.2.2 Strong theory

Analysis of wave amplitudes has brought into seismology several spectacular ideas from mathematics, plasma and material physics, optics, electromagnetism, and engineering. It would be fair to say that theoretical advances drive the attenuation studies, leading them to revealing subtle properties of the Earth's materials. Unfortunately, this prevalence of theory also somewhat erases the boundaries between the observations and theoretical predictions. Models become imprinted in the process of observation, and discussions of data measurements are often model driven. In global seismology, attenuation models tend to be more mathematical than physical, which causes additional risks for their misapplication.

Mathematical models of seismic attenuation need to be understood very carefully and sometimes critically. In particular, one should stay clear from using literal and intuitive associations based on the similarities in terminology. For example, as we will see below, the notion of the "quality factor" in seismology is not exactly the same as its namesake in mechanics. The "correspondence principle" is not the same as in quantum mechanics, and it is not guaranteed to work in all cases. Also, as shown below, the ties of "visco-elasticity" to the theories of viscosity and elasticity are not as strong it might appear from its name.

The use of the concept of scattering is another important case of intuitive connotation with a certain type of model. In seismic coda studies, "scattering" is usually understood as the presence of random heterogeneities within the subsurface. However, in its generic form, the "scattering approximation" is simply a way for solving the differential wave equations by presenting the solutions as small perturbations of some known "background" states (see Section 3.6). With the background being poorly known, the scattering approach may yield a highly non-unique characterization of the types and origins of such perturbations. For example, Q_s is typically interpreted as caused by random heterogeneities within the volume of the lithosphere. However, most of coda scattering should in fact occur near the surface, and the inverted Q_s may therefore be unrelated to small-scale heterogeneities, but correspond to the lithospheric structure and free-surface topography. This creates an ambiguous and sometimes confusing terminology. In the following, we will try keeping our terminology from leaning toward such intuitive associations.

1.2.3 Danger of model-driven argument

Empirical data fit is commonly used to justify attenuation models, which often become elaborate. Unfortunately, our natural attitude toward theoretical models is typically positive (*i.e.*, we would prefer to prove them), but they are often not that well constrained by the data. Analysis of such models requires breaking the loop of logical argument resembling the classic chicken-or-egg problem. In attenuation data analysis, the following line of argument can often be discerned: 1) assume a simple, theoretically tractable model of wave propagation; 2) add some intricate internal parameters of the medium, such as the frequency-dependent Q or local memory in strain-stress relations; 3) by using these parameters, achieve the attenuation data fit; and 4) based on this fit, conclude that both steps (1) and (2) "work." However, it is also clear that had (1) been selected differently, then (2) might change or disappear. Examples of such recursive inferences include justifications of the various "practical" geometrical-spreading models, inferences of rheological relaxation from the predictions of visco-elastic models, and separation of the anelastic and scattering attenuation.

It is quite difficult to pinpoint a single weak spot in the circular argument above. Generally, such problems arise from using limited datasets to corroborate permissive theoretical models. In the world of over-parameterized models, it would be significantly more important to find models which "don't work" than those that fit the data. Identification of such models would restrict the space of available solutions and reveal the true physical parameters of the Earth. Such a minimalistic approach is taken in this book.

1.3 Critique of the visco-elastic approach

The concept of Q is in the heart of the visco-elastic theory, and conversely, visco-elasticity strongly supports the key properties of Q. Hereafter, the term "visco-elastic" is used in the specific sense of the broadly accepted mathematical theory which describes anelasticity by extending the partial differential equations of the elastic field (*e.g.*, Carcione, 2007; Borcherdt, 2009). This theory will be reviewed in Chapter 2. At this point, let us note its key principles listed in Table 1.1. While scrutinizing the concept of Q, we will have to question all of these principles.

The reasons for questioning the principles of visco-elasticity are simple. This theory is only a phenomenological model describing *the fact* of attenuation

TABLE 1.1 Comparison of the visco-elastic and present approaches.

Concept	Visco-elastic theory	Present approach
Basic medium property	Q	Attenuation coefficient
Mechanical principle	Differential wave equations	Lagrangian (macroscopic) or microscopic mechanics
Strain-stress relations	Local phenomenological creep functions	Arising from Euler-Lagrange equations
Cause of attenuation	Retarded elastic moduli (creep)	Many factors
Source of energy dissipation	Elastic energy	Specific mechanisms; related to kinetic energy
Relation of attenuation to medium velocity	Fundamental: $Q^{-1} = -\tan(\arg V)$ (correspondence principle)	No specific relation
Geometrical spreading	Not considered	Part of attenuation coefficient
Frequency-dependent Q	Common in theory; often found in data	Likely in theory; not found in most datasets

without explaining its *causes*. According to this theory, attenuation results from the strain lagging behind stress during the loading-unloading cycle, which is described by "creep," or "memory" behavior encoded in the retarded moduli. However, this theory contains no actual mechanism supporting this memory. This model only explains the attenuation caused by material creep; however, creep is *not the only* mechanism to produce attenuation. On the other hand, *any* physical mechanism of attenuation would lead to the phase lags in the simple wave solutions which are usually studied in visco-elastodynamics. The material-creep model represents only an extrapolation of the empirical models constructed to describe the quasi-static laboratory experiments with rock samples.

Visco-elasticity describes the ability of the medium to dissipate the elastic energy by using a single parameter, which is the imaginary part of the complex modulus or of the seismic-wave velocity. However, a single parameter cannot be sufficient for describing the broad variety of physical processes causing attenuation. This model is *too simple* and basically considers only attenuation which behaves similarly to the velocity. Instead, or possibly in addition to such a heuristic theory, a rigorous physical description the dynamics of wave propagation and attenuation is required.

Therefore, being an elegant mathematical theory able to reproduce the attenuation data, visco-elasticity may not necessarily correspond to the complete reality of seismic wave propagation within the Earth. In particular, in heterogeneous media (*i.e.*, practically in all cases of interest), significant deviations from the visco-elastic paradigm can be found, as shown in Chapters 2 and 6. Its key hypothesis of a common factor, namely the frequency-dependent Q, being responsible for both attenuation and dispersion of seismic waves in the range of frequencies from the Chandler wobble ($\sim 10^{-8}$ Hz) and creep ($\sim 10^{-6}$–10^{-2} Hz) to sonic and even ultrasonic frequencies ($\sim 10^{5}$ Hz) is extremely bold but likely to skip over much of the tremendous complexity of physical processes taking place within this frequency range.

1.4 Attenuation-coefficient approach

As also recapitulated in Table 1.1, the approach to seismic attenuation taken in this book differs from the one adopted in visco-elastic theory. Our approach is conservative, data-driven, and based on recognition of multiple factors causing dissipation and scattering of the elastic-wave energy, by contrast to the single factor (Q^{-1}) included in the visco-elastic constitutive relations. This leads to reconsidering the entire methodology, from the selection of fundamental mechanical principles and observable quantities, to analyzing the role of geometrical-spreading corrections and frequency-dependent Q in interpretation (Table 1.1).

In brief, our approach consists in using the general frequency-dependent attenuation coefficients instead of the Q-factor based form (1.10). Considering, for example, the temporal attenuation coefficient, χ, and denoting its zero-frequency limit by $\chi|_{f=0} = \gamma$, one can write,

$$\chi(f) = \gamma + \kappa f . \qquad (1.14)$$

This form effectively replaces the two-parameter model (Q_0, η) with a different pair of parameters (γ, κ), which remain finite when $\chi|_{f=0} \neq 0$. For a notation reminiscent of the familiar Q terminology, the dimensionless parameter κ can be transformed into the "effective quality factor,"

$$Q_e = \frac{\pi}{\kappa} . \qquad (1.15)$$

The important difference between the attenuation-coefficient model (1.14) from Q-based one (1.10) is that χ is stable with respect to the uncertainties of the background geometrical spreading, whereas Q is not. Parameterization (1.14) encourages measurement of the zero-frequency attenuation γ, whereas the use of eq. (1.10) leads to endless and futile discussions about the accuracy of modeled G_0. As we will see, model (1.14) is applicable to a variety of seismic waves and frequency bands, both in theory and practical data observations, and it often leads to interpretations which are dramatically different from those of the frequency-dependent Q model. Several examples of such observations are given in Chapter 6.

1.5 Looking for physical properties

It is important to clearly differentiate the properties of an elastic medium from those of the waves propagating in it. Such differentiation may become obscure when using the concept of Q and visco-elasticity. In mechanics, the Q factor is a property of a system capable of stationary oscillations. In seismology, such systems are only represented by whole-Earth oscillations and resonant devices used in some lab measurements (for more on this, see Chapter 3). In such cases, Q values are well justified, but relevant only for the complete systems and not necessarily for the rock of which they are composed. These Q values also cannot be viewed as dependent on the oscillation frequencies. For propagating waves, Q^{-1} values are usually heuristically attributed to the medium, but justified only by considering plane- or spherical-wave solutions (*e.g.*, Aki and Richards, 2002), that is again determined by some specific "measurement devices". Plane or spherical waves only exist in uniform media, in which assigning the resulting Q^{-1} to the medium is trivial but not convincing. In such cases, $\alpha(f)$ is nearly proportional to f, and this proportionality compensates the artificial factor λ^{-1} in eqs. (1.9). However, this proportionality only holds for such simple cases of perfectly known geometrical spreading and is absent, for example, for all surface waves.

As we see, unlike α and χ, the quality factor is a phenomenological attribute of some specific wave rather than an *in situ* medium property. It is correct to say "Q of a resonant bar," "normal-mode Q," "coda Q," or "Q of a 30-Hz plane P-wave". However, expressions "Q within the lower crust" or "variations of Q within the Earth" are, strictly speaking, meaningless. Unfortunately, this second type of usage is quite widespread and often forms the basis for interpreting attenuation measurements. On the other hand, by contrast to Q, α and χ represent at least viable approximations for the local dissipation properties of the propagating medium, if not full solutions.

Although the transformation of the attenuation coefficients into Q (*e.g.*, eqs. (1.9)) may seem natural for wave-like processes, its underlying assumptions are neither trivial nor entirely innocent. Unless a singularity in $Q(f)$ near $f = 0$ is assumed, the very use of Q instead of the more general χ eliminates the possibility of the zero-frequency attenuation coefficient being non-zero, *i.e.*, $\gamma \neq 0$ in eq. (1.14). As shown in Chapter 4, such $\gamma \neq 0$ results from variations of geometrical spreading or reflectivity, and values of $\gamma > 0$ are commonly observed in short-period seismological data. In the quality-factor picture, positive γ values lead to spurious frequency dependences of Q, which appear to be frequently observed.

Another fundamental reason why Q should not be automatically viewed as a medium property is in the ambiguity of the concept of "scattering-attenuation quality," denoted Q_s^{-1} above. Typically, Q_s^{-1} is determined assuming a uniform background, and when the background structure becomes known more accurately, Q_s^{-1} decreases. Thus, Q_s^{-1} can be viewed as representing the "structure not accounted for in the selected background model". Although such Q_s^{-1} leads to correct predictions of amplitude decays, it is still a subjective property related to the interpreter's views and knowledge. As mentioned above, the term "scattering" here should not be understood as necessarily caused by random perturbations within the medium. If interpreted as random perturbations[4], Q_s^{-1} may lead to erroneous models when considered in the more complete context of seismological and geological data. For example, crustal velocity gradients and reflectivity, including that of the Moho, play the key roles in shaping the local-earthquake codas and in the S-wave envelope broadening. Their interpretations in the traditional sense (*e.g.*, Aki, 1980) and descriptions by a featureless "Q_s^{-1}" quantity suggests models of random heterogeneities uniformly distributed within large volumes of the lithosphere. Taken beyond the narrow context of attenuation studies, this is a grossly incorrect view ignoring the key elements of the Earth's structure. Therefore, to define a Q correctly, we need to accurately account for the effects of the background structure first. As it will be shown in Chapter 5, once this is done, the resulting Q (Q_e in eq. (1.15)) will in fact represent a property of the attenuation coefficient.

The above observations are just two illustrations of the uncertainty of the notion of the "medium Q". In interpreting the Q values arising from various measurements, it is important to keep in mind what type of quantity is being measured. In their Chapter 3, Bourbié *et al.* (1987) summarized a number of such measurements and noted that although most of them can be described by the corresponding visco-elastic models, there is little agreement between the resulting values of Q. Further discussions of the meaning of Q, and also of the relationships of its constituents Q_s and Q_i to the geometrical spreading continue in Chapter 2.

1.6 Key statements

Before proceeding to the detailed argument, let us summarize the key general statements of this book. First, methodologically, we reveal that **the concept of *in situ* Q** commonly used for describing the Earth-medium ability to attenuate seismic waves lacks physical basis and consistency, and it should be replaced with the attenuation coefficient. As a replacement to the apparent Q, we introduce the "**geometrical attenuation**," γ, and the "**effective Q_e**" parameters, which have simpler and clearer physical properties. We further illustrate their use on a number of model and data examples.

---

[4] This seems to be often done in the literature.

Second, we emphasize the **importance of small uncertainties in theoretical models** used to measure seismic attenuation. Such uncertainties most often take the form of **variations in the geometrical spreading**, but they may also be caused by near-surface reflectivity or scattering within the specimens or apparatuses used in lab attenuation measurements. Small but commonly occurring uncertainties in the models may bias and even overturn the observations of attenuation. In particular, a frequency-dependent Q is nearly always suspect of representing inaccurate background models.

Third, we present an **extensive critique of the visco-elastic theory** in application to measurement, modeling, and inverting for seismic attenuation. As will be shown, this theory correctly describes several known end-member solutions, but falls short of capturing the full physics of elastic-wave energy dissipation within the Earth.

Another methodological result relates to understanding the role of elastic scattering in observations. Of the three medium properties controlling the seismic amplitudes (geometrical spreading, scattering, and anelastic absorption), only two can be consistently resolved from attenuation data. We show that **scattering is indistinguishable from the other two properties in the data**, and it should therefore be merged with them and abandoned from data analysis. This shifts the emphasis of attenuation inversion from separation of scattering and anelastic attenuation to differentiating between the effects of the structure (γ, both large- and small-scale) and effective attenuation (Q_e, which includes the effects of anelastic dissipation and small-scale scattering).

Based on revising a number of published data examples, recognition of only a **frequency-independent** Q_e within the Earth may be the most important single result of this study. Frequency-independent Q may appear too simple and even disappointing from the viewpoint of supporting elaborate relaxation rheologies and scattering within the Earth's mantle and crust. However, it is important to clearly realize the limits of observational evidence and differentiate between the truly constrained quantities and artifacts resulting from inaccurate theoretical assumptions. As demonstrated in this book, model-independent analysis reveals spatially variable geometrical spreading and attenuation, but at same time suggests no need for a frequency dependence of Q. Therefore, frequency-dependent seismic-wave dissipation and scattering may not be nearly as pervasive as it is currently thought.

Note that a constant Q represents a significantly stronger constraint than the traditional frequency-dependent models. Instead of vague clauses contingent on unrealistic assumptions typical in frequency-dependent Q arguments (often reading as "… assuming a uniform and isotropic background, flat free surface, single scattering, and the geometrical spreading of $1/r$, Q increases with frequency as $f^{1.07}$ …"), the formulations given below consist of **specific, data-driven statements and quantitative estimates**. Frequency-independent attenuation within the Earth should certainly be much easier to understand, model, relate to geology, and use. If any deviation from this model is detected, it should have a significantly greater constraining power than the pervasive, permissive, yet inherently enigmatic frequency-dependent Q.

In regard to several observations of attenuation and lab measurements discussed in Section 3.9 and Chapter 6, the conclusions of this book may appear "embarrassingly simple." On one hand, the concept of attenuation coefficient emphasized here is not novel and had been used in seismic attenuation studies. On the other hand, this book shows that in several types of observations, the results change dramatically if not tailored to the ambiguous concept of "medium Q." When conducted in the form of the attenuation coefficient, the interpretations become simple, assumption-free, quantitative, and robust.

Within a broad range of frequencies, the **observed attenuation coefficient shows a piecewise-linear behavior,** often corresponding to different waves sampling different depths. The geometrical factor, γ, acquires the key significance and becomes related to geological structures, tectonic ages, or other specific conditions of the experiments. This parameter can even be predicted from a completely independent waveform modeling based entirely on the lithospheric structure. The effective Q_e is related to the physical state of the propagating medium and shows comparable values for different types of waves and observation areas. However, all this also comes at the cost of revising some of the results which have long been considered as well established.

Finally, the new view also leads to **predicted attenuation levels** that are **quantitatively different from the existing solutions**. This result is specifically demonstrated by numeric modeling of Love waves within the Earth's mantle. Unlike the existing model, the new solution conserves the total energy of the propagated and dissipated wavefields. The forward kernels for attenuation are principally different from those for phase velocity, which are commonly used in inversion today. This difference in the kernels, as well as the 10–20% difference in the predicted apparent Q, should have a significant impact on 1-D and 3-D inversion for mantle attenuation.

Chapter 2

Principles of Approach

Don't walk on the slope — you will wear down your boots on one side.

Kozma Prutkov, Fruits of Reflection (1853-54)

Before discussing the concept of Q and the attenuation-coefficient formulation, we need to formulate some general principles to guide our approach. As mentioned in the Chapter 1, there exist several approaches to seismic attenuation, ranging from "conventional" manipulation of experimental data to intuitive physical interpretation and further, to highly mathematical, postulate-driven visco-elastic theories. In this chapter, we compare three groups of general theoretical approaches that can be used as the basis of attenuative continuum mechanics:

(i) **Homogenization theory**, as used for deriving macroscopic laws by averaging the detailed dynamics at the microscopic level. In principle, this method should apparently be the most complete and adequate for approaching our problem. However, this approach is only briefly summarized here because it requires much better understanding of the microscopic physics of attenuation than is available now.

(ii) **Lagrangian formulation** and **Hamilton variational principle**, which describe the medium as a macroscopic continuum. This method is emphasized throughout this book. Fundamental physical principles such as the conservation of energy and utilization of the key symmetries are most important in this method. This approach is still not broadly used in seismology. To the author's knowledge, its only application is Biot's (1956) macroscopic theory of waves in fluid-saturated porous media. On the other hand, this theory also represents apparently the only case in which wave attenuation is successfully explained from basic mechanical principles.

(iii) **Visco-elastic theory**, which is currently broadly used in seismology. Its principles are used in modeling, interpreting, and in tomographic inversions for seismic attenuation at all scales. Practically all software for modeling seismic waves with attenuation also utilize this approach. Nevertheless, throughout this chapter, we carry out a systematic comparison of this approach to the Lagrange method and reveal the assumptions underlying this approach. As it will be shown, the visco-elastic approach might encounter serious problems when applied to waves in heterogeneous media.

A note on terminology is appropriate before comparing the above approaches. In the following, we use term "attenuative medium" for all cases in which elastic-energy dissipation is present. By contrast, the term "visco-elastic" and its derivatives are reserved specifically for the Q-based, axiomatic, mathematical theory using the elastic constants extended to incorporate memory parameters. Similarly, the term "internal friction" below is no longer directly associated with a "quality factor" (Q^{-1}), as, for example, in Aki and Richards (2002). Instead, this term is used to denote a variety of energy-dissipation processes caused by relative movements within the medium.

The principal question we will try answering in this chapter is whether we should describe the elastic energy dissipation in the medium as caused by "imperfect elasticity" or by "internal friction." The Lagrangian model shows that these two notions are not equivalent, and principally differ in attributing the dissipation to the elastic or kinetic energies, respectively. Several lines of argument suggest that "internal friction" provides the correct physical picture, whereas "imperfect elasticity" is inspired by analogies and established by mathematical postulates. At the same time, the "internal friction" model reveals the complexity of the attenuation process and shows that the problem of wave propagation in attenuative medium may still be far from being solved.

The "imperfect-elasticity" view has long become standard in earthquake seismology and it is unequivocally supported by the visco-elastic theory. Mathematically, it provides an extremely powerful view of Q as a true property of the medium representing the argument of its complex-valued wave velocity. However, physically, this picture is most concerning, considering that many factors control energy dissipation within the Earth, such as fracturing, fluid content, saturation, viscosity, porosity, permeability, tortuosity, properties of "dry" friction on grain boundaries and faults, and distributions of scatterers (Bourbié et al., 1987). Most of these factors are only remotely, if at all, related to the seismic velocities or elastic parameters.

The fundamental difference between the two approaches is that the Lagrangian approach studies the functional forms of the fundamental physical entities, such as the potential and kinetic energies; whereas, the visco-elastic method only uses the differential equations of motion. As shown below, visco-elastic constitutive equations can be viewed as heuristic generalizations of several simple wave solutions (such as plane waves), low-frequency experiments (such as creep measurements), and mechanical analogs. Such basis is insufficient for constructing a physical theory. We argue that the Lagrangian method is by far more trustworthy in application to realistic physical problems. We will try providing the links to the Lagrangian mechanics whenever possible.

As it appears, although the visco-elastic theory is self-consistent and complete, its underlying stress-strain relations have only been tested in simple uniform-medium cases. In a heterogeneous medium, parts of its volume are subjected to conditions that are never encountered in a uniform medium. For example, in a surface wave, its velocity V, frequency ω, and wavenumber k are not related by the relation $V = \omega/k$, which is characteristic for the uniform-medium case and from which the visco-elastic strain-stress law is inferred. This example is elaborated on in Section 6.1.2. Therefore, the visco-elastic constitutive equation needs to be carefully analyzed when applied to non-uniform media.

Another interesting indication of a failure of the visco-elastic model arises from extending the concept of the acoustic impedance to attenuative media. The visco-elastic theory predicts the imaginary part of the impedance which is *opposite* compared to the explicit solution. Thus, predictions of visco-elastic models may significantly deviate from reality in heterogeneous attenuative media.

The Lagrangian model illustrated below suggests that energy loss in a seismic wave occurs by means of dissipating its kinetic, and not elastic energy. This shows that the elastic moduli of the medium should remain real valued, and each dissipation mechanism needs to be modeled specifically, by using the appropriate parameterizations, dissipation functions, and differential equations. Quasi-static elastic relaxation certainly exists, but it can be described by the corresponding internal variables and dynamic mechanisms rather than by phenomenological memory parameters attributed to the elastic moduli. No common Q^{-1} parameter arises from this model, although the attenuation coefficient may exist under significantly more realistic approximations (Chapter 5).

In summary of this chapter, phenomenological analogies and analytic extrapolations on which the visco-elastic approach is based yield numerous insights into wave processes in media with energy dissipation. Nevertheless, their predictions may significantly deviate from reality in important practical cases, and therefore a description based on fundamental mechanical principles needs to be sought. Lagrangian continuum mechanics using realistic multi-phase models based on the specific physics of dissipative processes provide all the necessary tools for such a description.

2.1 Energy

When constructing a theory of energy dissipation, the first quantity that needs to be defined is the energy. The concept of energy plays a unique role in theoretical physics. Energy is not merely a quantity that is preserved in a class of processes called "conservative." Much broader than a simple conservation law, *all equations of motion* of any mechanical system, no matter how complex it is, are encoded in the *functional form* of a single function, called "the Lagrangian," which usually combines the *kinetic and potential energies* of the system. By taking partial derivatives of these functional forms, all equations of motion are generated. By utilizing various symmetries of the potential- and kinetic-energy forms, a complete set of conservation laws (such as the conservation of total energy and momentum) is derived. Once again, the definition of the kinetic and potential energies is crucial and at the same time completely sufficient for describing any physical system.

The kinetic and potential energy functions in mechanics are defined from the fundamental principles of time and translational invariance, linearity, presence of certain degrees of freedom, and types of interactions. These properties are defined for elementary systems, which are more basic than the spring-dashpot combinations commonly used as analogs in visco-elasticity (*e.g.*, Bland, 1960). For example, the simplest elementary system is the free mass m, which only possesses the kinetic energy, $E_k = \dfrac{m\dot{x}^2}{2}$, where $\dot{x}$ is its velocity. For another such system, a particle is subjected to an external force, and the potential energy is $E_p = U(x)$ regardless of the actual movement. The compete 1-D single-body mechanics is embedded in the combination of these two energy functions, $L = E_k - E_p$. It is important to see that E_k and E_p describe the behavior of the system even for trajectories which do not represent viable solutions, for example, those for which $\dot{x} \neq \dfrac{dx}{dt}$. Such trajectories are very important in wave mechanics.

The fundamental role of energy in wave mechanics can be further seen from the concept of the Hamiltonian operator. Using the Dirac notation conventional in quantum mechanics, the time evolution of a field $|\psi\rangle$ is described by the Hamiltonian, $\hat{H}$:

$$\frac{d}{dt}|\psi\rangle = -i\hat{H}|\psi\rangle. \tag{2.1}$$

For harmonic waves, $\hat{H}$ is diagonal: $\hat{H}|\psi\rangle = E|\psi\rangle$, where E is the energy of the system. Thus, energy "operator" is responsible for the time evolution of the system: $|\psi\rangle = e^{-i\hat{H}t}|\psi_{t=0}\rangle$.

The visco-elastic theory is constructed differently. In this theory, the energy density function cannot be uniquely defined (see Section 2.1 in Carcione, 2007). Visco-elasticity bypasses the variational principles and starts directly from a constitutive law (*i.e.*, equation of motion). The notion of "stored" energy is introduced only after this law is formulated, and typically only for harmonic processes. The stored energy is not partitioned into the kinetic and potential parts; instead, it is defined so that its time derivative corresponds to the dissipated power, but the functional form corresponds to the elastic energy (*ibid.*, or Section 1.2 in Borcherdt, 2009). In consequence, the cause of energy dissipation is described as "imperfect elasticity" (Anderson and Archambeau, 1964). However, as one can see, this energy definition is constructed backwards — by designing a suitable energy function for a known harmonic-wave solution instead of deriving such solutions from the variational principle.

The absence of consistent energy functions is a fundamental flaw of visco-elasticity. Without a variational dynamic principle, visco-elasticity is more of a model of known decaying-wave processes than a way to discover the true mechanics of attenuation. This model works well and allows the explanation of the behaviors of complex mechanical and electrical systems in engineering; however, in geophysics, we need a model reaching closer to the inner workings of the physical processes within the Earth. This subject is further discussed in Section 2.5.

2.1.1 Internal friction or imperfect elasticity?

From the basic analogies to mechanical and electrical systems, energy dissipation is usually related to velocities. For example, in electricity, current I represents an analog to the velocity, and the power dissipated on resistance, R, equals $-\delta E/\delta t = RI^2$. We can denote this quantity by $2D$, with $D = \frac{1}{2}RI^2$ being the dissipation function. The corresponding "force" of resistance is the voltage drop, which is proportional to I and equals the derivative of D in respect to I:

$$U = -RI = -\frac{\partial D}{\partial I}. \tag{2.2}$$

Note that D is functionally close to the electric-circuit equivalent of the kinetic energy, $E_k = \frac{1}{2}LI^2$, where L is the inductance.

In mechanics, viscous friction is also given by a similar equation:

$$\mathbf{f}_D = -\zeta \dot{\mathbf{r}}, \tag{2.3}$$

where $\dot{\mathbf{r}}$ is the velocity, and ζ is the constant of viscous friction. For a laminar-shear flow in a Newtonian viscous fluid, the shear stress corresponding to (2.3) is:

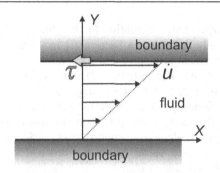

$$\tau = -\nu\rho\frac{\partial \dot{u}}{\partial y}, \qquad (2.4)$$

where $\dot{u}$ is the velocity along axis X (Figure 2.1), ν is the coefficient of kinematic viscosity, and ρ is the fluid density (Streeter and Wylie, 1981). The corresponding energy-density dissipation rate is:

FIGURE 2.1
Velocity and traction in viscous flow (eq. (2.4)).

$$\frac{\delta E}{\delta t} = \tau\dot{u} = -\nu\frac{\partial}{\partial y}\left(\frac{\rho\dot{u}^2}{2}\right) = -\nu\frac{\partial E_k}{\partial y},$$

$$(2.5)$$

where E_k is the kinetic energy density. From these examples, we see that energy dissipation in both electricity and fluid mechanics is closely associated with the *kinetic energy*. This is not surprising, because viscous friction dissipates energy by means of parts of the medium moving relative to each other.

2.2 Fundamental constraints

Two fundamental constraints on the attenuation coefficient exist regardless of any specific models. The first of these constraints comes from the conservation of energy, implying that the relative energy-dissipation rate should be proportional to the attenuation coefficient. Although in heterogeneous media this constraint is difficult to quantify, it can be used to assess the validity of solutions for α, as illustrated in a surface-wave example in Section 6.1.2. Another key constraint is imposed by causality, which requires that, in the most general case, non-zero attenuation must be associated with phase-velocity dispersion, and *vice versa*. This principle places some constraints on the frequency-dependence of α and phase velocities, and it is sometimes used as a general argument for a frequency-dependent Q within the Earth.

2.2.1 Causality

Physically, the Kramers-Krönig causality relation is easy to understand. Consider, for example, a seismic wave recorded at one point within the medium. Let us chose the time of the infinite-frequency wave onset at this point as the time origin, $t = 0$. Causality then requires that no energy arrives before this time, *i.e.*, $u(t) = 0$ for all $t < 0$. Therefore, in the time domain, causality removes "a half" of the values of $u(t)$ and consequently, in the frequency domain, also a half of the complex values of $u(\omega)$ must be removed. This condition can be expressed in several ways; in particular, we can require that $u(\omega)$ is analytic in the upper half-plane of ω. This analyticity means that the values of $u(\omega)$ for $\text{Im}\,\omega > 0$ are uniquely determined by its values at the real ω axis. Alternately, this condition can be expressed as the real part of $u(\omega)$ being uniquely determined by its imaginary part, and *vice versa* (see Appendix 1):

$$\begin{cases} \text{Re}\,u(\omega) = \dfrac{2}{\pi} P \displaystyle\int_0^\infty d\omega' \dfrac{\text{Im}\,u(\omega')}{\omega'^2 - \omega^2}, \\[4mm] \text{Im}\,u(\omega) = -\dfrac{2\omega}{\pi} P \displaystyle\int_0^\infty d\omega' \dfrac{\text{Re}\,u(\omega')}{\omega'^2 - \omega^2}, \end{cases} \tag{2.6}$$

where P denotes the Cauchy's principal values of the integrals.

The above argument is valid for any real-valued process $u(t)$ and therefore for any wave propagation in an arbitrary heterogeneous medium. In a uniform medium, additional useful forms of this causality condition can be expressed in terms of the wavenumber $k(\omega)$, phase velocity $V(\omega) = \omega/k(\omega)$, and attenuation coefficient $\alpha(\omega)$. A detailed discussion of this case is given by Aki and Richards (2002, p.167–169). Similar to eq. (2.6), this result relates the imaginary part of the wavenumber (α) to its real part (k), and *vice versa*:

$$\begin{cases} k(\omega) = \dfrac{\omega}{V_\infty} + \dfrac{\omega}{\pi} P \displaystyle\int_{-\infty}^\infty d\omega' \dfrac{\alpha(\omega')}{\omega' - \omega}, \\[4mm] \alpha(\omega) - \alpha(0) = -\dfrac{\omega}{\pi} P \displaystyle\int_{-\infty}^\infty d\omega' \dfrac{k(\omega') - \omega'/V_\infty}{\omega'(\omega' - \omega)}, \end{cases} \tag{2.7}$$

where V_∞ is the infinite-frequency limit of phase velocity, and $\alpha(0)$ is the "subtraction constant" for α. Note that this subtraction constant has the meaning of the zero-frequency attenuation coefficient (Nussenzveig, 1972). Both parameters V_∞ and $\alpha(0)$ can also be viewed as regularization constants for the integrals in eqs. (2.7), which are otherwise divergent.

Note that expressions (2.7) here are presented in the form of $k(\omega)$ and $\alpha(\omega)$ functions, unlike the $V(\omega) - \alpha(\omega)$ relations in Aki and Richards (2002, p. 169). The present form emphasizes the symmetry between the real and imaginary components of the wavenumber vector, and also shows that the Kramers-Krönig identity constrains the components of the wavenumber, and not the phase velocity. Therefore, causality only constrains the parameters of *solutions* to the wave equation. Parameters $k(\omega)$, $V(\omega)$, and $\alpha(\omega)$ above represent properties of a wave in a uniform medium, and they should not be automatically equated to the local properties of the medium, as it is often implied.

To conclude the discussion of causality, note that it is expressed by non-trivial integral equations only in the frequency domain. If the equations of motion are formulated in the time domain, causality is automatically satisfied if all interactions are local and instantaneous (as in the Lagrangian models below) or retarded (as in the visco-elastic model).

2.2.2 Phase-velocity dispersion

Causality relations (2.7) show that if some wave experiences attenuation ($\alpha > 0$), it must also exhibit the phase-velocity dispersion, and *vice versa*. In principle, these equations allow expressing the phase-velocity spectrum if attenuation values are known at all frequencies. However, the integrals in eqs. (2.7) typically converge very slowly near the limits of $\omega' \to \infty$ and $\omega' \to \omega$, and therefore much of the information required for using these expressions to predict either $k(\omega)$ or $\alpha(\omega)$ lies in the region of unphysically high frequencies.

The case of $\alpha \propto \omega$ is of particular interest, because there exists good evidence for it in both observations and theory (see below). However, in this case, the integral in the first equation in (2.7) is divergent and needs to be regularized. Such regularization can be done, for example, by assuming that $\alpha \approx \alpha_0 \omega$ within the seismic frequency band but flattens out at some high frequencies $|\omega| \gg 1/\alpha_1$ (Azimi *et al.*, 1968),

$$\alpha(\omega) = \frac{\alpha_0 \omega}{1 + \alpha_1 \omega}. \tag{2.8}$$

From (2.7), the corresponding phase slowness is:

$$\frac{1}{V(\omega)} = \frac{1}{V_\infty} - \frac{2\alpha_0}{\pi(1 - \alpha_1^2 \omega^2)} \ln(\alpha_1 \omega) \approx \frac{1}{V_\infty}\left[1 - \frac{2\alpha_0 V_\infty}{\pi} \ln(\alpha_1 \omega)\right]. \tag{2.9}$$

Denoting, in accordance with the definition of Q in eq. (1.9), $2\alpha_0 V_\infty = Q^{-1}$, we obtain the velocity dispersion law:

$$V(\omega) \approx \frac{V_\infty}{1 - \dfrac{1}{\pi Q} \ln(\alpha_1 \omega)}. \tag{2.10}$$

Equation (2.10) shows the general logarithmic phase-velocity increase with frequency in the presence of attenuation, which is supported by many attenuation models (*e.g.*, Carcione, 2007). However, its predictive power is very limited, and it is important to keep in mind the sensitivity of this law to the limits of $\alpha(\omega)$

assumed at $\omega\to0$ and $\omega\to\infty$. Parameters α_1 and V_∞ in (2.10) essentially represent arbitrary constants on which, however, the resulting values of phase velocities may depend very strongly.

Thus, causality conditions definitely require velocity dispersion in the presence of attenuation, yet the exact form of this dispersion should be determined from the specific models. Normally, any mechanical system possessing a time-domain (Lagrangian) description would satisfy the causality. Similarly, convergence of the Kramers-Krönig integrals requires a frequency-dependent attenuation $\alpha(\omega)$ at least at the very high and very low frequencies. For practical purposes, this requirement is not strong enough, because the low-frequency cut-off below which Q should decrease can be estimated to be about 10^{-99} Hz for $Q \geq 30$ (Futterman, 1962). Thus, the causality principle neither justifies the use of parameter Q (because it only appears as a replacement for parameter α_0 in the uniform-space problem) nor enforces any definite frequency dependence of Q within the seismological frequency band.

2.3 Homogenization approach

Broadly, homogenization theory deals with the derivations of equations for averages of solutions with rapidly varying coefficients. It tries obtaining macroscopic ("homogenized," or "effective") equations for systems with fine microscopic structure. The term "microscopic" here applies to physical laws governing the mechanisms at the scale of medium heterogeneity, whereas "macroscopic" refers to the scale of the observations. In the problem of wave attenuation, "microscopic" therefore refers to the scale of mineral grain, pore space, or fractures, whereas "macroscopic" corresponds to the scales of the wavelength and crust/mantle structures.

Homogenization methods can be subdivided into three groups. The first of them is based on some kind of averaging process, in which a microscopic problem is solved at the level of some elementary cell. Such a cell can contain an isolated heterogeneity, such as a scatterer or a fluid-filled channel. From a solution of such elementary problem (e.g., by using Born scattering), one can infer mean values of some "observables" (such as the stresses, strains, or scattering amplitudes) and relate them to macroscopic parameters of the cell (such as stresses or heterogeneity scale lengths). Once such functions are found, the actual heterogeneous medium can be replaced with a fictitious homogenous medium, which is described in terms of these macroscopic parameters. This procedure was followed by Biot (1956) in characterizing the flow of a fluid in a porous medium. This is also generally the approach taken in the classic single- or multiple-scattering local-earthquake coda model by Aki (1969, 1980).

The second group of homogenization methods represents the heterogeneous medium as a periodic repetition of some microscopic elementary structure, thereby imposing a periodicity on the solution. Once a solution in such a periodic medium is found, its spatial period is considered as tending to zero with respect to the macroscopic scale (Bensoussan *et al.*, 1978), and the macroscopic laws are obtained. This approach was used to generalize Darcy's law for fluid flows in non-steady state conditions. A drawback of this method is in that it only constrains the general form of the macroscopic law, whereas the averaging methods above also provide quantitative analytical estimates of the observables (Bourbié *et al.*, 1987).

The third approach to homogenization is appropriate when the dynamic equations governing the system are formulated through some variational principle, such as the Hamilton-Lagrange principle. In such cases, microscopic properties of the system can be accounted for by using some specific forms of solutions at short scale lengths, which are parameterized macroscopicly (*e.g.*, amplitudes of heterogeneities) and further combined into the macroscopic functional being optimized. This approach could be quite general and suitable for numerical computations. The Rayleigh-Ritz solution for Love-wave attenuation in Section 6.1.2 is derived in this spirit, although it is still quite limited in the depth of its microscopic level.

Overall, it appears that the degree of today's understanding of the microscopic physics of seismic attenuation within the Earth is still below the level required by homogenization approaches. As we argue throughout this book, although averaging methods are broadly used to model scattering (*e.g.*, Aki, 1980), their results may often be biased by inaccurate treatments of the macroscopic background. The available knowledge about the microscopic structure is limited, and even some of the basic facts (such as the existence of a "medium Q") need to be established from interpreting the attenuation measurements. In this environment, semi-phenomenological, macroscopic descriptions based on well-understood physics appear to be the most appropriate.

2.4 Lagrangian approach

The Lagrangian formalism is known for its generality and power, and represents the principal tool of theoretical mechanics, classical and quantum field theory, and the theory of elasticity (Landau and Lifshitz, 1986). Unfortunately, this approach appears to be still incompletely utilized in global seismology. In advanced texts on theoretical seismology, such as by Dahlen and Tromp (1998), this description and the associated group of variational methods are often presented as mostly mathematical tools. Here, we emphasize the value of this method for establishing the physical principles governing the dynamics of a system. We only overview the key ideas of the approach in relation to our problem of deciding whether wave attenuation can be best viewed as caused by dissipation of the kinetic or elastic (potential) energy, or be a totally independent entity.

In Lagrangian mechanics, the dynamics of any system, such as the elastic field, is described by a function of some generalized coordinates $\mathbf{q}$ and their time derivatives $\dot{\mathbf{q}}$:

$$L(\mathbf{q},\dot{\mathbf{q}}) = E_k - E_p, \tag{2.11}$$

where E_k and E_p are usually the kinetic and potential energies. Vector $\mathbf{q}$ consists of any parameters describing the field (e.g., local displacements, their Fourier amplitudes, or Rayleigh-Ritz coefficients in Section 6.1.2). Importantly, note that $\mathbf{q}$ and $\dot{\mathbf{q}}$ are treated as independent variables. The corresponding Euler-Lagrange equations of motion are given by:

$$\frac{d}{dt}\left(\frac{\partial L}{\partial \dot{q}_i}\right) - \frac{\partial L}{\partial q_i} = 0, \, i = 1,2, \ldots \tag{2.12}$$

In the presence of energy dissipation, these equations are modified by adding the generalized dissipative force Q_i^D (do not confuse with quality parameter Q),

$$\frac{d}{dt}\left(\frac{\partial L}{\partial \dot{q}_i}\right) - \frac{\partial L}{\partial q_i} = Q_i^D, \tag{2.13}$$

which is a partial derivative of some dissipation function D [also called "pseudo-potential;" not to be confused with the "specific dissipation function" Q^{-1} used by Bland (1960), Anderson and Archambeau (1964) and others] in respect to $\dot{\mathbf{q}}$:

$$Q_i^D = -\frac{\partial D}{\partial \dot{q}_i}. \tag{2.14}$$

If, as it is often the case, function D is quadratic in $\dot{\mathbf{q}}$, it can be interpreted as the rate of energy dissipation as a function of generalized velocities.

Equation (2.13) shows that the attenuation is caused by external friction forces which *are not included* in the Lagrangian. The effect of friction can nevertheless be described by an energy-like function D (eq. (2.18)), and this function is similar to the kinetic, but not to the elastic energy.

2.4.1 Linear oscillator

The damped linear oscillator is the simplest system exhibiting energy dissipation in an oscillatory process. Most characteristics of attenuative processes, including the definition of Q, arise from either explicit inferences or implicit analogies to this system. The concept of the linear oscillator represents a mathematical abstraction belonging to the field of "theoretical mechanics" (Landau and Lifshitz, 1976), however, its realizations are pervasive in virtually every area of physics and engineering, such as mechanics, electrodynamics, optics, and quantum mechanics.

In theoretical mechanics, a linear oscillator represents the simplest dynamic system described by only two parameters, which are its mass, m, and the natural frequency, ω_0. The parameters describe the functional forms of the kinetic and potential energies, respectively. These functional forms are quadratic in both the displacement, $\mathbf{r}$, and velocity, $\dot{\mathbf{r}}$, and the Lagrangian is:

$$L(\mathbf{r},\dot{\mathbf{r}}) = \frac{m}{2}\dot{\mathbf{r}}^2 - \frac{m\omega_0^2}{2}\mathbf{r}^2 .$$

(2.15)

Because this Lagrangian is explicitly independent of time, the "generalized energy,"

$$E = \dot{\mathbf{r}}\frac{\partial L}{\partial \dot{\mathbf{r}}} - L = \frac{m}{2}\dot{\mathbf{r}}^2 + \frac{m\omega_0^2}{2}\mathbf{r}^2 ,$$

(2.16)

is preserved, and the oscillator's movement consists of infinite oscillations at frequency ω_0 near the point $\mathbf{r} = 0$.

In order to introduce energy dissipation in this system, an additional "damping," or viscous-friction force is required. Using the two oscillator parameters, coefficient ζ in the equation of viscous friction (2.3) can be expressed through a dimensionless constant ξ by $\zeta = \xi m\omega_0$, so that the damping force becomes:

$$\mathbf{f}_D = -\xi m\omega_0\dot{\mathbf{r}} .$$

(2.17)

Consequently, the corresponding dissipation function (called Rayleigh) is:

$$D = \zeta\frac{m}{2}\dot{\mathbf{r}}^2 = \xi\omega_0 E_k .$$

(2.18)

Note that as above, the dissipation function is proportional to the kinetic energy.

With the above Lagrangian and dissipation functions, the equation of motion becomes:

$$m\ddot{\mathbf{r}} = -m\omega_0^2\mathbf{r} - \xi m\omega_0\dot{\mathbf{r}} .$$

(2.19)

For $\xi \ll 0$, its general solutions are $\mathbf{r}(t) = \mathrm{Re}[\mathbf{A}\exp(-i\omega'_0 t)]$, where:

$$\omega'_0 \approx \omega_0 \left(\pm 1 - \frac{i\xi}{2} \right), \tag{2.20}$$

and $\mathbf{A}$ is an arbitrary complex-valued amplitude. Quantity ω'_0 can be interpreted as the "complex frequency," and because $\mathbf{r}$ is real-valued, it is sufficient to consider only the solution with positive $\mathrm{Re}\omega'$. The imaginary part of ω_0' describes the logarithmic amplitude decrement of the free oscillation with time. Similar complex frequencies arise when describing the attenuation of the Earth's normal modes (Chapter 6 in Dahlen and Tromp, 1998) and in lab attenuation measurements using standing waves (see Section 3.9.2).

With the use of the complex frequency, eq. (2.19) takes the form of the free-oscillator eq. (2.18),

$$m\ddot{\mathbf{r}} = -m\omega_0'^2 \mathbf{r}. \tag{2.21}$$

If an external force $f(t)$ is added to the right-hand side of eq. (2.19), then:

$$A(\omega) = \frac{f(\omega)}{m}\Lambda(\omega), \tag{2.22}$$

where the shape of its frequency-domain response to $f(\omega)$ is determined by the complex frequency alone,

$$\Lambda(\omega) = \frac{1}{\omega_0'^2 - \omega^2}. \tag{2.23}$$

Thus, the displacement-amplitude response to a harmonic force exhibits a spectral peak centered at ω_0, with parameter ξ controlling its relative width (Figure 2.2).

Expression (2.23) is often used to describe the dissipation spectra near resonance. However, note that in practical observations, it is the *absolute* width of the spectrum that is directly measured, and this absolute width equals the temporal attenuation coefficient, $\chi = \omega_0 \xi/2$ (Figure 2.2). In terms of χ, eq. (2.23) reads:

$$\Lambda(\omega) = \frac{1}{\omega_0^2 - 2i\omega_0\chi - \omega^2}. \tag{2.24}$$

This form allows measuring χ in the observed power spectra by fitting, for example, the following normalized density function to the observed power spectrum:

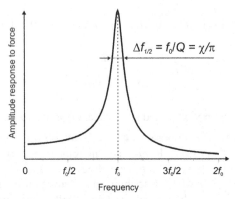

FIGURE 2.2
Definition of quality factor for a linear oscillator. In a steady-state oscillation, Q measures the relative width of the resonance peak: $\Delta f = f_0/Q$, measured at the level of $1/\sqrt{2}$ of the maximum amplitude. Example with $Q = 10$ shown.

$$\left|\tilde{\Lambda}(\omega)\right|^2 = \frac{1}{\left(\dfrac{\omega_0^2 - \omega^2}{2\omega_0\chi}\right)^2 + 1}. \tag{2.25}$$

2.4.2 Attenuative elastic medium

Similarly to the linear oscillator, the mechanics of an elastic medium can be best understood by using the Lagrangian formalism. For an isotropic elastic medium, the Lagrangian density is:

$$L = \frac{\rho}{2}\dot{u}_i\dot{u}_i - \left[\frac{\lambda}{2}\varepsilon_{kk}\varepsilon_{ll} + \mu\varepsilon_{ij}\varepsilon_{ij}\right], \tag{2.26}$$

where summation over all pairs of repeated indices is assumed. Parameters λ and μ are the Lamé elastic constants, and ρ is the mass density. Note that both the displacements and velocities are viewed as independent variables in function (2.26). According to the Hamilton variational principle, all equations of motion for the field result from the stationarity of the Hamiltonian action with respect to field perturbations. This principle can be expressed as:

$$\delta S = 0, \tag{2.27}$$

where the action is:

$$S[u] = \int dt \int L dV ,\tag{2.28}$$

and $\int dV$ denotes integration over the entire volume of the field. All differential equations of motion corresponding to (2.27) are then given by:

$$\frac{\delta S}{\delta u} \equiv \frac{\partial L}{\partial u} - \partial_n \left[\frac{\partial L}{\partial(\partial_n u)} \right] = 0 ,\tag{2.29}$$

where $\partial_n = \left(\frac{1}{\partial t}, \frac{1}{\partial x}, \frac{1}{\partial y}, \frac{1}{\partial z} \right)$, with $n = 0$ for time and $n = 1, 2, 3$ for the spatial variables.

Formulations (2.27) to (2.29) are significantly more powerful than the differential equations of motion used in visco-elastodynamics. For example, they readily show that, in the most general elastic case, harmonic oscillations at different frequencies do not interact with each other. To see this, note that all model parameters in eq. (2.26) are time-independent. Such mechanical systems are called "conservative" and can be shown to preserve the total energy. In a conservative system, let us express the time dependences of all functions in eq. (2.26) through their corresponding Fourier components,

$$b(t) = \frac{1}{2\pi} \int d\omega e^{i\omega t} b(\omega),\tag{2.30}$$

where b denotes any component of u_i or ε_{ij}. The action integral (2.28) then becomes:

$$S = \frac{1}{(2\pi)^2} \int dt \int d\omega \int d\omega' \int dV e^{it(\omega-\omega')} \left[\frac{\omega\omega'}{2} \rho \dot{u}_i^* \dot{u}_i - \left(\frac{1}{2} \lambda \varepsilon_{kk}^* \varepsilon_{ll} + \mu \varepsilon_{ij}^* \varepsilon_{ij} \right) \right],\tag{2.31}$$

where all u and ε values are now in the frequency domain. The time integral here gives the Dirac delta function:

$$\int dt e^{it(\omega-\omega')} = 2\pi\delta(\omega-\omega'),\tag{2.32}$$

and consequently the action takes a convenient frequency-domain form,

$$S = \frac{1}{2\pi} \int d\omega \int dV \left[\frac{\omega^2}{2} \rho u_i^* u_i - \left(\frac{1}{2} \lambda \varepsilon_{kk}^* \varepsilon_{ll} + \mu \varepsilon_{ij}^* \varepsilon_{ij} \right) \right].$$ (2.33)

This form contains no factors combining different frequencies, showing that all frequency harmonics participate in S independently and do not interact.

In the presence of energy dissipation, the dynamics of the elastic field becomes significantly more complex. A system of type (2.26) does not dissipate energy, and dissipation requires impurities, multi-phase flows, or fractures, which introduce additional degrees of freedom and parameters in the Lagrangian. The problem of finding these parameters for the Earth's crust and mantle and constructing the corresponding variational principles is far from being solved. Below, we only summarize some suggestions and key ideas relevant to subsequent discussions. For more complete treatments of elastic-energy dissipation in realistic porous media, see Bourbié et al. (1987) and Carcione (2007).

Hypothetical example

For a simple conceptual example, consider the following hypothetical form of D constructed similarly to the elastic energy (2.26), but using velocities instead of displacements:

$$D\{\dot{\mathbf{u}}\} = -\int \xi \left(\frac{\lambda}{2} \dot{\varepsilon}_{ii} \dot{\varepsilon}_{jj}^* + \mu \dot{\varepsilon}_{ij} \dot{\varepsilon}_{ij}^* \right) d^3\mathbf{r},$$ (2.34)

where ξ is a small attenuation constant. With such D, the P and S waves are not coupled by attenuation. For a plane P wave in a uniform medium, its displacement u can be used as the generalized coordinate q, and the homogeneous equation of motion (2.12) becomes:

$$\ddot{u} - V_m^2 u'' - \xi V_m^2 \dot{u}'' = 0.$$ (2.35)

As expected, from its plane-wave solution propagating in the direction of axis Z:

$$u_i(z,t) = A \exp\left(-i\omega t + i\tilde{k}z\right),$$ (2.36)

the dispersion relation gives a positive $\text{Im}\,\tilde{k}$:

$$\tilde{k} = \frac{\omega}{V_m \sqrt{1 - i\omega\xi}} \approx \frac{\omega}{V_m}\left(1 + \frac{i\omega\xi}{2}\right).$$ (2.37)

Comparing this equation to expression (2.119), we see that the spatial attenuation coefficient $\alpha = \operatorname{Im} \tilde{k}$ is proportional to ω^2, and the corresponding Q factor decreases with frequency:

$$Q = \frac{1}{\xi\omega}. \tag{2.38}$$

Similar dependences of α and Q on frequency are found in Biot's (1956) theory of porous saturated media outlined in the next section.

Following the visco-elastic practice, the wave phase velocity can also be rendered as complex-valued,

$$\tilde{V}_{phase} = \frac{\omega}{k'} = V_m \sqrt{1 - \frac{i}{Q}}. \tag{2.39}$$

In agreement with the correspondence principle (Section 2.5.4), the last two terms in eq. (2.35) can be lumped together in a complex-valued squared material $\tilde{V}_m$:

$$V_m^2 \rightarrow \left(\tilde{V}_m\right)^2 = V_m^2 \left(1 - i\xi\omega\right), \tag{2.40}$$

with the corresponding modifications to λ and μ. However, although this transformation makes $\tilde{V}_m$ equal V_{phase}, it does not extend to boundary conditions (see Section 2.7.1) and should not be interpreted as a fundamental principle and generalized to non-uniform media.

Saturated porous medium

Another instructive example, and apparently the only one well-developed among those relevant to seismology, is the case of fluid-saturated porous medium. We only briefly outline the general sense of this theory here, following Bourbié et al. (1987, p.69–81). The main point to note is that in order to produce anelastic effects, at least a two-phase medium, such as porous rock saturated with fluid, is required. By relative movements of the two phases, energy is dissipated, and various other kinematic and dynamic effects are produced.

In Biot's (1956) theory of saturated porous rock, the dimensions of the macroscopic elementary volumes are assumed to be small compared to the wavelength. This is the typical case encountered in seismology (see Figure 1.1 on p. 5) and thus it is sufficient to consider only quadratic terms in the volumetric density of the kinetic energy,

$$E_k = \frac{\rho_u}{2} \dot{u}_i \dot{u}_i + \frac{\rho_w}{2} \dot{w}_i \dot{w}_i + \rho_{uw} \dot{u}_i \dot{w}_i \,, \tag{2.41}$$

where $\dot{w} = \dot{U} - \dot{u}$ is the filtration velocity (*i.e.*, velocity of the fluid $\dot{U}$ relative to the rock matrix $\dot{u}$). The kinetic matrix-fluid coupling parameters ρ above are determined by the porosity and permeability of the rock. For example, ρ_u in this expression is the average density,

$$\rho_u = \rho = (1 - \phi) \rho_s + \phi \rho_f \,, \tag{2.42}$$

where ρ_s and ρ_f are the densities of the matrix and fluid, respectively, and ϕ is the porosity.

Note that in porous media, the density factor is modified in the presence of energy dissipation. The relative movement of pore fluids causes an additional inertial coupling force (Bourbié *et al.*, 1987, p.71),

$$F_{inertial} = \rho_f (1 - a) \ddot{u}_i \,, \tag{2.43}$$

where $a \geq 1$ is the "tortuosity" parameter, and ρ_f is the pore fluid density. This force is proportional to the acceleration and effectively modifies the inertial property (*i.e.*, density ρ) to $\rho + \rho_f(1-a)$. Therefore, the modification of density discarded by Anderson and Archambeau (1964) in their early attenuation model is actually relevant (in this sense) for saturated porous rock.

Similarly to E_k, the dissipation function is quadratic in filtration velocity,

$$D = \frac{\eta}{2\kappa} \dot{w}_i^* \dot{w}_i \,, \tag{2.44}$$

where η is the fluid viscosity, and κ is the absolute permeability, which depends on the geometry of the pores among other factors. For P waves, the displacements of rock matrix and fluid can be expressed through the corresponding scalar potentials, which can be combined in a two-component vector field,

$$\begin{pmatrix} \mathbf{u} \\ \mathbf{U} \end{pmatrix} = \begin{pmatrix} \nabla \Phi_1 \\ \nabla \Phi_2 \end{pmatrix} = \nabla \mathbf{\Phi} \,, \tag{2.45}$$

where ∇ denotes the gradient operator. Similarly to eq. (2.35), the Euler-Lagrange equations (2.12) in respect to $\mathbf{\Phi}$ are:

$$\mathbf{R} \nabla^2 \mathbf{\Phi} - \mathbf{A} \dot{\mathbf{\Phi}} - \mathbf{M} \ddot{\mathbf{\Phi}} = 0 \,, \tag{2.46}$$

where **R** and **M** are the rigidity and mass matrices, respectively,

$$\mathbf{R} = \begin{pmatrix} \lambda_f + 2\mu + M\phi(\phi - 2\beta) & M\phi(\beta - \phi) \\ M\phi(\beta - \phi) & M\phi^2 \end{pmatrix}, \tag{2.47}$$

$$\mathbf{M} = \begin{pmatrix} \rho + \phi\rho_f(a - 2) & \phi\rho_f(1 - a) \\ \phi\rho_f(1 - a) & a\phi\rho_f \end{pmatrix}, \tag{2.48}$$

and **A** is the damping matrix,

$$\mathbf{A} = \begin{pmatrix} b & -b \\ -b & b \end{pmatrix}. \tag{2.49}$$

In these expressions, λ_f is the second Lamé modulus of a closed system (i.e., corresponding to the case of fluid content being constant), and parameter M has the meaning of pressure that needs to be exerted on the fluid in order to increase the fluid content by a unit value at iso-volumetric conditions ($\varepsilon_{kk} = 0$). Parameter b is defined as $b = \phi^2/K$, where $K = \kappa/\eta$ is the hydraulic permeability, or mobility of the fluid within the rock. Parameter $\beta \in [0,1]$ quantifies the proportion of the apparent macroscopic volumetric strain caused by the variations in fluid content, which can be illustrated by the expression of the strain-stress relation,

$$\sigma_{ij} = \left(\lambda_f - \beta^2 M\right)\varepsilon_{kk}\delta_{ij} + 2\mu\varepsilon_{ij} - \beta p\delta_{ij}, \tag{2.50}$$

where p is the mean fluid pressure.

For our comparison to the visco-elastic theory, it is important to see that the above strain-stress relation: 1) is "instantaneous," i.e., containing no "memory" mechanism or variables and 2) it is not constructed in some ad hoc manner, but arises from the elastic-energy part of the Lagrangian:

$$\sigma_{ij} = \frac{\partial E_{el}}{\partial \varepsilon_{ij}}, \text{ and } p = \frac{\partial E_{el}}{\partial \xi}, \tag{2.51}$$

where ξ is the volumetric fluid content within pores. The selected form of stress tensor (2.50) corresponds to the most general case of isotropy, implying that E_{el} depends on ξ and on only two invariants of the strain tensor, which are $I_1 = \varepsilon_{kk}$ and $I_2 = 2(\varepsilon_{ij}\varepsilon_{ij} - I_1^2)$. Invariant I_1 corresponds to the dilatation of the medium, whereas I_2 describes shear. Considering small perturbations, E_{el} can be approximated by a quadratic form in ε_{ij} and ξ, and therefore the most general form of potential energy is:

$$E_{el} = \frac{1}{2}\left(\lambda_f + 2\mu\right) I_1^2 + \mu I_2 + M\xi^2 - \beta MI_1\xi, \tag{2.52}$$

which leads to eq. (2.50). This quadratic form also must be "positive-definite," which places three additional constraints on its coefficients (Biot, 1962):

$$\mu \geq 0, \quad M \geq 0, \quad \text{and} \quad \lambda_f - \beta^2 M + \frac{2}{3}\mu \geq 0. \tag{2.53}$$

The single-phase (and therefore elastic) case corresponds to $\xi = 0$ and $\beta = 0$.

In the absence of attenuation (*e.g.*, for $\eta = 0$, or $\xi = 0$ and $\beta = 0$), matrix $\mathbf{R}^{-1}\mathbf{M}$ in (2.46) is positive definite and symmetric, and consequently it has two real positive eigenvalues corresponding to the squares of two P-wave velocities, $V_{P,1}^2$ and $V_{P,2}^2$. The corresponding eigenfunctions have the forms of plane waves with wavenumbers equal $\omega/V_{P,1}$ and $\omega/V_{P,2}$, respectively. In the presence of attenuation, these wavenumbers (but not velocities) become complex. The resulting attenuation coefficients can be expressed through the normalized functions $\tilde{\alpha}_P\left(\omega/\omega_B\right)$ for both the slow and fast P waves:

$$\alpha(\omega) = \frac{\omega_B}{V_P}\tilde{\alpha}_P\left(\frac{\omega}{\omega_B}\right), \tag{2.54}$$

and a similar expression for S waves. Note the appearance of Biot's reference frequency ω_B, which depends on the filtering medium:

$$\omega_B = \frac{\phi}{\rho_f \mathrm{K}}. \tag{2.55}$$

This frequency may be somewhat analogous to the heuristic relaxation frequencies commonly used in visco-elastodynamics. Bourbié *et al.* (1987, p.79) noted that ω_B can be extremely variable, ranging from 30 kHz to 1 GHz for water. Nevertheless, this frequency is always very high compared to seismic frequencies.

Similarly to eq. (2.54), the resulting phase velocities are given by:

$$V_P(\omega) = \frac{\sqrt{\lambda_f + 2\mu}}{\rho}\tilde{V}_P\left(\frac{\omega}{\omega_B}\right) \quad \text{for } P \text{ waves, and} \tag{2.56a}$$

$$V_S(\omega) = \frac{\sqrt{\mu}}{\rho} \tilde{V}_S\left(\frac{\omega}{\omega_B}\right) \quad \text{for } S \text{ waves.} \quad (2.56b)$$

Functions $\tilde{\alpha}_P(\omega/\omega_B)$, $\tilde{\alpha}_S(\omega/\omega_B)$, $\tilde{V}_P(\omega/\omega_B)$, and $\tilde{V}_S(\omega/\omega_B)$ were tabulated for different combinations of parameters by Bourbié *et al.* (1987, p.76–78). At the zero-frequency limit, attenuation disappears [$\tilde{\alpha}_{P,S}(0) = 0$], and the velocities are the lowest [$\tilde{V}_{P,S}(0) = 1$]. With increasing frequencies, both attenuation and velocities monotonously increase, approximately as ω^2.

An important conclusion from eqs. (2.54) and (2.56) is that the zero-frequency limits of both velocities and attenuation coefficients equal the corresponding limits of zero dissipation within the medium. The zero-dissipation case occurs when the system is closed, *i.e.*, when fluid movement is not allowed (*e.g.*, by zero permeability). This observation supports our conjecture that the "geometrical" (dissipation-free) limit can be studied by considering the zero-frequency limit in the observed amplitudes.

Expressions above were only given to illustrate the flavor of the Lagrangian macroscopic theory of wave attenuation and dispersion in two-phase media. As one can see, the description is based on rigorous mechanics, with several physical parameters affecting wave propagation (M, p, ξ, μ, λ_f, ρ_f, ϕ, β, a, κ, and η) revealed and precisely defined. This prepares us for the discussion of visco-elastodynamics, in which all of this variety of detail is replaced by a heuristic association of dissipation with only two visco-elastic moduli. Although at this point it seems clear that such an approach should be very limited in application to real media, it is nevertheless predominant in the theory of the solid Earth, and therefore deserves careful consideration.

2.5 Visco-elastic approach

The linear visco-elastic theory is based upon the popular, but intuitive interpretation of attenuation as "imperfect elasticity" (Anderson and Archambeau, 1964). To describe such imperfect elasticity, Boltzmann's (1874) concept of memory (*Nachwirkung*; after-effect) is invoked, according to which the stress at any point in the medium depends upon strain at the same point, but at all preceding times. Similarly, the strain at any given point also records a memory of the stress history at that point. Because the response of such a medium is not instantaneous, the elastic moduli are replaced with time-dependent functions characterizing the anelastic behavior of the medium. The strength of this dependence equals zero in the future, is the greatest for perturbations in the recent past, and weakens with time.

With the intuitive interpretation of material memory, it is nevertheless important to not take this concept too literally. Looking at this strain-stress relationship differently, one can see that this memory results simply from the linearity of the strain-stress solutions, which require, for example, that strain represents a convolutional response to the stress history. Such solutions can be reproduced by several types of mechanical models consisting of springs and dashpots, as discussed in most texts on visco-elasticity (*e.g.*, Bland, 1960; Carcione, 2007). However, such systems nevertheless do not contain any real memory mechanisms. The memory arises by summarizing the general properties of wave solutions, which: 1) are given by causal, attenuating harmonic waves in the time domain and 2) can be linearly superimposed on each other. Therefore, all constitutive equations reduce to multiplications in the frequency domain, and consequently, they become convolutions in the time domain. Thus, the visco-elastic formulation can be viewed as a *phenomenological extrapolation of the empirical strain-stress relations* found in creep and other experiments rather than a model of the actual mechanics of the attenuating medium.

Mathematically, the retarded linear response of, for example, strain ε to stress σ is often described by the "creep function," $\phi(t)$, giving the time-dependent ε resulting from application of a step-function stress $\sigma = \sigma_0 \theta(t)$:

$$\varepsilon(t) = \frac{\sigma_0}{M_U}\left[1 + \phi(t)\right]. \qquad (2.57)$$

Here, M_U is the "unrelaxed" (initial-state) modulus, and $\phi(0) = 0$ and $\phi(t)|_{t\to\infty} = \text{const}$. The "relaxed" (static, final-state) modulus then becomes $M_R = M_U/[1 + \phi(\infty)]$. The form of this creep function is only constrained by causality $[\phi(t) = 0$ for $t < 0]$, but is otherwise free. This freedom can be exploited in order to reproduce the observed strain-stress histories (in creep measurements) or attenuation spectra (in seismic measurements). Consequently, numerous visco-elastic creep-function models were developed to describe various types of attenuation observations.

2.5.1 Strain-stress model

By generalizing the elementary linear response to a step-function stress (2.57), the strain can be presented as a convolutional response to the preceding stress-rate history, $\dot{\sigma}(\tau)$ (Dahlen and Tromp, 1998, p. 194-195):

$$\varepsilon(t) = \int_{-\infty}^{t} J(t-\tau)\dot{\sigma}(\tau)d\tau, \qquad (2.58)$$

and conversely, the stress can also be written as a response to the strain-rate history,

$$\sigma(t) = \int_{-\infty}^{t} M(t-\tau)\dot{\varepsilon}(\tau)d\tau,$$ (2.59)

where $J(t)$ and $M(t)$ are the generalized time-dependent compliance and visco-elastic modulus, respectively. Function $M(t)$ represents the response of the stress to a step-function strain $\varepsilon = \varepsilon_0 \theta(t)$, which is equivalent to a delta-function perturbation of strain rate, $\dot{\varepsilon}$,

$$\frac{\sigma(t)}{\varepsilon_0} = \int_{-\infty}^{t} M(t-\tau)\dot{\theta}(\tau)d\tau = \int_{-\infty}^{t} M(t-\tau)\delta(\tau)d\tau = M(t).$$ (2.60)

For $J(t)$, there is a similar meaning of a response to stress rates at the preceding times. Time convolutions correspond to multiplication in the frequency domain, and therefore, the frequency-domain modulus equals the ratio of stress to strain rate,

$$M(\omega) = \frac{\sigma(\omega)}{\dot{\varepsilon}(\omega)}.$$ (2.61)

Note the symmetry between treating stress as caused by strain variations in eq. (2.59) and the inverse relation in eq. (2.58). This symmetry is typical for the visco-elastic theory, which treats all field variables as linearly related to each other by convolutional time-domain laws. Generally, it arises from generalizing the linear relations between the frequency-domain representations of the strain and stress. If one only looks for a linear relation between these quantities, the question about their cause-and-effect relationships does not arise — strain can be viewed as caused by the stress or its time derivative, or *vice versa*. This symmetry is emphasized in Zener's (1948) generalized stress-strain law known as the standard linear solid,

$$\sigma + \tau_\sigma \dot{\sigma} = M_R(\varepsilon + \tau_\varepsilon \dot{\varepsilon}),$$ (2.62)

where M_R denotes the relaxed (static) modulus. This relation is satisfied by many physical mechanisms that were proposed to explain attenuation. Parameters τ_ε and τ_σ are the relaxation time constants describing the delay in $\varepsilon(t)$ responding to a step-function stress $\sigma = \sigma_0 H(t)$, and *vice versa*; these parameters must satisfy $\tau_\varepsilon > \tau_\sigma$. The unrelaxed modulus corresponding to the instantaneous response $\dot{\sigma}/\dot{\varepsilon}$ at $t = 0$ equals $M_U = M_R \tau_\varepsilon / \tau_\sigma$. Note that the case of $\tau_\varepsilon = \tau_\sigma$ is degenerate in the sense that M_U becomes equal to M_R (and therefore there is no relaxation), and quantities σ and ε become linearly related and decoupled from the corresponding time derivatives.

By using eq. (2.61), the frequency-dependent modulus M can be expressed through $\phi(t)$,

$$M(\omega) = \frac{M_U}{1+\dot{\phi}(\omega)}, \tag{2.63}$$

where $\dot{\phi}(\omega)$ is the Fourier transform of the time derivative of the creep function. Assuming that modulus $M(\omega)$ is constructed so that it diagonalizes the wave operator and the equation of motion becomes $\rho\omega^2 u = M(\omega)\partial^2 u/\partial x^2$, we obtain a fundamental solution $\exp(-i\omega t + ikx)$, with complex wavenumber $k(\omega) = \omega\sqrt{\rho/M(\omega)}$. Therefore, the phase velocity for such a wave is (assuming that $\mathrm{Im}M(\omega) \ll \mathrm{Re}M(\omega)$):

$$V(\omega) = \frac{\omega}{\mathrm{Re}\,k(\omega)} \approx \sqrt{\frac{\mathrm{Re}\,M(\omega)}{\rho}}, \tag{2.64}$$

and the attenuation factor (eq. (**1.9**)) turns out to be independent of ρ,

$$Q^{-1}(\omega) = \frac{2\,\mathrm{Im}\,k(\omega)}{V(\omega)} \approx -\frac{\mathrm{Im}\,M(\omega)}{\mathrm{Re}\,M(\omega)}. \tag{2.65}$$

Thus, in visco-elastodynamics, the frequency-domain Q^{-1} is *defined* as the ratio of the imaginary and real parts of the complex modulus (Anderson and Archambeau, 1964).

2.5.2 Absorption peak

A generalized rheology called the "standard linear solid" is often used for fitting the observed attenuation spectra. This rheology is defined by the following creep function:

$$\phi(t) = \left(\frac{\tau_\varepsilon}{\tau_\sigma} - 1\right)\left(1 - e^{-t/\tau_\varepsilon}\right). \tag{2.66}$$

From the above equations, its complex modulus $M(\omega)$ equals (Figure 2.3):

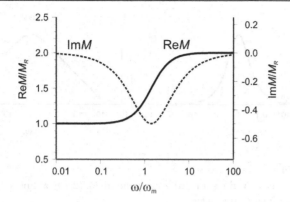

FIGURE 2.3

Real and imaginary parts of $M(\omega)$ for a standard linear solid with $\tau_\varepsilon/\tau_\sigma = 2$. Frequency ω_m corresponds to the position of the peak in Q^{-1}.

$$M(\omega) = M_R \frac{1 - i\omega\tau_\varepsilon}{1 - i\omega\tau_\sigma}. \tag{2.67}$$

For waves in such a medium, the $Q^{-1}(\omega)$ and phase velocity $V(\omega)$ are:

$$Q^{-1}(\omega) = \frac{\omega(\tau_\varepsilon - \tau_\sigma)}{1 + \left(\dfrac{\omega}{\omega_m}\right)^2}, \text{ and} \tag{2.68}$$

$$V(\omega) = \frac{V_U}{\sqrt{1 + \dfrac{(V_U/V_R)^2 - 1}{1 + \left(\dfrac{\omega}{\omega_m}\right)^2}}}, \tag{2.69}$$

where $\omega_m = 1/\sqrt{\tau_\varepsilon\tau_\sigma}$, and the "relaxed" and "unrelaxed" limits for V are denoted $V_R = \sqrt{M_R/\rho}$ and $V_U = \sqrt{M_U/\rho}$, respectively. Therefore, the standard linear solid exhibits an absorption peak at $\omega = \omega_m$, with $Q^{-1} \propto \omega$ for $\omega \ll \omega_m$, and $Q^{-1} \propto \omega^{-1}$ for $\omega \gg \omega_m$ (Figure 2.4). Note that the peak attenuation value $Q_m = Q(\omega_m)$ is only controlled by the separation between τ_ε and τ_σ. It is therefore instructive to replace both of the relaxation-time constants in the expression for $Q^{-1}(\omega)$ by Q_m and ω_m:

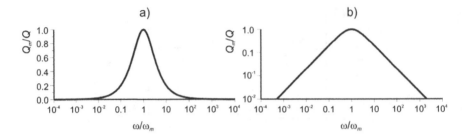

FIGURE 2.4
Absorption peak for the standard linear solid in eq. (2.70): a) linear scale
in Q^{-1} and b) logarithmic scale.

$$Q^{-1}(\omega) = 2Q_m^{-1} \frac{\omega/\omega_m}{1+\left(\dfrac{\omega}{\omega_m}\right)^2}. \tag{2.70}$$

This expression shows that the shape of the absorption peak is completely fixed. In order to produce the desired absorption spectra, this standard shape is only replicated and scaled by ω_m in frequency and by Q_m^{-1} in amplitude. The phase-velocity spectrum $V(\omega)$ follows $\mathrm{Re}M(\omega)$ and monotonously increases with frequency from V_R at $\omega = 0$ to V_U at $\omega \to \infty$ (Figure 2.3).

2.5.3 Phenomenological character

In eqs. (2.69) and (2.70), it is important to see that both $V(\omega)$ and $Q(\omega)$ for the standard linear solid are determined entirely by the "observable" characteristics V_R, V_U, Q_m, and ω_m. Therefore, relaxation times τ_ε and τ_σ can be viewed as fictitious parameters selected so that the retarded strain-stress system exhibits the specified attenuation peak at frequency ω_m:

$$\tau_\varepsilon = \frac{1}{\omega_m}\left[Q_m^{-1} + \sqrt{Q_m^{-2}+1}\right], \text{ and } \tau_\sigma = \frac{1}{\tau_\varepsilon \omega_m^2}. \tag{2.71}$$

This representation is suitable for seismic-wave applications, in which Q_m is measured but relaxation times are inferred. Conversely, in quasi-static creep experiments, τ_ε represents a primary quantity, and Q_m is derived from it (Figure 2.5):

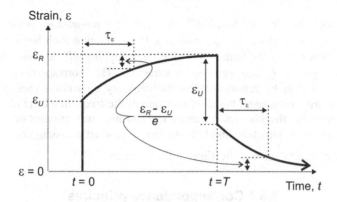

FIGURE 2.5

Schematic of rock creep measurements (the elastic after-effect). Stress is applied at time $t = 0$ and released at $t = T$. The principal measured quantity are the relaxation time τ_ε and the relaxed and unrelaxed strains ε_R and ε_U, respectively.

$$Q_m^{-1} = \frac{\left(\tau_\varepsilon \omega_m\right)^2 - 1}{2\tau_\varepsilon \omega_m}. \tag{2.72}$$

Note that in these cases, the values of ω_m (similarly to Lomnitz's constant a; Section 2.5.6), and consequently of Q_m, are quite uncertain.

Another important observable characteristic of an attenuating process is the attenuation coefficient. For visco-elastic systems, the energy density function is non-unique (see Section 2.1 in Carcione, 2007) and consequently there exists no unique definition of the attenuation coefficient. Nevertheless, a practical functional definition can again be obtained from the shape of the resulting wave solution. The spatial attenuation coefficient should equal the imaginary part of the complex wavenumber, which is:

$$\alpha = \operatorname{Im} k = \left(\operatorname{Re} k\right)\tan\left[\arg\left(k\right)\right] = -\frac{1}{2}\left(\operatorname{Re} k\right)\tan\left[\arg\left(M\right)\right] = \frac{\left(\operatorname{Re} k\right)Q^{-1}}{2}. \tag{2.73}$$

Therefore, $\alpha = \pi Q^{-1}/\lambda$ and $\chi = \omega Q^{-1}/2$, as expected.

Thus, all the essential parameters of the solution are independent of the relaxation-time constants. It appears that the "internal construction" of the standard linear solid, and likely other similar relaxation mechanism may in fact be dropped from consideration, and the solution only understood as an elementary *phenomenological* "basis function," or "*ansatz*" possessing some desired properties (*i.e.*, sets of V_R, V_U, Q_m, and ω_m). By combining such basis functions, practically arbitrary (only limited to rising and dropping off no faster than $\omega^{\pm 1}$) attenuation spectra $Q^{-1}(\omega)$ can be constructed. The corresponding velocity dispersion can then be estimated from the causality constraints (Section 2.2.1) (once again, assuming zero attenuation outside of the frequency band of interest). Most importantly, the physical mechanisms causing such attenuation spectrum $Q^{-1}(\omega)$ may be completely unrelated to the visco-elastic assumptions, such as "imperfect elasticity" or modifications of the Lamé moduli.

2.5.4 Correspondence principle

In the derivation of eqs. (2.65) and (2.64), V and $(-VQ^{-1}/2)$ were interpreted as the real and imaginary parts of a complex phase velocity, respectively:

$$\tilde{V} = \frac{\omega}{k} = V - \frac{iV}{2Q}. \tag{2.74}$$

Note that V and Q in this equation represent apparent quantities, *i.e.*, related to the specific solution of the wave equation. However, when considering a uniform-medium case, these quantities equal the corresponding medium properties, such as those given by the right-hand sides of eqs. (2.64) and (2.65). By generalizing this observation, one can *postulate* that in the general case, the effect of attenuation can also be described by a negative complex argument of medium velocity,

$$V_{\text{anelastic}} = V_{\text{elastic}} \left(1 - \frac{i}{2Q} \right). \tag{2.75}$$

This postulate is known as the *correspondence principle* (Bland, 1960).

The correspondence principle provides a very powerful way for solving problems of attenuative elasticity by extrapolating the differential equations and results from elastic-wave theory (Aki and Richards, 2002, p.174–75). By virtue of its close ties to the visco-elastic theory, this principle opens numerous possibilities for numerical modeling and inversion for Q^{-1}, which can be thereby done even at the incredible spatial resolution of cross-well reflection imaging (see Chapter 12 in Wang, 2008).

Although derived from the retarded strain-stress relations discussed previously, justification of the correspondence principle does not require any physical model of the medium at all. Transformation (2.75) is phenomenological and arises simply from interpreting the attenuation coefficient α in the wave-amplitude decay law $\exp(ikx-\alpha x)$ as a positive imaginary part of the complex wavenumber, $\tilde{k} = k + i\alpha$. Such amplitude decay could also be achieved by any other model of energy dissipation. Thus, the correspondence principle is not related to the visco-elastic theory, but is entirely due to a hypothesis equating the phase velocities found in some special cases with the general properties of the medium.

The correspondence principle is unquestionably valid only for wave propagation in homogenous media, because, only in such media does there exist an elastic parameter directly corresponding to the phase velocity, such as $V_S = \sqrt{\mu/\rho}$ for S waves. This allows attributing the imaginary shift in the wavenumber to the phase velocity, and further — to μ, thereby justifying the retarded strain-stress model above. However, in a heterogeneous medium, the phase velocity does not equal the material velocity at any point, and the reasons for introducing a complex-valued material μ and velocities remain questionable. As shown below, when extrapolated to heterogeneous media, principle (2.75) leads to incorrect results in several important cases.

2.5.5 Causality

Because both $J(t)$ and $M(t)$ and are real-valued functions and presumed to be causal, as described by the integration limits in eqs. (2.59) and (2.58), their frequency-domain components must satisfy the Kramers-Krönig relations (see Section 2.2.1 and Appendix 1). Therefore, for example, for $M(\omega)$, we have:

$$\begin{cases} \operatorname{Re} M(\omega) = \dfrac{2}{\pi} P \int_0^\infty d\omega' \dfrac{\operatorname{Im} M(\omega')}{\omega'^2 - \omega^2}, \\ \operatorname{Im} M(\omega) = -\dfrac{2\omega}{\pi} P \int_0^\infty d\omega' \dfrac{\operatorname{Re} M(\omega')}{\omega'^2 - \omega^2}, \end{cases} \tag{2.76}$$

where P denotes the Cauchy's principal value. Consequently, for the phase velocity and Q^{-1}, we can write:

$$\begin{cases} V^2(\omega) = \dfrac{-2}{\pi} P \int_0^\infty d\omega' \dfrac{V^2(\omega')Q^{-1}(\omega')}{\omega'^2 - \omega^2}, \\ Q^{-1}(\omega) = \dfrac{2\omega}{\pi V^2(\omega)} P \int_0^\infty d\omega' \dfrac{V^2(\omega')}{\omega'^2 - \omega^2}. \end{cases} \tag{2.77}$$

The second of these expressions allows deriving $Q^{-1}(\omega)$ from the complete spectrum of phase-velocity dispersion, although the inverse relation is not so straightforward in this form.

The importance of the Kramers-Krönig relations is in emphasizing the fact that a medium with attenuation must also exhibit velocity dispersion and *vice versa*. These relations are also often used for setting the phase-velocity spectra corresponding to known or desired attenuation spectra. However, because the relations only mean causality and the integrals converge slowly, in practical cases, these equations rarely produce highly definitive constraints.

2.5.6 Observational basis

Having outlined the visco-elastic theory, now let us consider what support for the above local-memory model is available from observations. Two types of such support have been advanced, one from studies of creep in rock samples, and another from *in situ* seismological Q observations.

Most of the experimental evidence for transient strain-stress relations such as (2.58) and (2.59) come from quasi-static rock-creep measurements (Figure 2.5), which only overlap with the lowest end of the seismic frequency spectrum. Creep experiments clearly indicate relaxation-type behavior of most materials. In such experiments, a step-function stress $\sigma = \sigma_0 \theta(t)$ is typically applied, and a near-instantaneous ("elastic") response $\varepsilon_e = \sigma_0/M_U$ is recorded, followed by a slow increase of deformation whose rate slows down with time. Describing such observations led to isolating the creep in the form of function $\phi(t)$ shown in eq. (2.57).

Summarizing his observations of creep in magmatic rocks under low stresses, Lomnitz (1956) proposed the following logarithmic law for the creep function:

$$\phi(t) = \begin{cases} 0 & \text{for } t < 0, \\ q \ln(1+at) & \text{for } t \geq 0, \end{cases} \qquad (2.78)$$

where parameter a was interpreted as some fundamental frequency. The inverse of this frequency, $t_E = 1/a$, can be viewed as the relaxation time during which the transition from the elastic- to creep-type deformation occurs. This parameter was poorly constrained from the data and selected so that $t \gg t_E$ within the observation time range. Its role is in reconciling the logarithmic creep $\phi(t) \approx q \ln t$ within the creep observation time range (from about 30 s to about 10 weeks) with the initial value $\phi(0) = 0$ at the beginning of the experiment. Different functional forms for $\phi(t)$ were proposed by Michelson (1917, 1920), Griggs (1939), Cottrell (1952), Jeffreys (1958), and others.

To determine whether the creep law (2.57) applies to the observations at seismological frequencies, some knowledge of the value of its relaxation time is critical. Lomnitz (1956) used $a = 1000$ Hz, but this parameter could also be as high as the vibration frequency of a vacancy in the crystal lattice, which could be 10^4–10^{10} Hz (Lomnitz, 1957). However, for both of these estimates, all seismological frequencies are extremely low, $f << a$, which means that all real-Earth observations belong to the transient creep regime.

However, once we had set out on a critical evaluation of the postulates of the visco-elastic theory, let us also pose some questions as to the empirical creep law above. First, should parameter a in eqs. (2.78) be definitely associated with rock rheology? Note that only parameter q is reliably determined from the data, but a arises as a regularization of the singularity in the logarithmic creep law $\phi(t) \propto \ln t$ for $t \to 0$. Such logarithmic law reflects the fundamental observation that the creep rate decreases inversely proportionally to the time elapsed from the stress onset: $\phi(t) \propto t^{-1}$. However, for $t < t_E$, we expect a linear dependence, $\phi(t) \propto t$, and the quasi-static creep dependence cannot be established earlier than when the apparatus equilibrates itself. Therefore, this regularization parameter a should lie within the range of the natural oscillation frequency of the *measurement apparatus*, which is likely to be about ~100–10,000 Hz. Thus, Lomnitz's fundamental frequency a should indeed be high compared to seismic frequencies, but it might not be related to the vibration frequencies or relaxation mechanisms operating at the microscopic level within an *in situ* rock.

Second, with $at >> 1$ and q in the ~10^{-4}–10^{-2} range, Lomnitz's (1956) strain-stress relation reduces to:

$$\varepsilon(t) \approx \frac{\sigma_0}{M_U}\left[(1+q\ln a)+q\ln t+\frac{q}{at}\right] \approx \varepsilon_0 + q_M\sigma_0\ln t, \qquad (2.79)$$

with at least ~0.01% accuracy, where $q_M = q/M_U$, and

$$\varepsilon_0 = \frac{\sigma_0}{M_U} + q_M\sigma_0\ln a\cdot \qquad (2.80)$$

This expression shows that the role of parameter a is to correct the assumed value of σ_0/M_U in order to match the observed intercept of the straight-line creep curve ($\ln t, \varepsilon$) coordinates. Experimental data only show this straight line (Lomnitz, 1956), from which q_M can be reliably measured. The transition to the elastic limit $\varepsilon(0) = \sigma_0/M_U$, in which M_U equals the "elastic" μ, seems to be not measured but *inferred*. If we allow the elastic value of M_U to be uncertain, a reasonable approach to its measurement could be to take $M_U = \sigma_0/\varepsilon_0$ in eq. (2.79). With this elastic limit, Lomnitz's law (2.79) would reduce to Grigg's (1939) logarithmic creep relation. Note that one does not need to worry about the formal divergence of eq. (2.79) in the limit of $t \to 0$, because physically, the creep regime becomes established only after some period of quick "elastic" equilibration of the rock sample.

M_U appears to be a difficult quantity to measure, and its equivalence to the elastic shear modulus (μ) measured in other experiments may be difficult to justify. In particular, M_U represents the infinite-frequency limit of M_U (Figure 2.3). According to the same model (2.78), M_U equals the ratio of the time derivatives $\dot{\sigma} / \dot{\varepsilon}$ during the initial, very fast ($t \ll t_E$) stage of the experiment, which is extremely difficult to record and analyze. Inferring values of M_U from those measured in quasi-static experiments should run into the same problem of measuring not the elastic stage but the beginning of creep. In addition, stress distributions in torsional experiments may depend on the shapes of the samples and their uniformity. The above uncertainty in the determination of M_U and its effect on subsequent interpretation looks very similar to the trade off of the frequency-dependent Q with the assumed geometrical spreading, which will be discussed in Chapter 3.

Third, the distinction between the "elastic" and "creep" deformation stages in creep experiments also pose an interesting question: during which of these two stages does the energy dissipation predominantly occur? To answer this, consider the schematic diagram of creep regimes in Figure 2.6 (Griggs, 1940). During periodic weak loading and unloading, an elementary volume within the medium passes through the elastic stage (labeled OA in Figure 2.6), transient creep (AP), elastic unloading (PQ), transient relaxation creep (PR), and finally returns to the undeformed state. The "elastic" deformations (highlighted by gray) occur much faster than creep, and because of high velocities, they are likely to cause the greatest energy dissipation. This argument (although somewhat speculative as is the creep-wave analogy itself) suggests that elastic-module relaxation caused by creep may not be the main reason for seismic wave attenuation. Viscous and grain-boundary friction, as well as other effects related to particle velocities, should be the most important in producing the elastic-wave energy dissipation.

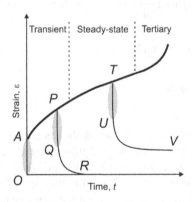

FIGURE 2.6

Schematic creep curve showing the transient (primary), steady-state (secondary), and tertiary creep regimes. Gray ellipses indicate the intervals of high velocities and interpreted energy dissipation.

Finally, with the surface- and body-wave conditions, there exists no experimental data directly verifying the strain-stress memory relations. Creep and relaxation at seismic frequencies are inferred only implicitly through the interpreted quality factor Q and the use of the visco-elastic models described above. Thus, if we question the applicability of the visco-elastic theory and the existence of an *in situ* Q (see Chapter 3), the idea of local strain-stress memory within the mantle also loses its support.

2.5.7 "Visco-elastic" form of Newtonian mechanics

Visco-elasticity only requires the linearity of equations of motion with respect to the field variables and their time derivatives, but does not imply any specific mechanical properties of the medium. To illustrate this point, let us look at a somewhat absurd re-formulation of Newton's second law, closely following the strain-stress model in Sections 2.5.1 and 2.5.2. As it turns out, virtually any particle motion fits into the visco-elastic formalism. While going through this example, note once again that such an effect is only possible because of the focus on *describing the solution* rather than *analyzing the construction of the dynamic equations* themselves.

Consider the Newton's equation of motion for a single particle:

$$\ddot{x}(t) = \frac{f(t)}{m}, \tag{2.81}$$

where x is its coordinate, m is the mass, and f is the force. This equation is linear in x, its time derivatives, and f. Consequently, its solution can be written as a superposition of responses to step-function forces applied at times t_0, $f(t) = f_0 \theta(t - t_0)$. For a single step function, such response is:

$$x(t) = \phi(t - t_0), \tag{2.82}$$

where $\phi(t)$ is the "creep function" analogous to the one in eq. (2.57),

$$\phi(t) = \theta(t) \frac{f_0}{2m} t^2. \tag{2.83}$$

In eq. (2.82), we dropped the homogenous part of the solution $x(t) = x_0 + V_0 t$, to make x similar to the elastic deformation ε. The step-function force $f_0 \theta(t - t_0)$ can be further expressed to a delta-function "force rate," $\dot{f}(t) = f_0 \delta(t - t_0)$; this gives:

$$x(t) = \int_{-\infty}^{t} J(t - \tau) \dot{f}(\tau) d\tau, \tag{2.84}$$

where $J(t)$ can be called the "retarded compliance function," similarly to the one in eq. (2.58),

$$J(t) = \frac{\theta(t)}{2m} t^2.$$

(2.85)

Conversely, the equation of motion (2.81) can be read in the opposite direction, as the force resulting from acceleration:

$$f(t) = m\ddot{x}(t).$$

(2.86)

This gives the force as a "retarded response to velocity,"

$$f(t) = \int_{-\infty}^{t} M(t-\tau)\dot{x}(\tau)d\tau,$$

(2.87)

where $M(t)$ is the "retarded modulus," as in eq. (2.59):

$$M(t) = m\frac{\delta(t)}{t},$$

(2.88)

because $\dot{\delta}(\tau) = \delta(\tau)/\tau$.

In the frequency domain, eq. (2.81) yields a harmonic solution:

$$x(\omega) = -\frac{f(\omega)}{m\omega^2},$$

(2.89)

and also $\dot{f}(\omega) = -i\omega f(\omega)$. Therefore, the frequency-domain "compliance" equals:

$$J(\omega) = \frac{x(\omega)}{\dot{f}(\omega)} = \frac{1}{im\omega^3}.$$

(2.90)

Similarly, the frequency-domain "complex modulus" is:

$$M(\omega) = \frac{f(\omega)}{\dot{x}(\omega)} = -im\omega.$$

(2.91)

If we now add a viscous force to eq. (2.81):

$$\ddot{x}(t) = \frac{f(t) - \zeta\dot{x}(t)}{m},$$

(2.92)

the frequency-domain solution (2.89) modifies to,

$$x(\omega) = -\frac{f(\omega)}{m\omega^2 + i\omega\zeta} \; ; \qquad (2.93)$$

$J(\omega)$ and $M(\omega)$ become,

$$J(\omega) = \frac{1}{\omega^2(-\zeta + im\omega)} \quad \text{and} \qquad (2.94)$$

$$M(\omega) = \zeta - im\omega \cdot \qquad (2.95)$$

Finally, from eq. (2.65), the frequency-dependent Q^{-1} can also be formally defined as:

$$Q^{-1}(\omega) = -\frac{\operatorname{Im} M(\omega)}{\operatorname{Re} M(\omega)} = \frac{m\omega}{\zeta} \cdot \qquad (2.96)$$

However, this result shows the "dissipation" as inversely proportional to ζ/m, which is opposite to what we might expect. This contradiction occurs because the system is in fact non-oscillatory.

In summary, a complete set of visco-elastic parameters can be derived for a single-body problem in mechanics, which nevertheless has no real relevance to the problem of attenuation in elastic media. As this example shows, the essence of the retarded-moduli description consists of exploiting the linearity of the relations between strain, stress, and all of their time derivatives. This linearity allows defining $M(\omega)$ and $J(\omega)$ as ratios of these quantities. Nevertheless, these ratios only represent attributes of the solutions to eq. (2.81) written in the form of convolutional integrals. They do not mean that Newton's second law includes memory properties or physically relates the force to velocity, as it might appear, for example, from eq. (2.87). Visco-elasticity is therefore not a true mechanism, but a *mathematical model* of it, similar to the dashpots and springs commonly used for its own illustration.

2.5.8 "Visco-elastic" linear oscillator

The key principles of the "visco-elastic" approach can be illustrated by using a damped mechanical oscillator example. As shown in Section 2.4.1, the complex-frequency view gives an adequate picture of the process, and the complex-valued moduli are not needed for describing the decaying oscillations.

However, following the visco-elastic practice, one can also define a complex-valued spring constant,

$$\tilde{k} = m\omega_0'^2 \approx k(1-i\xi).$$ (2.97)

This spring constant is analogous to the elastic modulus in continuum mechanics. Its phase rotation due to attenuation $\arg \tilde{k} = -\xi = Q^{-1}$ also equals the one attributed to the visco-elastic moduli, similar to an oscillator without energy dissipation. Due to proportionality $A \propto f$ in the frequency domain, they are related by convolution (2.59) in time domain.

Interestingly, the linear oscillator also represents a "standard linear solid." This can be verified by considering the following equation of motion in time domain:

$$m\ddot{x} = f(t) - f_{VE},$$ (2.98)

(now 1D for simplicity), where the "visco-elastic force" is:

$$f_{VE} = -m\omega_0^2 x - \xi m\omega_0 \dot{x}.$$ (2.99)

Now, consider this equation for a constant external force $f(t) = f_0$ at $t > 0$. Its solution with initial conditions $x(0) = 0$ and $\dot{x}(0) = 0$ reads:

$$x(t) = \frac{f_0}{m\omega_0^2}\left[1 + \left(\frac{\xi}{2}\sin\omega_0 t - \cos\omega_0 t\right)e^{-\frac{\xi}{2}t}\right].$$ (2.100)

Substituting this solution into eq. (2.99), we obtain, after some re-arrangement, the following identity equivalent to eq. (2.62):

$$f_{VE} + \tau_f \dot{f} = M_R(x + \tau_x \dot{x}),$$ (2.101)

where

$$\tau_f = \frac{1}{\xi\omega_0}, \quad \tau_x = \frac{1+\xi^2}{\xi\omega_0}, \text{ and } M_R = -m\omega_0^2.$$ (2.102)

These expressions give a consistent analog to an elastic-wave attenuation process and show how the complex-valued moduli and relaxation spectra arise. However, note that these complex moduli only exist due to the existence of an oscillatory solution with $\omega = \omega_0$. Equation (2.101) is only similar to (2.62) because it also results from a harmonic solution (2.100). In general, this derivation shows that standard-linear solid-type equations can likely be constructed for any process described by combinations of trigonometric and exponential functions in the time domain.

Along with the similarities to the oscillator, an important difference of the propagating-wave case is that it has no "natural" oscillation frequency arising from the wave equation. Without special relaxation mechanisms, the propagating medium is non-resonant, and all of its oscillations are of a purely forced nature. Energy dissipation occurs per unit volume or travel path and can be measured by attenuation coefficient α. In order to obtain a Q value, α has to be converted to Q by using eq. (1.9) and introducing a frequency or wavelength dependence similar to the one shown in eq. (1.13).

2.5.9 Relation to Lagrangian description

As mentioned above, no Lagrangian formulation exists for the visco-elastic approach, because the latter has no unique definition for the elastic energy density (*e.g.*, Carcione, 2007, p.52). This appears to be a fundamental deficiency of the visco-elastic method, particularly in looking for its roots in mechanics and testing its solutions for preserving the energy balances during attenuation processes.

Nevertheless, let us now look at the same problem from another angle and try constructing a true dynamic (Lagrangian) system which would exhibit a visco-elastic (standard linear solid-type) behavior. Consider, for simplicity, a 1-D dynamic system, such as the linear oscillator above. Its dynamics are described by its Lagrangian function, $L(q, \dot{q})$, where q is the generalized coordinate, and $\dot{q}$ is the generalized velocity. Coordinate q corresponds to ε, and generalized force $f = \partial L / \partial q$ – to σ in the standard-linear solid law (2.62). Therefore, we are now trying to generalize eq. (2.62) to:

$$\frac{\partial L}{\partial q} + \tau_f \frac{d}{dt}\left(\frac{\partial L}{\partial \dot{q}}\right) = -M_R\left(q + \tau_q \dot{q}\right), \qquad (2.103)$$

and find the corresponding constants M_R, t_f, and t_q. The negative sign in front of M_R is selected here because σ is actually assumed to act in the opposite direction relative to ε in eq. (2.62). The time derivative in the left-hand side of this equation equals $\dfrac{d}{dt}\left(\dfrac{\partial L}{\partial \dot{q}}\right) = \dfrac{\partial L}{\partial t} + \dfrac{\partial^2 L}{\partial \dot{q}^2}\dot{q} + \dfrac{\partial^2 L}{\partial q \partial \dot{q}}\ddot{q}$. To simplify our analysis, consider only stationary Lagrangians with the generalized momentum functionally independent of the coordinate, *i.e.* those having $\frac{\partial L}{\partial t} = 0$ and $\frac{\partial^2 L}{\partial q \partial \dot{q}} = 0$. Equation (2.103) then becomes:

$$\frac{\partial L}{\partial q} + \tau_f \frac{\partial^2 L}{\partial q^2} \dot{q} = -M_R \left(q + \tau_q \dot{q} \right). \tag{2.104}$$

Note that the above equation only contains the derivative of L with respect to the generalized coordinate, and therefore it only involves the potential (elastic) energy. If we want this equation to be satisfied as a constitutive law, *i.e.*, for *functionally independent* q and $\dot{q}$, then we need to require that:

$$\frac{\partial L}{\partial q} = -M_R q, \text{ and } \frac{\tau_q}{\tau_f} = -\frac{1}{M_R} \frac{\partial^2 L}{\partial q^2}, \tag{2.105}$$

which means that:

$$L = E_{kin}\left(\dot{q}\right) - \frac{1}{2} M_R q^2 \text{ and } \tau_q = \tau_f, \tag{2.106}$$

where $E_{kin}\left(\dot{q}\right)$ is an arbitrary kinetic-energy function.

Two interesting conclusions follow from the above analysis:

1) The standard-linear solid law can be derived solely from the quadratic form of the potential-energy function.

2) If the strain-stress relation (2.103) is viewed as a dynamic principle (*i.e.*, with functionally independent q and $\dot{q}$), then the relaxation times *must be equal*, $t_q = t_f$. This means that relaxation is actually absent from this model.

Thus, the Lagrangian model appears to disallow the standard linear solid-type strain-stress relation. This result can be understood by noting that we only considered a single generalized coordinate q, which was supposed to be equivalent to the bulk strain in some creep experiment. However, creep apparently occurs because of the existence of additional, internal variables that can slowly accommodate when stress is applied or removed. Such variables correspond, for example, to the orientation of mineral grains or dislocations in crystalline lattices, or fluid distributions within pores (for reviews of creep mechanisms, see Evans and Dresden, 1991; Evans and Kohlstedt, 1995). Figure 2.7 shows a possible schematic model of such creep mechanism. Under constant axial (or shear) stress,

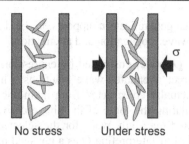

No stress   Under stress

FIGURE 2.7
Conceptual creep model. If the rock contains anisotropic elements such as
elongated mineral grains (ellipses), their orientation may gradually change
under stress, causing deformation increasing with time.

Brownian motion would cause preferential re-orientation of the grains, which
should occur in the direction of decreasing potential energy (more precisely,
enthalpy) and increasing strain. Detailed discussions of such models are beyond
the scope of this book. However, three points should be noted for our discussion:

1)   Such systems are well within the scope of Lagrangian continuum
     mechanics and homogenization theory. A simple example of such a
     model is given in Appendix A2.

2)   Although based on thermodynamics and likely causing some dissipation
     of mechanical energy into heat, such re-orientation (and therefore creep)
     is not the only cause of elastic-wave energy dissipation as it may appear
     from the visco-elastic constitutive law. Relative movements of grain
     boundaries (Figure 2.7) could be a more significant factor for energy
     dissipation.

3)   The difference between the present model and the visco-elastic retarded-
     strain model is in that the creep is explained by its present-time state and
     non-linearity, rather than by a linear retardation of the response in time.
     The system shown in Figure 2.7 has no "memory," although
     thermodynamic relaxation of grain orientations should lead to the
     phenomenological Boltzmann's eqs. (2.58) and (2.59).

2.5.10 Critical summary

In this section, we examined the visco-elastic constitutive equations trying
to clarify their relations with the fundamental physical principles. Our overall
conclusion is that the visco-elastic model represents an generalization of two
empirical observations (creep in quasi-static observations and Q within the
seismological frequency band) of the *in situ* constitutive laws for the Earth's

material. Moreover, this generalization appears to be lacking support in the fundamental physics of wave propagation and attenuation.

The visco-elastic Q model postulated by expression (2.65) "works" in the sense of producing correct attenuated-wave solutions. This could be expected, because this model is actually constructed from harmonic waves decaying with time. However, recapitulating the logic of the above derivation, we can see that this model is heuristic and may only work for the selected cases for which it is designed. The justification of interpreting Q as a physical property of the medium is *entirely* based on considering the following simple problem:

(i)   Assume a uniform isotropic medium with complex modulus $M(\omega)$;

(ii) Solve for complex-valued plane- or spherical-wave phase velocity, $\tilde{V}(\omega) = \sqrt{M(\omega)/\rho}$, where ρ is the density. Its complex argument consequently equals $\arg \tilde{V}(\omega) \approx -Q^{-1}(\omega)/2$.

(iii) From the dispersion relation, $\tilde{k}(\omega) = \omega/\tilde{V}(\omega)$, the corresponding complex argument of k becomes $\arg \tilde{k}(\omega) \approx Q^{-1}(\omega)/2$, and therefore the spatial attenuation coefficient is $\alpha(\omega) = \operatorname{Im}\tilde{k}(\omega) = \dfrac{\pi}{\lambda Q^{-1}(\omega)}$, as expected (compare to eqs. 1.5 and 1.9).

Such *a posteriori* justification of a theoretical quantity (here, the complex modulus $M(\omega)$) by referring to the resulting plane-wave solution still does not mean that this property is meaningful in the more general cases. In addition, points (i) to (iii) above have several general problems:

1) The definition of $M(\omega)$ (2.65) is designed *for only two moduli* corresponding to the known fundamental-wave solutions. Thus, for longitudinal waves, M_U is taken to equal $\lambda + 2\mu$, and for shear waves, $M_U = \mu$ (*e.g.*, Dahlen and Tromp, 1998; p.195). Such separation is only appropriate in a *uniform medium*, in which P and S waves are independent. In a heterogeneous medium, these wave modes are not separable and interact with each other at every point (such as in a Rayleigh wave). Therefore retaining the use of the corresponding moduli at every point represents only a hypothesis.

3) On the contrary, the two most fundamental moduli, λ and μ (which constitute the elastic energy in the Lagrangian; see eq. 2.26), are not properly honored in visco-elastic formulation.

4) Derivation (i) – (iii) above is also valid only for *uniform media*. In a heterogeneous medium, both expressions for expressions for $V(\omega)$ and $k(\omega)$ become incorrect. For example, in a surface wave, there is a single

phase velocity, but $\sqrt{M(\omega)/\rho}$ is variable and may not equal $V(\omega)$ at any point.

The ability to fit a simple exponential amplitude decay law exp(-αx) for a plane or other simple wave is still an insufficient argument for assigning complex values to such fundamental physical quantities as the elastic moduli. This approach may be successful in only a narrow range of models, such as the uniform-media cases and phenomenological quasi-static creep. For example, as shown in Section 6.1.2, predictions of Love wave attenuation based on expression (2.65) violate the energy balance and result in over-estimated levels of Q^{-1}. This model also predicts wrong polarities of reflections from attenuation contrasts, as shown in the next section. On the other hand, any other energy dissipation model outlined in Section 2.4.2 could surely achieve the same amplitude decay with x. As also shown in Section 2.4.2, the elastic moduli should remain real, and dissipation treated by the specific, corresponding mechanisms.

In summary, the generalized visco-elastic model (2.59) satisfactorily describes a variety of field problems, yet its significance in relation to the fundamental medium properties should not be overstated. Visco-elastic models can be viewed as working mathematical tools rather than an end in themselves. In different experimental environments (wave propagation, forced harmonic oscillations, or free vibrations), different values of Q arise from relaxation models, and they cannot be reduced to each other (for a detailed overview of this point, see Chapter 3 in Bourbié et al., 1987). For example, Biot's (1956) theory for porous saturated rock, or scattering theory clearly lead to the spatial attenuation factors α (eq. 1.5). Although these factors can be converted into Q by using eq. (1.9), such Q's are different from those determined in the experiments with resonant bars. We will continue this discussion in Chapter 3.

2.6 Equipartitioning of energy

For surface waves and normal modes, the dispersion equation relating the oscillation frequencies to the corresponding wavenumbers or normal-mode numbers can be established from the energy equipartition relation. This states that for any traveling wave, the potential energy, E_p, equals its kinetic energy, E_k (Aki and Richards, 2002, p.284). For a stationary standing wave, such as the normal-mode oscillation, energy has a different spatial distribution, but the time-averaged elastic energy also equals the average kinetic energy. In the latter case, the equipartition relation can be simply related to the conservation of energy.

The energy equipartition relation arises from applying the Hamilton principle, or the corresponding equations of motion, to harmonic oscillations. When harmonic dependences on time, $e^{-\omega t}$, are only considered, all time derivatives reduce to multiplications by $-i\omega$, and the equations of motion become reduced to an eigenvalue problem $E_k = E_p$, where E_k is usually proportional to ω^2. For normal modes, this eigenvalue problem is formulated as the Rayleigh's principle and becomes the key tool for finding both the normal-mode frequencies and attenuations (see Dahlen and Tromp, 1998; p.109–117). Its application for finding the dispersion relation can generally be reduced to three steps: 1) assume a harmonic wavefield with frequency ω; 2) determine the corresponding spatial distributions of amplitudes; and 3) use relation $E_k = E_p$ to establish the value of ω for the given mode.

2.6.1 Application to *SH* waves

For a simple example, consider an *SH* surface-wave field of form in a layered, isotropic, and lossy medium in Cartesian coordinates. Assuming that model parameters depend only on the depth, z, and the field is harmonic in time, the most general form of displacement within the wave propagating in the direction x is given by:

$$u_y(x,z,t) = l_1(z)\psi(x,t),$$

(2.107)

where $\psi(x,t) = e^{-i\omega t + ikx}$, $l_1(z)$ is the amplitude of the mode of interest, and k is the wavenumber. Note that similarly to the time dependence, $\psi(x,t)$ is also taken as a harmonic function of x, because the problem is translationally invariant in this direction. Our problem now consists in finding the relation of ω to k.

From the Lagrangian density (2.26), the time-averaged kinetic energy density is:

$$\langle E_k \rangle = \left\langle \frac{1}{2}\rho\dot{u}_i\dot{u}_i^* \right\rangle = \frac{uu^*}{4}\rho\omega^2 l_1^2,$$

(2.108)

where complex conjugation (denoted by the asterisk) is used to account for the complex-valued wavefield amplitudes in eq. (2.107). The corresponding average elastic energy density is:

$$\langle E_p \rangle = \left\langle \frac{1}{2}\lambda\varepsilon_{kk}\varepsilon_{nn}^* + \mu\varepsilon_{ij}\varepsilon_{ij}^* \right\rangle = \frac{uu^*}{4}\left[\mu kk^* l_1^2 + \left(\frac{dl_1}{dz}\right)^2 \right],$$

(2.109)

and the total energy:

$$\tilde{E} = \int_0^\infty \langle E_k + E_{el} \rangle \, dz = \frac{1}{2}\left(\omega^2 I_1 + kk^* I_2 + I_3 \right),$$

(2.110)

where the energy integrals are:

$$I_1 = \frac{uu^*}{2} \int_0^\infty \rho l_1^2 dz \, , \; I_2 = \frac{uu^*}{2} \int_0^\infty \mu l_1^2 dz \, , \text{ and } I_3 = \frac{uu^*}{2} \int_0^\infty \mu \left(\frac{dl_1}{dz}\right)^2 dz \, . \quad (2.111)$$

By equating the kinetic and potential energies, the equipartition relation is obtained:

$$\omega^2 I_1 = kk^* I_2 + I_3 \, . \quad (2.112)$$

In the absence of attenuation, k in this equation is real and can be determined for any given ω. With attenuation, this equation still remains valid, but an additional principle is needed for resolving the resulting modifications in both $\text{Im} k$ and ω. Two approaches to this problem are discussed in Section 6.1.2.

2.6.2 Application to normal modes

When using the normal-modes picture for whole-Earth oscillations, the equipartition principle is similar, but wavenumbers, k, are replaced with spherical-harmonic numbers and the spectrum of ω becomes discrete. For a given mode, eq. (2.112) becomes (see Dahlen and Tromp, 1998; p.116):

$$\omega^2 T = U = U_K + U_\mu + U_g + U_\psi \, , \quad (2.113)$$

where T can be viewed as the total kinetic energy in an oscillation of unit amplitude, U is the corresponding total potential energy, and U_K, U_μ, U_g, U_ψ are its parts related to the effects of dilatational and shear deformations, gravity, and centrifugal acceleration, respectively. In the presence of attenuation, and unlike in (2.112), it is convenient to view ω as complex valued, with $\chi = \text{Im}\,\omega$ being our temporal attenuation coefficient.

However, allowing complex-valued ω in eq. (2.112) brings up an important problem, namely that T or U become complex valued as well. In visco-elasticity, this requirement is put into the foundation of the theory by making the bulk and shear moduli of the medium, and consequently U_K and U_μ, complex. However, as already explained, we regard this as a violation of the most fundamental principle of mechanics, which requires that the energy functionals must be real-valued. Looking closely into its derivation (Dahlen and Tromp, 1998, p.109–117), one can see that the equipartition equation (2.113) originates from the total Lagrangian of the system of the following form:

$$L(u, \dot{u}) = T\dot{u}^2 - Uu^2 \, , \quad (2.114)$$

where u is the normal-mode displacement. This Lagrangian does not contain attenuation, which needs to be added separately, by using some specific dissipation function as described in Section 2.4.

Using complex U instead of the dissipation functions takes us out of the range of the Hamilton variational principle, on which the entire dynamical model is based. Therefore, similarly to the surface-wave case, we need to look for other ways for introducing attenuation in eq. (2.113). This problem appears to be still unsolved, which is discussed in Chapter 8.

2.6.3 Relation to visco-elasticity

As shown above, the equipartition relation contains practically everything needed for finding the surface-wave and normal-mode frequency spectra. When extended to complex parameters $\alpha = \operatorname{Im} k > 0$ (for surface waves) or $(-\chi) = \operatorname{Im} \omega < 0$ (for normal modes), this relation also produces the attenuation coefficients. This shows that the equipartition relation is inherently close to visco-elasticity.

The intimate connection of the equipartition relation to visco-elasticity is through its focus on harmonic-wave solutions. When harmonic solutions $u \propto e^{-\omega t}$ are considered without looking at the original Lagrangian function in the time domain, the time derivatives $\dot{u}$ disappear, and the kinetic and potential energies become functionally indistinguishable. Nevertheless, because it follows from the Hamilton principle, the equipartition equation correctly predicts the $\omega(k)$ relation *for real ω and k.* Shifting the ω or k into the complex planes allows extending this relation and producing formally decaying solutions. However, in these solutions, the kinetic ($E_k = T\dot{u}^2 \propto \omega^2 T u^2$) and potential ($E_p \propto k^2 u^2$, because of the spatial derivatives in U) energies are intermixed and complex valued. Both E_k and E_p in fact contain contributions from a dissipation function which is omitted from the formulation.

Thus, similarly to the rest of the visco-elastic paradigm, the extension of the equipartition relation to the cases with attenuation is heuristic and phenomenological. Equipartition allows constructing solutions which decay in time and/or space, and in which the complex E_k and E_p are equal. However, this relation bypasses the discussion of the physical causes of attenuation, and attributes the effects of the unspecified dissipation function to E_k and E_p.

2.7 Anelastic acoustic impedance

An interesting test for the visco-elastic model arises from generalizing the concept of the acoustic impedance to attenuative media. In elasto-dynamics, the impedance (often denoted Z) is constructed so that the reflection coefficient r_{12} from a boundary of two media is solely determined by the ratio of their impedances Z_1 and Z_2:

$$r_{12} = \frac{\dfrac{Z_2}{Z_1} - 1}{\dfrac{Z_2}{Z_1} + 1}. \qquad (2.115)$$

For P waves without attenuation at normal incidence, the acoustic impedance equals the product of density (ρ) and wave velocity (V) of the medium:

$$Z = \rho V. \qquad (2.116)$$

Reflectivity is among the principal causes of attenuation, and therefore it is not surprising that Z should be related to attenuation. By solving the wave boundary conditions on a welded contact of two attenuative media, Lines *et al.* (2008) noted that attenuation contrasts cause phase-rotated reflections. Morozov (2009e) pointed out that such reflectivity can be expressed by incorporating the conventional Q factor into the acoustic impedance:

$$Z = \rho V \left(1 + \frac{i}{2Q} \right), \qquad (2.117)$$

and using formula (2.115). A problem arises, however, when comparing this expression to the correspondence principle (2.75); because of the negative sign of $\mathrm{Im}\,V$, the correspondence principle gives an opposite sign of $\mathrm{Im}\,Z$ compared to the exact expression (2.117). Thus, the correspondence principle should be used with caution in this type of problem.

To derive the correct expression (2.117), consider a plane harmonic P wave normally incident on a welded contact of two media (Figure 2.8a). In the presence of attenuation, its scalar potential is:

$$\varphi(\mathbf{x},t) = A \exp(-i\omega t + i\mathbf{k}\mathbf{r} - \alpha\mathbf{r}) \equiv A \exp(-i\omega t + i\tilde{\mathbf{k}}\mathbf{r}), \qquad (2.118)$$

where ω is the angular frequency, $\mathbf{k}$ is the wavenumber vector, and α is the attenuation vector. Complex-valued vector $\tilde{\mathbf{k}} = \mathbf{k} + i\alpha$ is the effective wavenumber including both the propagation and spatial attenuation effects. For simplicity, consider a homogeneous wave with parallel $\mathbf{k}$ and α. The spatial attenuation coefficient α can then be related to $\mathbf{k}$ through the spatial quality factor Q:

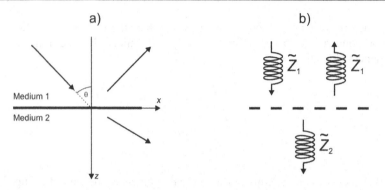

FIGURE 2.8.
Definition of anelastic acoustic impedance. a) Model geometry
and notation and b) analogy to the equilibrium of elastic springs.

$$\alpha = \frac{\mathbf{k}}{2Q}. \qquad (2.119)$$

From the expression for *P*-wave potential (2.118) in each of the two media,
we obtain:

the displacement,

$$u_i(\mathbf{r},t) = \partial_i \varphi(\mathbf{r},t) = i\tilde{k}_i \varphi(\mathbf{r},t); \qquad (2.120)$$

velocity,

$$\dot{u}_i(\mathbf{r},t) = \omega\tilde{k}_i \varphi(\mathbf{r},t); \qquad (2.121)$$

strain,

$$\varepsilon_{ij}(\mathbf{r},t) = \partial_i u_j(\mathbf{r},t) = -\tilde{k}_i\tilde{k}_j \varphi(\mathbf{r},t); \qquad (2.122)$$

and stress,

$$\sigma_{ij}(\mathbf{r},t) = -\left(\lambda\delta_{ij}\tilde{k}_n\tilde{k}_n + 2\mu\tilde{k}_i\tilde{k}_j\right)\varphi(\mathbf{r},t); \qquad (2.123)$$

where λ and μ are the Lamé constants, and δ_{ij} is the Kronecker symbol (identity-
matrix element). Now let us only consider the normal-incidence case. From eqs.

(2.121) and (2.123), we can relate the traction and velocity boundary conditions by a single factor Z:

$$\sigma_{zz}(\mathbf{r},t) = -Z\dot{u}_z(\mathbf{r},t),$$ (2.124)

where Z is the acoustic impedance (see Appendix A3),

$$Z = \frac{(\lambda+2\mu)\tilde{k}}{\omega} = \rho V\left(1+\frac{i}{2Q}\right),$$ (2.125)

as in eq. (2.117).

Equation (2.125) shows that Z depends on Q similarly to the wavenumber, and not as the complex-valued phase velocity (2.75); therefore, $\mathrm{Im}Z \geq 0$. In the presence of a contrast in Q, r_{12} becomes complex valued and leads to phase-shifted reflections. For example, for small contrasts in ρV and Q^{-1}, eq. (2.115) gives:

$$r_{12} \approx \frac{\delta\ln Z}{2} \approx \frac{\delta\ln(\rho V)}{2} + i\frac{\delta(Q^{-1})}{4},$$ (2.126)

showing that reflections from positive attenuation contrasts are delayed in phase by $\delta\phi = \arg(r_{12}) \approx \arctan[\delta(Q^{-1})/\delta\ln(\rho V)/2]$.

Note that, although the reflections in some cases appear advanced in time (Figure 2.9), their causality is disturbed by neither positive nor negative phase shifts in eqs. (2.125) and (2.126). Because these phase shifts are frequency independent, the corresponding group velocity delay $d(\delta\phi)/d\omega = 0$. Nevertheless, correct signs of phase delays are important for determining the shapes of wavelets reflected from the boundary. In particular, reflections from pure attenuation contrasts (Figure 2.9d) change polarities after switching from the explicit to the correspondence-principle formulation.

2.7.1 Boundary conditions

The reason for the impedance discrepancy resulting from the correspondence principle is that in reality, $V_{\mathrm{anelastic}}$ in eq. (2.75) represents the phase velocity and not the material wave speed V_m, which is merely a combination of elastic parameters[5]. As mentioned above, the material and phase velocities equal each other only in a uniform, isotropic, boundless, and attenuation-free medium. Phase velocity is related to the complex wavenumber as $\tilde{V} = \omega/\tilde{k}$, and

---

[5] Such as $V_m = \sqrt{(\lambda+2\mu)/\rho}$ for P waves.

consequently its imaginary part is negative when $\mathrm{Im}\,\tilde{k} > 0$. However, in heterogeneous structures, a single phase velocity corresponds to a distribution of V_m, and analytical relations between them hardly exist in the general case. Assumptions of such analytical relations may produce elegant, but erroneous solutions.

In wave mechanics, the description of the displacement field consists of two parts: 1) the wave equations, or the corresponding Lagrangian and 2) the boundary or radiation conditions. For a harmonic wave with known Q, the wave equations are correctly transformed by the correspondence principle (Section 2.5.4). However, when applied to the boundary conditions (see eq. (A3.4) in Appendix A3), this transformation makes λ and/or μ in formula (2.125) complex-

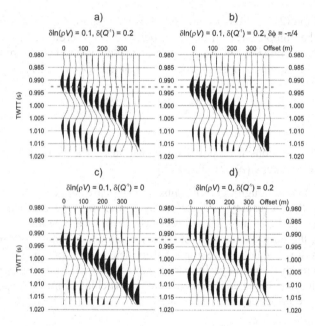

FIGURE 2.9.
Numerical models of a P-wave reflection from 1-km depths for a range of source-receiver offsets by using the *reflectivity* method (Fuchs and Müller, 1971): a) with positive velocity/density and attenuation contrasts; b) same, with an additional -45° phase rotation to compensate the effect of $\delta(Q^{-1})$; c) with no attenuation contrast; and d) with a pure attenuation contrast. Overburden velocities are $V_P = 2$ km/s and $V_S = 1$ km/s, and density $\rho = 2$ g/cm³. Velocity and attenuation contrasts across the boundary are indicated in the labels. Dashed gray line indicates the position of the near-offset peak in plot c).

valued. The stress,

$$\sigma_{ij} = \lambda \varepsilon_{kk} \delta_{ij} + 2\mu \varepsilon_{ij}, \tag{2.127}$$

acquires an additional phase delay $[-2\tan^{-1}(Q^{-1}/2)]$ relative to the elastic case, as can be seen, for example, from relation $\lambda + 2\mu = \rho V^2(Q)$ for P waves. This delay exceeds the attenuation-related $\tan^{-1}(Q^{-1}/2)$ phase advance of ε_{ij} with respect to u_i, reverses the phase of AI in eq. (2.117), and leads to incorrect phases of reflections from attenuation contrasts. The assumption "… it is understood that the boundary conditions for the two problems are identical …" (Bland, 1960, p.96) appears to be unviable with complex-valued elastic moduli.

The sensitivity of the boundary conditions to the correspondence-principle transformation does not just affect the reflection/transmission and AI problems considered above. In modeling the surface waves in a layered Earth, this transformation, applied to the layers with contrasting Q, shifts the phases of the boundary-value equations related to stress. Consequently, the eigenfunctions of the wavefield similarly modify themselves to the reflection coefficients in the acoustic impedance problem (Section 2.7). The resulting distributions of the kinetic and potential energies also change, leading to predictions of frequency-dependent Q^{-1} different from those given by Anderson and Archambeau (1964) (Section 6.1.2). Thus, it appears that the correspondence principle may only reliably apply to wave-equation solutions in boundless uniform media. Unfortunately, such cases are of little practical use in seismology.

The issue of boundary condition is significantly more difficult, but it can generally be understood as follows. For a contact of two media, the differences in their elastic energies plus the presence of the contact itself lead to some kind of "surface tension" on the boundary, which can be described by the corresponding kinetic and potential energies. Similarly, along with the volume dissipation functions of type (2.44), dissipation functions should also be attributed to the boundaries. Such functions would describe, for example, fluid filtration across the contact of two materials (Bourbié et al., 1987). Therefore, boundaries should also possess some specific attenuation properties. Two distinct approaches to setting these properties were considered:

1) In the visco-elastic approach, the boundary is also "visco-elastic," because the equations of boundary conditions are formulated in terms of the same two complex moduli, which are made responsible for attenuation.

2) In the approach by Lines et al. (2008) and discussed here, boundary conditions are assumed to be "elastic," as well as the Lamé modules. The dissipation attributed to the contact is set equal to zero. In the absence of information about the specific properties of the contact, this appears to be a simple and reasonable approach.

Thus, as above, the visco-elastic approach offers a compact and internally consistent approach to boundary conditions, yet this approach appears too special and unlikely to suit the conditions in real media. The second, our approach, may also be insufficiently specific at present, but at least it leaves room for developing a more complete model in the future, when the contact energies between mantle interfaces become better understood.

2.8 Frequency dependence of attenuation coefficient

All results presented in this book can be explained without invoking frequency-dependent attenuation mechanisms. Because the attenuation coefficients are not guaranteed to tend to zero as $f \rightarrow 0$, the leading attenuation effect should come from the frequency-independent parts of α and χ. If this effect is disregarded (which is unfortunately often done), $Q(f)$ becomes unstable and strongly variable at low frequencies.

In addition to being non-zero, frequency-independent $\alpha|_{f \rightarrow 0}$ and $\chi|_{f \rightarrow 0}$ are also *positive* in many real-data cases. The physical causes of such positive frequency-independent attenuation coefficients seem to be quite fundamental and related to the real observation environments being more complex than assumed in the models underlying the measurements. For traveling waves, such an increased complexity leads to reflectivity and ray bending leading to positive and frequency-independent contributions to α and χ. For standing waves and free vibrations (such as in laboratory attenuation measurements), unaccounted-for scattering within the sample and within the measurement apparatus should also lead to (usually) positive energy dissipation.

The three key contributions to the attenuation coefficients that can be expected in the observations are schematically shown in Figure 2.10:

$$\chi = \gamma + \kappa f + \chi_{err} . \tag{2.128}$$

The first two terms in this expression can be viewed as the Maclaurin series for $\chi(f)$. The frequency-independent contribution, denoted γ, is often the most significant part, which is positive for most lithospheric datasets. On top of this constant level, a linear (usually also positive) $\chi(f)$ trend can be recognized. From the slope of this linear trend, a frequency-independent "effective" $Q_e = \pi/\kappa$ can be derived for comparisons to the conventional terminology. Resonances and other frequency-dependent effects cause oscillatory contributions χ_{err} in the vicinity of the general linear trend (Figure 2.10). However, when presented in the frequency-dependent Q form, these contributions become overwhelmed by the strong $Q \propto f$ trends and are difficult to discern.

As will be shown on several theoretical and data examples below (Section 3.9.1 and Chapters 5 and 6), $\chi(f)$ dependences shown in Figure 2.10 are

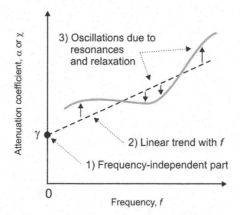

FIGURE 2.10

Three elements of the general dependence of the attenuation coefficients
on frequency expected in most observations.

encountered in most cases. Generally, the "static" term γ can be interpreted as
related to the residual geometrical spreading (Chapter 4). The linear term κf is
empirical. Note that models of saturated rocks (Section 2.4.2) suggest
dependences proportional to f^2, which are similar to (2.128) in its most important
aspect, which is the increase from a non-zero level at $f = 0$. In addition, parameter
κ in eq. (2.128) may vary with frequency; for example, Biot's case in Section
2.4.2 corresponds to $\kappa \propto f$. Our most important point here is that dependences of
the general type (2.128) should be considered *first* when analyzing the data
(Figure 2.10), without imposing restrictive constraints $\chi|_{f\to0} = 0$, which are implied
when the traditional Q-based interpretation is used.

Chapter **3**

The *Q* Paradigm

If ever asked: What is more useful, the sun or the moon, respond: The new moon. For the sun only shines during daytime, when it's light anyway, whereas the moon shines at night.

Kozma Prutkov, Fruits of Reflection (1853-54)

The seismological quality factor is a remarkable quantity. Although it is not directly measured and may not represent a local property of the medium, it is nevertheless the basis of almost all theoretical models of seismic attenuation. Its measured values are ambiguous and depend on a variety of theoretical assumptions, yet it represents practically the only way attenuation observations are presented. Its name suggests analogies with vibrating systems, and yet it is broadly used to describe non-oscillatory, transient processes, such as creep, pulse broadening, and scattering.

Starting to analyze this paradox in Chapter 1, we offered a general argument showing that Q is unlikely to be a true medium property common to many waves. In particular, the way Q is defined implies a built-in frequency dependence of $Q \propto f$, which can be overcome only in ideal cases of the geometrical spreading being absent and scattering being linear in frequency. Here, this analysis is continued by looking into the foundations of the Q paradigm. We review the theoretical arguments used to infer this parameter, and the measurement techniques commonly used for its estimation from various data types.

When discussing such a subtle subject as Q, it is important to clearly differentiate between its empirical value, such as its ability to explain amplitude-decay measurements, and its viability as a true material property. Successful phenomenological theories are abundant in the history of science — with pre-Copernicus geocentric cosmology among the best-known examples. Such theories were based on solid observational evidence and fitted those data well until it was discovered that they were based on incorrect astronomical principles. In some applied areas, such as in nuclear-test monitoring, only phenomenological data predictions may be important[6], and therefore even if inaccurate or incorrect, the Q-based empirical models may work just as well. For geocentric cosmology, medieval navigation is an example of such a successful applied field. In this sense, Q is perfectly valid as an empirical wave property (although better descriptions arise from the attenuation-coefficient model). If viewed as a true physical parameter of the Earth, the concept of Q deserves a fundamental criticism.

Continuing the above cosmological analogy, note that only some qualitatively new evidence, such as provided by the telescope, was able to disprove the erroneous phenomenology. Unfortunately, we still have no such dramatic evidence for truly vetting the theory of Q, and are therefore left to scrutinizing its foundations and checking how well they fit with the established principles of theoretical physics. As we will see in this chapter, the concept of "medium Q" is based on several mathematical conjectures, analogies, and assumptions. Despite their elegance, long history, and tremendous impact on attenuation studies, these assumptions nevertheless appear to fail.

The subject of attenuation of elastic processes in continuous media is very large and cannot be covered in the limited space of this book. Only several aspects of this subject are therefore discussed below:

(i) The significance and validity of the definitions of Q used in several different contexts, such as: interpretation, modeling, and visco-elastic theory. In particular, we will be interested in the impact of these definitions on the different types of attenuation measurements.

(ii) The key theoretical assumptions leading to accepting Q and visco-elastic models of Earth materials.

(iii) The general character of Q-based observations. Attenuation measurements differ from many other physical observations by their dependence on theoretical models. These "background" models are generally known to be inaccurate, yet it is rarely realized that this inaccuracy may jeopardize the very validity of the measurements. Q measurements also stand out by their reliance on a certain conventional routine, such as selecting the fiducial frequency $f_0 = 1$ Hz (eq. 1.13) for comparing measurement results in different

---

[6] The importance of full "physical" models is also well understood in nuclear-test monitoring, and such models are being developed.

areas, or using certain scattering- and geometrical-spreading models in coda measurements.

(iv) General causes of the theoretically-predicted and observed frequency dependences of Q and their roles in substantiating the Q concept.

The "scattering Q" represents a strong generalization of the quality-factor idea which is perhaps the most remote from its original use in mechanics. Even intuitively, it seems clear that scattering has little to do with the widths of spectral peaks near resonances, from which the notion of Q has emerged. Scattering Q also highlights most problems of Q, such as its difficulty in relating to the basic mechanics, sensitivity to the assumed background structure, trade-off with geometrical spreading, and the strongest frequency dependence[7]. For these reasons, we devote a special section to this quantity and also discuss its various aspects in relation to the specific subjects.

The main conclusion of this chapter is that the "wave Q" is a valid and useful phenomenological property closely related to wave observations. Such Q represents an "apparent," *i.e.*, observed quantity. By contrast, the *in situ*, or "medium Q" is only inferred theoretically by extrapolating such apparent attributes of simple wave solutions to the medium. Instead of relying on the mechanical properties of the propagating medium, this inference may often be based on unjustified or inaccurate assumptions. Consequently, the use of "medium Q" should generally be discouraged. These problems of methodological consistency become most apparent for the frequency-dependent Q, and also for the scattering Q.

3.1 Quality factor in mechanics

Before considering the seismological Q, let us briefly review the definition of the quality factor used in physics and engineering. The most important point here is that the quality, or Q factor is only defined for mechanical, electrical, and optical systems oscillating in a steady state near resonance. The quality factor is a dimensionless parameter describing how under-damped an oscillator or resonator is. Free oscillators with high quality factors ring longer, because they have low damping. Resonators with $Q < \frac{1}{2}$, $Q = \frac{1}{2}$, and $Q > \frac{1}{2}$ and are called over-damped, critically damped, or under-damped, respectively, depending on whether they exhibit an oscillatory behavior with time or not. Dampers keeping doors from slamming shut are examples of critically damped systems with $Q = \frac{1}{2}$, and atomic clocks reach values of $Q > 10^{11}$.

In spectral terms, Q equivalently characterizes a resonator's bandwidth relative to its center frequency f_0 (see Figure 2.2 on p. 30). For a linear oscillator

---

[7] Note that the frequency dependence of scattering $Q_s(f)$ often exceeds even the $\eta = 1$ limit set by the visco-elastic mechanics.

discussed in Section 2.4.1, quality factor Q is a measure of the half-width $\Delta f_{1/2}$ of the peak, measured at the level of a half of its power, which can be shown to equal ξ^{-1}:

$$Q = \frac{f_0}{\Delta f_{1/2}} = \frac{1}{\xi} = \frac{m\omega_0}{\zeta} \, . \tag{3.1}$$

For oscillations at frequency f_0 (but not at other frequencies), energy dissipation during one period δE is proportional to the peak energy, E_{max} (see Figure 2.2), and parameter Q^{-1}, describing the dissipation rate can be defined,

$$Q^{-1} = \left. \frac{\delta E}{2\pi E_{max}} \right|_{f=f_0} . \tag{3.2}$$

Thus, for mechanical and other resonant systems, the quality factor is merely a constant combining of the parameters of the oscillator. Consequently, there is no question of its frequency dependence. This factor increases with decreasing viscosity and increasing mass, and also with increasing spring constant, $k = m\omega_0^2$, of the oscillator. This factor is a property of the entire oscillator, and linking it solely to k would be neither necessary nor well justified from first principles.

3.2 Definitions of Q in seismology

Compared to mechanics, the Q factor in seismology is defined differently and is usually introduced as follows (e.g., Aki and Richards, 2002, p.162–163): if a volume of Earth's material is cycled in stress at frequency f, such as caused by a traveling or standing wave, then the energy loss in each cycle, δE, is proportional to the peak strain energy stored in that volume. Consequently, the dimensionless ratio of these quantities is taken as a measure of anelasticity,

$$Q^{-1} = \frac{\delta E}{2\pi E_{max}} . \tag{3.3}$$

For traveling or standing waves, this definition leads to the one given in formula (1.9) on p. 4.

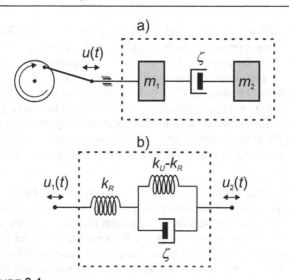

FIGURE 3.1

Two mechanical models for seismic wave attenuation (compare to Figure 1.1 on p. 5):

a)　Model of this book (Biot, 1956). The attenuating element (dashed box) is approximately uniform and participates in a forced oscillation of amplitude $u(t)$. The dissipation occurs because of the existence of at least two "phases" represented by masses m_1 and m_2.

b)　Visco-elastic model for a Zener (1948) material. The dissipating element is placed between other similar elements, and the attenuation occurs because of the *difference* between $u_1(t)$ and $u_2(t)$. Instead of the internally-moving elements m_1 and m_2, the dissipation is associated with the springs. This spring arrangement ensures that the "relaxed" spring constant is k_R and the "unrelaxed" one is $k_U < k_R$.

　　Despite its simplicity and similarity to eq. (3.2), there exists a serious problem with this definition. Let us pose the following question: does definition (3.3) actually say anything about the medium? Both quantities on the right-hand side of eq. (3.3) refer to some oscillatory process at frequency f. The seismic wave represents a non-resonant, purely forced oscillation (Figure 3.1a), in which there exists neither preferential frequency f_0 nor ringy behavior characteristic of weakly damped systems[8]. Although the energy ratio of eq. (3.3) can be taken at any

---

[8] The difference between the processes of wave propagation and linear oscillation can also be seen in the character of energy balance. For an elastic oscillator, *the sum* of the kinetic and potential energies is always constant, whereas in a wave, the kinetic and

frequency, it is still a property of the traveling or standing wave. This is the first example of inferring a property of a propagating medium from a characteristic of a wave solution in it. By analogy with eq. (3.2), it is *postulated* in eq. (3.3) that energy decay should occur proportionally to the number of oscillations. This postulate is incorrect because eq. (3.2) only applies to frequencies $f = f_0$, and, as it will be shown below, at other frequencies, this ratio does not equal Q^{-1} even for a linear oscillator. This postulate also breaks down in many practical observations: for example, Q is often found to be frequency dependent. Nevertheless, despite these problems, ratio (3.3) is still viewed as a physical quantity in most standard texts in seismology.

Another major flaw of definition (3.3) consists in the energy E being conventionally presented as the *strain* energy. From our discussion of Lagrangian mechanics in Chapter 2, our second key question therefore is: why strain and not the *kinetic* energy? This is the key point for understanding the physics of the attenuation processes. As described in Section 2.4, energy dissipation usually occurs due to relative movements of the parts of the mechanical system, and is described by viscosity-type forces and dissipation functions, which are close to the kinetic energy. In this case, the internal construction of the dissipating medium may be complex and variable (Figure 3.1a).

By contrast to mechanics, in the theory of seismological attenuation, energy dissipation is associated to "imperfect elasticity" and completely described by parameter Q^{-1} included in the visco-elastic modulus (*e.g.*, Anderson and Archambeau, 1964). The dissipation is thus attributed to the *deformation* of the elementary volume (Figure 3.1b) rather than to its internal movement. The reasons for such association are not easy to identify. One reason may be in the emphasis on equivalent mechanical models (Figure 3.1b) which often replace detailed mechanical descriptions of the medium. Such models lead to an elegant theory of visco-elasticity (Section 2.5) which greatly simplifies the resulting modeling and inversion for Q^{-1}. Nevertheless, this Q^{-1} is still not a physical property but only a phenomenological attribute.

The physical difference between the two models in Figure 3.1 can be seen in the following thought experiment. Assume that a volume of rock is enclosed in a rigid casing, so that it does not deform, and subjected to acceleration, for example, by oscillatory motion (Figure 3.2). Will the mechanical energy dissipate within this volume, which would cause, for example an increase in its temperature? In a real material, acceleration would cause relative movements of pore fluids, grains, and dislocations within the rock matrix, which would lead to mechanical energy loss (Figure 3.1a). By contrast, for the model in Figure 3.1b, there would be no energy dissipation. Because the sample would not deform, even time-dependent, visco-elastic moduli would lead to zero energy loss.

---

potential energy densities always *equal each other* (see Aki and Richards, 2002, p. 122). The energy in a wave propagates rather than oscillates.

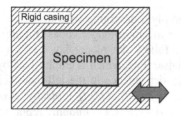

FIGURE 3.2
Thought experiment to
differentiate between the
two energy-dissipation
models in Figure 3.1. A
rock sample is placed into a
rigid casing and subjected
to motion.

Thus, definition (3.3) appears to arise from an intuitive association of harmonic waves with oscillators and equivalent mechanical models while ignoring the important differences between them. The oscillator or wave Q is further interpreted as a medium property, and the physical model of the wave-propagating medium is replaced with mathematical models which generalize certain basic observations. In the following, we will discuss some of these observations and models while trying to dissect the multi-faceted problem of Q in several aspects: 1) its phenomenological character, 2) theoretical background, 3) frequency dependence, and 4) relation to scattering and geometrical spreading.

3.2.1 Observed *versus in situ*

Like any other physical parameters, the observed ("apparent") Q needs to be clearly differentiated from the corresponding true physical quantities. The apparent Q is straightforward to derive from amplitude measurements. For example, for S-wave coda attenuation, coda Q_c^{-1} is usually obtained from velocity seismograms averaged within some selected coda window T by using the following expression:

$$P(t,f) \equiv \sqrt{\left\langle \left| \dot{u}(t,f) \right|^2 \right\rangle_T} \propto t^{-\nu} \exp\left[-\frac{\pi f}{Q_c(t,f)} t \right], \qquad (3.4)$$

where the power-law exponent ν lies within the range [½,1] depending on whether surface, diffusive, or body waves are viewed as the primary contributors to the coda (Sato and Fehler, 1998, p. 58). In recent years, most investigators preferred choosing $\nu = 1$ in coda studies. Although such selection was apparently motivated by the single-scattering body-wave model by Aki (1980), the use of a single value of ν is mostly important as a suitable convention for correlating results from different studies. The exponential decay factor in expression (3.4) describes the excess of the amplitude decay rate over this reference level of $t^{-\nu}$.

Expression (3.4) shows that Q_c is simply a descriptor of coda shape, with lower Q_c values meaning shorter, and higher meaning longer codas (Figure 3.3). As the actual coda shapes usually do not follow the model law (3.4), $Q_c(t,f)$ becomes a function of frequency and coda lapse time, which is emphasized by the notation in eq. (3.4). Because the frequency dependence in the left-hand side of this equation is usually much slower than $\exp(-af)$, $Q_c(t,f)$ typically strongly increases with frequency ($Q_c \propto f^{\eta}$ with values $\eta \approx 0.5$–1.2 commonly reported). In many cases, this increase nearly cancels the dependence of eq. (3.4) on f. Also, because the assumed t^{-v} law does not cover the entire coda time range, $Q_c(t,f)$ is often dependent on the lapse times over which the formula is applied.

Many problems arise in finding a physical meaning for Q_c above. Considering its dependence on f and t, and relation to the selected exponent v, it is even difficult to see why this particular functional form (3.4) was chosen for parameterizing the time-frequency dependence of coda amplitudes at all. The best reason apparently is that such amplitude dependence follows from Aki's (1969) single-scattering model, and this model is often used to interpret coda shapes and to present Q_c results. Note that once the model is introduced, the *apparent* Q_c in coda-envelope expression (3.4) transforms into the *true medium Q* in the model. This medium Q is typically explained by scattering on small random heterogeneities within the medium. The lapse-time and frequency dependences of the apparent Q_c are mapped into depth-dependent distribution densities of small heterogeneities and their frequency-dependent scattering amplitudes. In recent studies, 3-D spatial distributions of Q were even inverted by using procedures similar to migration (*e.g.*, O'Doherty *et al.*, 1997). Numerous extensions of this model to multiple scattering were also developed, as reviewed by Sato and Fehler (1998).

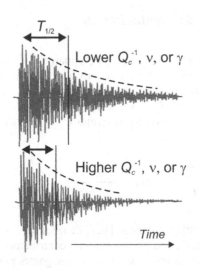

FIGURE 3.3
Schematic seismograms showing codas with lower and higher attenuation. Higher apparent attenuation means shorter duration of the coda $T_{1/2}$, which can be described by larger values of Q_c^{-1}, v, or γ.

However, the connection of the apparent Q_c to the corresponding *in situ* Q hinges entirely on the assumed theoretical model for the background. As discussed in Section 1.2.1, data constraints used to support parameterization (3.4) are weak and can be satisfied by other models. Therefore, the coda-amplitude data fit does not "prove" the selected scattering model. Models allowing detailed theoretical treatment (such as Aki's single-scattering model) are usually very simple and based on single-mode and uniform- and isotropic-space approximations. Such approximations are grossly inaccurate in describing the second-order amplitude effects, such as those of Q^{-1}. For example, the increase of the apparent Q_c with frequency can be explained by values of ν being slightly higher than 1, or $\gamma > 0$ in eq. (1.14) (Figure 3.3), and its variations with lapse time — by crustal reflections contributing at the intermediate distances (usually 30–90 km). Heterogeneities do not have to be distributed throughout a large volume of the lithosphere and can be concentrated near the surface, as most geological and seismic evidence suggests[9]. Therefore, the resulting "medium Q" resulting from them needs to be interpreted with great caution. Despite its presentation as a function of depth and heterogeneity scale-lengths, such scattering model may in fact describe not much more than the observed coda decay rate in (3.4).

3.2.2 Empirical *versus* theoretical

As most researchers agree, the general purpose of using the concept of "medium Q" is for describing the ability of a unit volume of the medium to dissipate the elastic-wave energy. This is how the interpreters usually understand Q^{-1} within the Earth, associating its increased values with unconsolidated rock, elevated temperatures, presence of fluids, or heterogeneity that causes scattering. Frequency dependence of Q plays the key role in such intuitive interpretations. For example, in local-earthquake coda-wave studies, frequency-dependent Q is routinely interpreted in terms of scale-lengths of heterogeneities (*e.g.*, Aki and Chouet, 1975), which are often variable with depth. In global observations, the observed frequency dependence of Q gives rise to the absorption-band hypothesis, which attributes this dependence to relaxation mechanisms operating at microscopic scales within the mantle (Anderson *et al.*, 1977).

However, for theorists, the *in situ* Q is often a different quantity. Because there exists no single parameter describing the ability of the medium to dissipate the elastic energy, various phenomenological proxies were proposed (Bourbié *et al.*, 1987). In particular, in formal visco-elastodynamics, Q^{-1} is indeed rigorously *defined* as a local property of the medium, equal to the argument of the complex elastic modulus in the frequency domain (Anderson and Archambeau, 1964). Such Q^{-1} leads to elegant mathematical models of wave propagation and energy dissipation (*e.g.*, Borcherdt, 2009). Nevertheless, this still does not mean that complex elastic moduli are indeed present within the Earth and that such Q^{-1} is related to any geologically meaningful properties.

---

[9] Alternate models of local-earthquake coda based on the attenuation-coefficient concept are given in Sections 5.6.4 and 6.8.

Seismic attenuation is never measured directly, but inferred from a variety of observations by using theoretical models. Such models are never sufficiently accurate — merely because the actual Earth structure is variable and practically never known with accuracy sufficient for isolation of the attenuation effects. At the same time, compared to the geometrical-spreading compensation factor G, the effect of attenuation is relatively small and practically entirely lies within the region of instability of eqs. (1.5) to (1.9). As data examples below show, only ~10% of the errors in the geometrical-spreading compensation factor G eliminate the entire frequency-dependent Q signatures and often change the Q values (commonly compared at 1-Hz frequency and denoted Q_0) by 20–30 times (see Chapter 6).

Considering the general weakness of constraints arising from attenuation observations (*e.g.*, the inability to differentiate between the χ- and Q-based models; see Figure 1.2 on p. 7), preferred theoretical models gain *de facto* precedence over the observations. In other words, attenuation data processed with a frequency-dependent Q in mind usually support such assumed $Q(f)$ models. Frequency-dependent Q models are very permissive and capable of fitting the available data in many ways. Thus, it appears that the key to understanding the existence of Q needs to be sought in the theory. Having said this, note that the data nevertheless *do play* a most important role in testing the truthfulness of parameters revealed by this theoretical analysis.

3.2.3 Axiomatic *versus* physical

Theoretical formulations of the dynamics of deformation in the literature can be subdivided into two general groups. First, in the **"axiomatic" approach of linear visco-elasticity** discussed in Section 2.5, which is broadly used in theoretical global seismology (Dahlen and Tromp, 1998), a rigorous mathematical theory is constructed by starting from the constitutive law (2.59) or similar. Unlike rigorous theoretical mechanics, this theory does not use the Lagrangian method and Hamilton variational principle but starts from differential equations of motion and relies on dashpot-spring analogies for their illustration and support (*e.g.*, Carcione, 2007). Once the constitutive strain-stress relation is established, further description becomes entirely self-consistent and close to that of an elastic problem. Nevertheless, anelasticity also leads to several new types of solutions, such as inhomogeneous waves, which sometimes exhibit peculiar properties (Richards, 1984).

By contrast, the approach that can be called **"physical"** attempts to build a wave-propagation model by using traditional mechanics, which describes the energy dissipation by viscous flows or dry friction. Apparently because of its attention to fluids, this approach is more developed in exploration seismology (*e.g.*, Chapter 2 in Bourbié *et al.*, 1987). The medium is described by using the Lagrangian formulation (Section 2.4). Notably, when specific solutions to the wave equations are considered (*e.g.*, harmonic plane waves in homogeneous media), phenomenological complex moduli also arise from this treatment (Bourbié *et al.*, 1987; Carcione, 2007).

The "physical" approach is by far more preferable for unraveling the true mechanisms of wave propagation and attenuation. Lagrange formulation of mechanics is well known for its depth, power, and generality. Instead of "over-fitting" the data with permissive axiomatic models, it encourages looking for true physical constraints arising from the observations.

3.3 Four assumptions of the visco-elastic *Q* model

The key theoretical observation facilitating both the early 1-D (Anderson *et al.*, 1965) and modern 3-D attenuation inversions (Romanowicz and Mitchell, 2007) is that the Fréchet sensitivity kernels F_q relating the *in situ* surface-wave properties q of to the observed ones, q_{obs},:

$$q_{obs} = \int F_q(\mathbf{r}_{obs}, \mathbf{r}) q(\mathbf{r}) d^3\mathbf{r}, \tag{3.5}$$

are the same for q, taken equal to VQ^{-1}, or to the wave velocity V within the medium. This statement is closely related to the correspondence-principle interpretation of Q^{-1} as the negative complex argument of the medium velocity (Section 2.5.4). Physically, this equivalence can hardly take place, because many factors control the energy dissipation within the Earth, such as fracturing, fluid content and saturation, viscosity, porosity, permeability, tortuosity, properties of "dry" friction on grain boundaries and faults, and distributions of scatterers (Bourbié *et al.*, 1987). These factors can hardly be lumped together in a cumulative medium Q^{-1}, and their effects on the observed attenuation should hardly by similar to those of V. The above similarity of VQ^{-1} to V suggests that only some specific wave mode is in fact considered, and its properties are substituted for the properties of the medium.

Indeed, Q^{-1} in eq. (1.9) is not the type of quantity that can be uniquely attributed to any point in the medium, as can parameter $q(\mathbf{r})$ in eq. (3.5). Values of Q^{-1} are different for different waves (for example, *P*, *S*, and various inhomogeneous waves, see Borcherdt, 2009). This difference is attributed to the two elastic parameters of the medium, such as the bulk (K) and shear (μ) moduli (Anderson and Archambeau, 1964). However, let us ask ourselves, what properties of the *elastic moduli* led to their association with attenuation? In answering this question, four fundamental hypotheses can be recognized in the pioneering studies of the Earth's attenuation in the 1960s and 1970s:

(H1)   Visco-elastic moduli (retarded in the time domain, complex-valued in the frequency domain) can be used to write the anelastic equations of motion.

(H2)   Phase-velocity dependencies on the medium parameters can be analytically extrapolated into the complex plane in order to derive the attenuation properties (*e.g.*, Anderson and Archambeau, 1964). This conjecture directly led to the similarity of the forward-modeling kernels for velocity and Q^{-1} in eq. (3.5).

(H3)   The Q is expected to vary with frequency, and its power-law frequency dependence $Q = Q_0 f^\eta$ is often suitable for describing the observed (apparent) and also the material attenuation (*e.g.*, Aki and Chouet, 1975; Anderson and Given, 1982).

(H4)   Geometrical spreading can be "reasonably" accurately modeled theoretically, which allows correcting the observed amplitudes for it and inferring the frequency-dependent *in situ* Q.

These hypotheses form the basis of both attenuation modeling and measurements and are rarely questioned today. Nevertheless, all four of these assumptions appear to fail. The key assumption (H1), *i.e.*, the constitutive law for attenuation, was discussed in Section 2.5 and found lacking in its basic physical justification and lead to problems when considering heterogeneous media. Assumption (H3) can be viewed as only a convenient parameterization for Q; however, in conjunction with (H4), it leads to incorrect values and spurious frequency dependences of Q in cases of incompletely corrected geometrical spreading (Morozov, 2009a, 2009b, 2009c, 2009d). Assumption (H4) may be the most harmful, because it affects the very procedure of data measurement and presentation (Morozov, 2009a). As demonstrated by revisiting several key studies (Chapter 6), geometrical-spreading models are insufficiently accurate in most observational cases, and their corrections often eliminate the need for a frequency-dependent *in situ* Q.

The hypothesis of analyticity (H2) is closely related to the correspondence principle (Section 2.5.4). Analyticity (holomorphism) is a very strong condition on a complex function, stating that the derivatives of its imaginary part are related to those of the real part by the Cauchy-Riemann equations. Once such analyticity is assumed, the many degrees of freedom (V_S and all the attenuation parameters, or at least α) collapse to a single one (V_S), and the sensitivity kernels to $V_S Q_S^{-1}$ and V_S attain the same shapes in both 1D (Anderson *et al.*, 1965) and 3D (Romanowicz and Mitchell, 2007). Inversion for Q_S^{-1} or Imμ thus becomes closely related to velocity tomography. However, the observed quantities (such as the apparent Q or phase velocities) represent integral transforms of the corresponding subsurface parameters (for an example, see Section 6.1.2). The analyticity of these transforms with respect to their functional integrands is unlikely in the general case. For example, analyticity of the dependence of the phase velocity on the crustal or sedimentary-layer velocities can hardly be proven, and it is also not something that can be safely assumed for convenience. Finally, the existence of a numerically determined partial derivative (for example, $\partial V_{\text{phase}}/\partial V_S$ for a surface wave) still does not mean analyticity with respect to complex V_S. Thus, we need to avoid treating the phase velocity as a holomorphic function and viwing attenuation as an imaginary part of velocity. Instead, we must regard the velocities and attenuation coefficients as independent physical quantities, as described in Chapters 5 and 6.

3.4 Standard models for Q

Because Q^{-1} in visco-elasticity represents in fact the attenuation coefficient χ divided by the frequency, some properties of χ also apply to Q^{-1}. In particular, both χ and Q^{-1} are additive for multiple processes contributing to attenuation. Based on this idea, practically any form of $\chi(f)$ and $Q^{-1}(f)$ distribution can be achieved by combining narrow-band relaxation mechanisms with the appropriate intensities. As we saw in Section 2.5.3, the choice of the specific models and their relaxation parameters τ_ε and t_s is not important and may have little physical significance. If we need to construct a system with a certain $Q(f)$ spectrum, then the attenuation coefficient in the desired system should simply equal

$$\chi(f) = \pi f Q^{-1}(f).$$

This section summarizes two of the most commonly used frequency-dependent Q models that appear to best represent the nearly constant-Q observations made in many seismological studies: 1) the absorption-band type behavior, called the "nearly constant Q model" and 2) a model with Q strictly independent of the frequency (the "constant-Q model").

The **nearly constant Q model** was introduced by Liu *et al.* (1976) and is often cited today. This model reconciles the narrow-band absorption of the standard visco-elastic models with the observations of long-period Q's which usually slowly vary with frequencies. By superimposing multiple relaxation peaks corresponding to the standard linear solids with varying absorption frequencies, Liu *et al.* (1976) constructed a band of nearly constant Q. Figure 3.4 shows an example of such approximate arrangement using seven standard linear–solid peaks.

The **constant-Q model** represents an extension of the above model to an infinite set of standard-linear-solid elements. This model is described by three parameters: the reference frequency $1/t_0$, the phase velocity at that frequency, and the value of Q. Its complex modulus equals (Kjartansson, 1979):

$$M(\omega) = M_0 \left(\frac{i\omega}{\omega_0} \right)^{2\gamma}, \tag{3.6}$$

where $\omega_0 = 2\pi/\tau_0$, $M_0 = |M(\omega_0)|$ is the absolute value of the complex modulus at the reference frequency, and parameter γ is related to the Q factor:

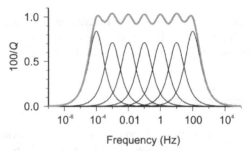

FIGURE 3.4

Frequency band of a near-constant Q^{-1} (gray) obtained by superimposing seven standard linear-solid–type dissipation bands (black lines).

$$\tan(\pi\gamma) = Q^{-1}. \tag{3.7}$$

The creep function corresponding to this complex modulus is:

$$\phi(t) = \begin{cases} 0 & t < 0, \\ \dfrac{1}{M_0\Gamma(1+2\gamma)}\left(\dfrac{t}{t_0}\right)^{2\gamma} & t \geq 0, \end{cases} \tag{3.8}$$

where Γ is the gamma function, and $M_0 = M(2\pi/\tau_0)$ is the complex modulus taken at the reference frequency of $1/t_0$ (Kjartansson, 1979).

Bourbié *et al.* (1987, p.125–128) compared the impulse responses of the constant-Q and the near-constant–Q models and found them to be practically the same for small γ, *i.e.*, for Q greater than about 10. Note that this threshold of Q is well below the levels normally considered in regional or global seismology. Bourbié *et al.* (1987) also concluded that these models are essentially equivalent, with the constant-Q model offering the convenience of fewer parameters. At the same time, in terms of physical validity, these two models appear to be equivalent to any other models. As argued throughout this book, Q^{-1} itself represents only a conventional proxy for the attenuation coefficients observed in the specific wave modes. Accordingly, various visco-elastic models can be viewed as mechanisms phenomenologically reproducing the corresponding $V(\omega)$ and $\chi(\omega)$ spectra.

3.5 Bulk and shear Q

The phenomenological mechanics of visco-elasticity, as applied to practical seismic observations, can be summarized by the following, somewhat accentuated narrative emphasizing the principal statements and assumptions (*e.g.*, Anderson *et al.*, 1965).

Seismic waves attenuate due to only two anelastic factors. According to the partitioning of strain into the dilatational and shear components, these factors are similarly associated with the corresponding bulk and shear moduli, and denoted by Q_K^{-1} and Q_μ^{-1} respectively. To infer a relation of these factors to the observable parameters, two cases of harmonic, elastic *P*- and *S*- body waves are first considered in a uniform space. For such waves, the corresponding phase velocities are:

$$V_P = \sqrt{\frac{K + \frac{4}{3}\mu}{\rho}} \text{ and } V_S = \sqrt{\frac{\mu}{\rho}}, \tag{3.9}$$

where K and μ are the bulk and shear moduli, respectively, and ρ is the density. Next, an attenuation is introduced, and the attenuation coefficients, α_P and α_S, for these waves are shown to be proportional to their respective wavenumbers, k_P and k_S (see Section 2.5.3). Therefore, the corresponding quality factors Q_P and Q_S are defined:

$$\alpha_{P,S} = \frac{Q_{P,S}^{-1}}{2} k_{P,S}. \tag{3.10}$$

These attenuation coefficients are then included in the complex wavenumbers,

$$\tilde{k}_{P,S} = k_{P,S} + i\alpha_{P,S}, \tag{3.11}$$

fom which complex phase velocity is defined in order to preserve the form of the elastic-case dispersion relation:

$$\tilde{V}_{P,S} = \frac{\omega}{\tilde{k}_{P,S}}. \tag{3.12}$$

By using the correspondence principle (Section 2.5.4), $\tilde{V}_{P,S}$ are then interpreted as true properties of the medium. Therefore, for weak attenuation, the *P*- and *S*-wave quality factors represent complex arguments of $\tilde{V}_P$ and $\tilde{V}_S$, respectively:

$$\arg \tilde{V}_{P,S} = -\arg \tilde{k}_{P,S} = -\frac{Q_{P,S}^{-1}}{2}. \tag{3.13}$$

Further, to explain the complex-valued $\tilde{V}_P$ and $\tilde{V}_S$, eqs. (3.9) are assumed to be valid in the anelastic case, and K and μ are considered as complex valued. Density ρ is still treated as real based on its intuitive connotation with "imperfect gravity". Therefore, the two body-wave parameters Q_P and Q_S become mapped into the complex-valued moduli of the medium,

$$\mu = \rho \tilde{V}_S^2 \text{, and } K = \rho \left(\tilde{V}_P^2 - \frac{4}{3} \tilde{V}_S^2 \right). \tag{3.14}$$

Finally, the complex arguments of the two new moduli are denoted as the desired bulk, Q_K^{-1}, and shear, Q_μ^{-1}, "quality-factor" values:

$$Q_K^{-1} = -\arg K \text{, and } Q_\mu^{-1} = -\arg \mu. \tag{3.15}$$

Up to this point, the derivation only represents a way to formally express the end-member P- and S-wave solutions (eq. (3.9)) in terms of K and μ. However, by using these complex moduli to construct the visco-elastic constitutive equations, one becomes able to solve problems to which the original problems did not apply, such as the surface waves and normal-mode oscillations. For example, for Rayleigh waves, the attenuation factor can be expressed as a linear combination of Q_K^{-1} and Q_μ^{-1}, or of Q_P^{-1} and Q_S^{-1},

$$Q_R^{-1} = mQ_P^{-1} + (1-m)Q_S^{-1}, \tag{3.16}$$

where m is a function of the Poisson's ratio, which was tabulated by Macdonald (1959). In another important example, the dispersion relations originally derived for plane-wave wavenumbers can also be attributed to the complex moduli (Dahlen and Tromp, 1998, p. 218):

$$\frac{K(\omega)}{K(\omega_0)} \approx 1 + \frac{2Q_K^{-1}}{\pi} \ln \frac{\omega}{\omega_0} \text{, and} \tag{3.17a}$$

$$\frac{\mu(\omega)}{\mu(\omega_0)} \approx 1 + \frac{2Q_\mu^{-1}}{\pi} \ln \frac{\omega}{\omega_0}. \tag{3.17b}$$

Assuming frequency-independent Q_K^{-1} and Q_μ^{-1}, such dependences of K and μ on frequency allow the corresponding time-domain modules, $K(t)$ and $\mu(t)$, to become causal, i.e., vanish at $t < 0$. This allows the use of these modules in the constitutive eq. (2.59) and completes the construction of the visco-elastic picture of the process.

In regard to the relative values of Q_K and Q_μ, it is usually believed that losses in pure compression are lower than losses in shear (Anderson *et al.*, 1965), and consequently Q_K is considered to be high. In practical inversions for mantle Q^{-1}, most of the model variability usually belongs to Q_μ, and Q_K^{-1} is set equal to zero where significant Q_μ^{-1} is present (see Dahlen and Tromp, 1998).

To recapitulate, the above idea of tracking the Q values to the bulk and shear moduli of the medium represents a generalization of the uniform-space, body-wave solutions, which allows unleashing the power of the axiomatic visco-elastic theory. However, despite its elegance, this is still a heuristic extrapolation of two simplest-case solutions with no real support from the fundamental theory. To see whether, and how, this model works, it is useful to compare it to Biot's (1962) model of porous saturated rock (Section 2.4.2), for which both the plane-wave V_P and V_S, and also the corresponding attenuation coefficients were derived from physical considerations, which were much more rigorous than the above analogies.

Figures 3.5 and 3.6 show the phase velocities and attenuation coefficients for the principal P and S waves in a porous saturated rock as functions of f/f_B, where $f_B = \omega_B/2\pi$ is Biot's characteristic frequency (2.55). All four of these dependences are strongly sensitive to the porosity of the rock, ϕ, and also to the ratio of its effective bulk modulus (*i.e.*, the total modulus of the matrix and fluid combined), K_0, to the bulk modulus of the solid matrix, K_s. In addition, both the velocities and attenuations strongly depend on f_B, which combines the properties of the filtering fluid. Thus, this model immediately shows a departure from the two-parameter attenuation model (3.15).

The quasi-static (zero-frequency) velocity limits in Figures 3.5 and 3.6, denoted V_P^0 and V_S^0, correspond to expressions (3.9) in which $K = K_f$ is the bulk modulus for a closed system in which no relative movement of the pore fluid occurs. These limits are used to normalize both the velocities and attenuation-coefficient values in these plots.

To continue with the derivation of the complex K and μ for Biot's model, note that these moduli are equal to neither K_0, K_s, or K_f, but *defined* by eq. (3.14). They represent combinations of ρ, V_P, and V_S rather than independent thermodynamic characteristics[10]. Therefore, their perturbations due to non-zero frequency can be expressed by combining the normalized $\delta V_{P,S}/V_{P,S}^0 \equiv \left(V_{P,S} - V_{P,S}^0\right)/V_{P,S}^0$ and $\alpha_{P,S}/a_{P,S}$ variations into complex phase velocities (Figures 3.5 and 3.6):

---

[10] Such as the thermodynamic definition for the bulk modulus: $K = -V\dfrac{dp}{dV}$, where p is the pressure, and V is the volume.

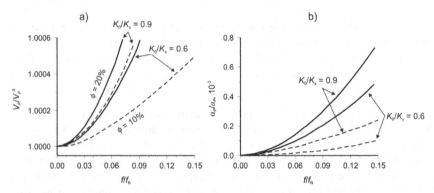

FIGURE 3.5
P-wave reslts from Biot's (1956) model of porous saturated rock.
a) velocity variation and b) attenuation coefficient. Solid lines
correspond to porosity $\phi = 20\%$ and dashed lines correspond to
$\phi = 10\%$. Results for two values of bulk-module ratios K_0/K_s are
shown. V_P values are normalized by the zero-frequency limit V_P^0,
and α_p values are normalized by $a_P = 2\pi f_B/V_P^0$.

$$\frac{\delta\mu}{2\rho V_P^2} = \left(\frac{V_S}{V_P}\right)^2 \left[\frac{\delta V_S}{V_S^0} - i\frac{f_c}{f}\frac{\alpha_S}{a_S}\right],$$
(3.18a)

$$\frac{\delta K}{2\rho V_P^2} = \left[\frac{\delta V_P}{V_P^0} - i\frac{f_c}{f}\frac{\alpha_P}{a_P}\right] - \frac{4}{3}\frac{\delta\mu}{2\rho V_P^2}.$$
(3.18b)

Figure 3.7 correlates the real and imaginary parts of these complex moduli. As we
see, δK and $\delta\mu$ depend on several parameters in complicated ways, and do not
appear to follow any simple frequency dependences. Note, in particular, that
$\mathrm{Im}\delta K > 0$ in this example, which directly contradicts the visco-elastic model
(3.15). Interestingly, Figure 3.7b suggests an approximate relation $\mathrm{Im}\delta K = -\mathrm{Im}\delta\mu$,
although its origin still appears unclear.

In summary of this section, the model in Figures 3.5 to 3.7 is certainly far
from giving a realistic picture of the process of attenuation within the mantle.
Nevertheless, this model illustrates a viable physical mechanism of seismic
attenuation by pore-fluid flow and is much better justified than the model given by
eqs. (3.9) to (3.15). It provides a specific illustration of our general statement that
only two Q-type parameters, Q_K^{-1} and Q_μ^{-1}, are insufficient for a consistent model
of seismic attenuation. It also shows that these parameters could be no more than
transformations of the phase-velocity and Q-factor values arising from two plane-
wave solutions.

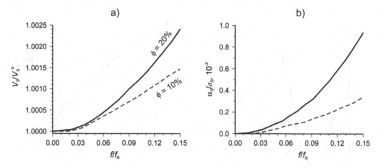

FIGURE 3.6
S-wave reslts from Biot's (1956) model of porous saturated rock.
a) velocity variation and b) attenuation coefficient in Biot's
model of porous saturated rock. Note that these quantities do
not depend on K_0/K_s. Lines and labels as in Figure 3.5. V_S values
are normalized by the zero-frequency limit V_S^0, and α_s values are
normalized by $a_S = 2\pi f_B/V_S^0$.

3.6 Scattering *Q*

The concept of scattering Q (denoted Q_s here) highlights most of the
problems of Q in general. Wave scattering in a random medium is the type of
process that is clearly unrelated to oscillations and resonances, and therefore
"quality factors" do not naturally arise in its descriptions. Scattering is also most
difficult to differentiate from geometrical spreading and anelastic attenuation, and
therefore Q_s is in the focus of the geometrical-spreading–$Q(f)$ trade-off problem.
In observations, Q_s typically shows the strongest frequency dependences.

To understand scattering Q, we need to look closely into the concept of
scattering itself, which is done in Section 4.5. At this point, it is important that
scattering is naturally described by the differential and total cross sections, the
mean free path ℓ, or medium turbidity g_0. All these quantities measure the
logarithmic decrements of wave amplitude with distance, and do not lead to the
notion of Q. Nevertheless, based on these parameters, a "scattering Q" factor can
still be introduced to provide a similarity to Q-based models of anelastic
attenuation (Aki, 1980):

$$Q_s = \frac{k}{g_0} = \frac{2\pi}{\lambda g_0}, \tag{3.19}$$

where k is the incident-wave wavenumber, λ is the corresponding wavelength, and
g_0 is the turbidity parameters describing the random heterogeneity of the medium.

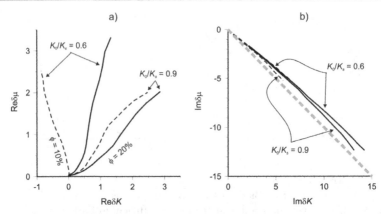

FIGURE 3.7

Variations of a) ReK compared to Reμ and b) ImK compared to Imμ, in Biot's year model of porous saturated rock. Labels as in Figure 3.5. Gray dashed line in plot b) indicates trend Im$\delta\mu$ = - ImδK. K and μ values shown in units of $2 \cdot 10^{-4}\rho V_P^2$.

However, this analogy does not appear particularly productive, as the resulting Q_s becomes dependent on the parameters of the incident wave and exhibits an excessive dependence on frequency. Even if using Q_s, the interpretation invariably relies on the underlying attenuation coefficients α or g_0.

3.7 Trade off with geometrical spreading

A problem with steep and positive frequency dependences of Q arises from the fact that such dependences may actually result from over-parameterized inversions. In particular, dependences $Q \propto f$ lie entirely within the uncertainty of attenuation-free geometrical spreading, and therefore they are likely to represent measurement artifacts.

A small error in geometrical spreading may often absorb the entire frequency-dependent Q signal. For example, consider two hypothetical measurements of the type $Q = Q_0 f^\eta$ conducted in two areas in which the geometrical spreading may differ by a factor of $f^{\delta\nu}$. For typical local-earthquake coda frequencies (~1–10 Hz) and observation times (~60 s), two observations of the apparent Q equal 200 and 600 in these areas can be interpreted in three different ways (Figure 3.8): 1) as a three-fold difference in frequency-independent $Q = Q_0$ (line $A \rightarrow B0$); 2) as a nearly constant Q_0 and a strong positive change in η ($A \rightarrow B1$); and 3) as an even greater difference in Q_0 and a negative variation of η ($A \rightarrow B2$).

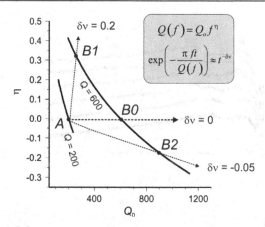

FIGURE 3.8
Comparison of hypothetical observations with apparent $Q = 200$ (point A) and $Q = 600$ (points $B0$, $B1$, and $B2$) using assumptions about the geometrical spreading being constant ($B0$) or variable ($B1$ and $B2$).

Not surprisingly, and possibly disappointing for researchers interested in relaxation-type rheologies, all geometrical-spreading measurements performed to date lead to virtual elimination of the observed frequency dependences of Q (see Chapter 6). Although considered as potentially frequency dependent, the measured Q values become constant after only ~10% geometrical-spreading corrections in these studies. This estimate arises from considering a geometrical-spreading correction in the attenuation-coefficient type form of $e^{-\gamma t}$. The typical values for γ in body- and coda-wave measurements are ~0.01 s^{-1} (Chapter 6). By equating this effect to an approximately equivalent $t^{\delta v}$, we obtain $\delta v \approx \gamma t / \ln t$, which gives $\delta v \approx$ 0.09 for a typical local-earthquake coda observation time $t \approx 30$ s. With the usual background value of $v = 1$, such geometrical-spreading variations should certainly be expected from, for example, diving and reflected waves within a heterogeneous lithosphere.

Without making a clear physical distinction between the geometrical-spreading and frequency-dependent Q, interpretation may become tricky and even erroneous. For example, Figure 3.9a shows codas of the Peaceful Nuclear Explosion (PNE) Kimberlite-3 recorded within the Siberian Craton. Within the conventional coda model proposed earlier (Morozov and Smithson, 2000), scattered-wave geometrical spreading was expected to be compensated by the scattering volume increasing with time, leading to $G_0 \approx 1$ in eq. (1.5). Consequently, coda-amplitude decay was originally interpreted as caused entirely by attenuation. At the same time, the time-domain log-amplitude slopes are practically independent of the frequency (Figure 3.9a), leading to $Q \approx 330 f^{0.85}$ in

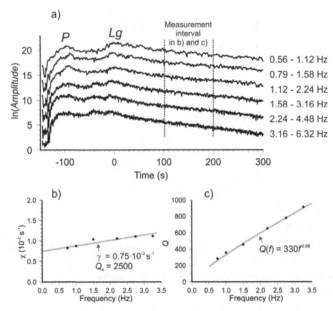

FIGURE 3.9

Coda attenuation data from PNE Kimberlite-3 in the Siberian Craton. Six stacked recordings within 660–760-km distances shown.

a)   Records filtered within several frequency bands (labeled). Regional phases and coda measurement intervals are indicated. Times are aligned at the L_g arrival, and log-amplitude envelopes are shifted in order to separate the curves. Note that the logarithmic amplitude decrements are almost independent of the frequency.

b)   Interpretation by using linear attenuation-coefficient expression (1.14);

c)   Traditional interpretation by using theoretical geometrical-spreading compensation, resulting in $Q(f)$ steeply increasing with frequency. Note that $Q_e \gg Q(f)$ within the measured frequency band.

eq. (1.9) for this explosion (with f measured in Hertz; Figure 3.9c). However, this was an erroneous result, which in fact indicated a failure of our initial model of the geometrical-spreading compensation. Such low Q_0 and high η are not expected for the stable Siberian Craton. The fact that crustal Q at 0.5–1 Hz cannot be as low as ~200–300 in this area was clear from previous observations of P_g traveling to over 1600 km along the PNE profiles in this area (for an illustration, see Section 6.8.3).

By contrast, if not transforming the observed coda slopes into Q, but interpreting them in the $\chi(f)$ form, we see that coda slopes are only weakly increasing with frequency (Figure 3.9b). This shows that that their origin is "geometrical," and the attenuation is low, as indicated in the labels in Figure 3.9b. Similar observations of steep $Q(f)$ dependencies which actually correspond to much higher Q and stronger geometrical spreading were found in many studies (Morozov, 2008, and Chapter 7). The very interesting question of why $Q(f)$ turns out to be typically increasing with frequency in most crustal and lithospheric observations can be explained by the effects of reflectivity within the upper crust. These effects are discussed in detail in Chapter 4.

The trade-off of Q values with the background geometrical spreading is well known (*e.g.*, Kinoshita, 1994). The common approach to resolving this trade-off is to adopt a standard convention for geometrical spreading, such as $G_0(r) = r^{-1}$ for body waves or $r^{-1/2}$ for frequency-domain surface waves. If the background geometrical spreading is considered perfectly known, the entire residual-amplitude signal is attributed to the frequency-dependent Q. However, although such conventions allow comparing the Q_0 and η values measured in different areas, these values become "apparent," *i.e.*, only useful for describing the seismic amplitudes. When used to characterize the subsurface, such Q_0 and η become subjective, *i.e.*, related to the interpreter's opinion about the mode of wave propagation. In this book, we are not satisfied with this approach and seek attenuation parameters that describe the subsurface objectively. Consequently we need to look for a way to unravel the $Q(f)$–geometrical-spreading trade-off.

Realizing that geometric spreading is variable, unknown, and affects the Q measurements, let us consider the following question: under what minimal conditions imposed on geometrical spreading can it be separated from Q^{-1} effects *empirically*? Considering that practically the only source of information for such separation is the frequency dependence of the attenuation coefficient, the answer is that the geometrical-spreading variation can be separated from attenuation only by approximating it as frequency independent. This is a much weaker constraint than assuming a fixed geometrical spreading, and this approach is taken here. This discussion will be continued in Chapter 4.

3.8 Frequency dependence

Almost every seismological journal contains articles presenting frequency-dependent attenuation. Values of Q reported in numerous publications often increase nearly proportionally to the frequency f or even faster[11]. The notion of frequency-dependent scattering or rheological Q (Aki and Chouet, 1975; Liu *et*

---

[11] For examples from the *Bulletin of the Seismological Society of America* from 2008 alone, see Castro *et al.* (2008), Kinoshita (2008), Morasca *et al.* (2008), Mukhopadhyay *et al.* (2008), Oth *et al.* (2008); and with somewhat weaker $Q(f)$ dependencies, Drouet *et al.* (2008) and Ford *et al.* (2008).

al., 1976) has been imbued in the minds of more than one generation of seismologists and is now accepted as common knowledge.

The principal argument in favor of the frequency-dependent Q is that it is theoretically possible. Theoretical models broadly show that, depending on the statistical properties of the scattering medium, frequency dependence of the elastic (scattering) Q_s can range from nearly frequency-independent Q_s (Frankel and Clayton, 1986) to $Q_s \propto f$ or steeper (Chernov, 1960; Dainty, 1981; Sato and Fehler, 1998). Theoretical intrinsic $Q_i(f)$ dependencies also range from near constant to nearly proportional to the frequency, as predicted by 'creep' or 'relaxation' rheological models (Jackson and Anderson 1970; Liu *et al.*, 1976). Such similarity of observable properties leads to the virtual impossibility of separating these quantities in the data without making various stringent assumptions, of which the key one is about the geometrical spreading. Although it is broadly known that common geometrical-spreading assumptions are often inadequate, they nevertheless do not break the validity of the theoretical $Q(f)$ models.

In addition to the well developed theory, in observational seismology, one also needs to test how well such assumptions and models relate to the Earth and to the measurement procedures, and whether the fact of frequency-dependent Q is indeed established. Such testing is often subdued in the use of elaborate inversion methods, curve-fitting, and in the use of model assumptions. Assumptions have become a part of not only theoretical treatments, but also of data measurements and interpretations. As shown in many examples in this book, frequency-dependent Q resulting from typical inversions can usually be traced back to inaccurate model assumptions.

Thus, there seems to exists a disconnect between the theoretical arguments for a *possible in situ* frequency-dependent Q within the Earth and the *actual* observations of $Q(f)$ from seismological data. Observational evidence for the frequency-dependent Q, such as advanced in support for the absorption-band concept (Doornbos, 1983), are usually limited to the apparent $Q(f)$ derived by using theoretical geometrical-spreading compensation. For example, as shown in Chapter 7, the band-like shape of the apparent $Q(f)$ within $\sim 3 \cdot 10^{-3}$–30 Hz can be explained by the Earth's layering, which affects the geometrical spreading in the short- and long-period waves in opposite ways. Generally, through the use of the uniform-space geometrical spreading or sometimes models for source spectra, attenuation interpretations appear to be biased toward the frequency-dependent Q.

3.8.1 Frequency-dependent *Q* of a linear oscillator

To understand the caveats of the Q definition and the significance of frequency-dependent seismological Q, it is instructive to consider a similar example in mechanics. For example, the Q value of a linear oscillator is a constant (eqs. 3.1 and 3.2), but if defined "as in seismology" (by eq. 3.3), it starts linearly increasing with frequency. This example shows that the frequency-dependent medium Q derived from seismic observations is largely controlled by its definition, and no special rheological or scattering properties may follow from such observations.

The above statement can be demonstrated as follows. If the oscillator (see eq. 2.15 and Figure 2.2 on p. 30) is driven by an external force with circular frequency $\omega = 2\pi f$, then the energy δE dissipated in one period $T = 1/f$ equals:

$$\delta E = \int_0^T \xi m \omega_0 \dot{r}^2 dt = \xi \omega_0 T E_k,$$ (3.20)

where E_k is the peak kinetic energy in that period.

For a linear oscillator, the quality factor is simply a constant parameter of the system denoted by $Q = \xi^1$ (eq. 3.1). However, if we use the "seismological" definition (3.3), the oscillator Q becomes:

$$Q = \frac{2\pi E_k}{\delta E} = \frac{\omega}{\omega_0 \xi},$$ (3.21)

i.e., it linearly increases with frequency, because δE increases with period (see eq. 3.20). This shows that dependences $Q \propto f$ should be expected when applying definition (3.2) even to the simplest vibrating mechanical system.

3.8.2 "Natural" frequency dependences of Q and t^*

When interpreting attenuation measurements, significant emphasis is usually placed on the frequency dependence of Q. The usual expectation appears to be to find a frequency-independent Q in "normal" media. As a result of this expectation, variations of Q with frequency are attributed to the effects of relaxation mechanisms or to the statistical distributions of heterogeneity scale lengths within the media. However, this is again only an intuitive hypothesis, and our argument above shows that it may be unjustified. Thus, it would be useful to specify what forms of frequency dependence of Q can the considered as "natural," *i.e.*, present in the most basic cases.

Whenever a $t^* \propto f^1$ or $Q \propto f$ behaviors are observed, their reasons should not be immediately sought in the frequency-dependent rheology or scattering. The main reason appears to be *the way these quantities are defined*. Recalling that $Q^{-1} \propto \chi(f)/f$ and considering the general attenuation-coefficient form $\chi(f) = \gamma + \kappa f$ (Section 2.8), we see that the first two leading terms in Q^{-1} should be f^{-1} (*i.e.*, $Q \propto f$) and a constant. Such dependences, or their combinations, are indeed found in many observations. Equation (3.21) shows that seismological Q is defined so that Q becomes proportional to f in all systems with frequency-independent energy dissipation rates, such as resulting from short-scale reflectivity or viscous friction. In a seismic wave with variable wave amplitude and period, the local energy density increases with amplitude, but the amount of energy dissipated in one period increases with both the amplitude and period. As a result, both the apparent Q^{-1} and t^* increase with the incident-wave period (Figure 3.10):

$$t^*(f,t) = -\frac{\ln \delta P(f,t)}{\pi f} = -\frac{\ln \delta P(f,t)}{\pi}T, \qquad (3.22)$$

and

$$Q^{-1} = -\frac{\delta E}{2\pi E} = -\frac{\langle P_{\text{diss}} \rangle}{2\pi E}T, \qquad (3.23)$$

where E is the peak energy during the wave-oscillation cycle, δE is energy dissipation per cycle, and $\langle P_{\text{diss}} \rangle \propto E$ is the mean dissipated power.

The most likely reason for the observed Q^{-1} and t^* increasing with wave periods is the effect of inaccurate geometrical spreading or some other type of background-model correction. For example, for all types of lithospheric body and surface waves from earthquake sources (see examples in Chapter 6), the geometrical spreading turns out to be slightly under-corrected by the conventional correction G_0 in eq. (1.1). As discussed in the next section, the residual positive geometrical spreading remaining after this correction gets propagated into the estimated apparent $Q(f)$ and leads to in the often-reported rapid increase of Q with frequency.

Thus, allowing for an imperfectly corrected–for geometrical spreading or short-scale scattering along the ray path, the most "natural" dependence we can expect is $Q^{-1} \propto f^1$. After these

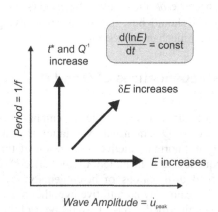

FIGURE 3.10

Directions of increasing peak energy (E), its dissipation per period (δE) and attenuation parameters (t^* or Q^{-1}) in a seismic wave with constant dissipation rate with time. Note that both t^* or Q^{-1} increase with period simply by virtue of their definitions as proportional to $\delta E/E$.

effects are accounted for, the next expected dependence is $Q^{-1} \approx \text{const}$[12]. Constant Q values are often observed in experiments in which the uncertainty of geometrical spreading is either accurately accounted for (such as for plane waves) or canceled by the measurement procedure (such as by using the spectral ratios). Note that once again, in the attenuation-coefficient formulation, the two "natural" dependences simply correspond to the leading terms in the Taylor series for $\chi(f)$. Also, looking at the physical mechanisms of attenuation, such as the effects of fluids and scattering (Chapter 5), we can also expect $\chi(f)$ to increase faster than $\propto f$ as well, which would correspond to Q increasing with frequency.

By contrast to Q or t^{*}, Q_e is defined differently, and approximately as:

$$Q_e = \frac{1}{\pi} \frac{\partial \chi}{\partial f} = -\frac{1}{2\pi} \frac{\partial^2 (\ln E)}{\partial f \partial t}.$$ (3.24)

Therefore, to explain Q_e, we only need to explain the increase of the relative energy dissipation rate, $\left(-\partial \ln E / \partial t\right)$, with frequency.

Phenomenologically, the increase of dissipation rate with frequency can be modeled by "retarded elasticity," as it is done in the visco-elastic theory. However, if we do not accept such local memory in a mechanical system, we need to look for a simpler, physical explanation. A natural explanation arises from considering the wave equation. For example, in our hypothetical model (2.35), we have:

$$\ddot{u} - V_m^2 u'' - \xi V_m^2 \dot{u}'' = 0.$$ (2.35 again)

Because of the presence of the second time derivative in its first term, the resulting spatial attenuation coefficient always increases with frequency as ω or faster:

$$\text{Im} \, k' = \omega \, \text{Im} \left(\frac{1}{V_m \sqrt{1 - i\omega\xi}} \right).$$ (2.37 modified)

This shows that Q_e should be positive in media with energy dissipation. At the same time, once linear or faster increases in χ and α with frequency can be expected, their frequency-independent ("geometrical") parts also cannot be ruled out, and therefore the general $\chi(f) = \gamma + \kappa f$ dependence should always be considered.

---

[12] This argument is done in terms of Q^{-1} being represented by a Maclaurin series in f.

3.9 Measurement methods

Q^{-1} is commonly measured by several methods: from the relative width of the resonance peak ($\Delta f_{1/2}/f_0$, measured at the level of $1/\sqrt{2}$ of the maximum displacement amplitude); from the phase lag of the strain behind the stress (δ); or from the logarithmic decrement of its amplitude decay (χ/f_0). The first two of these methods use harmonic signals, and measure Q indirectly, often by invoking the visco-elastic theory. For small levels of dissipation, these measures should be related as follows:

$$Q^{-1} = \frac{\Delta f}{f_0} = \tan \delta \cdot$$

(3.25)

The third measure uses transient (decaying) signals, and the derived attribute directly corresponds to the logarithmic amplitude decrement,

$$Q^{-1} = \frac{\chi}{\pi f_0} \cdot$$

(3.26)

These three types of Q are not automatically equivalent and require substantial empirical corrections, and mathematical modeling for reconciling. In the transient-signal methods of (3.26), the attenuation coefficient χ always represents an intermediate product, and therefore they are relatively easy to analyze for frequency-independent contributions in χ. By contrast, the methods of (3.25) do not measure χ but derive it from the properties of the resonance peaks of the sample. Most importantly, in both of these cases, non-zero values of the frequency-independent part of χ may lead to biased Q^{-1} estimates.

Measurement of the elastic- or acoustic-wave attenuation is technically difficult, even in the laboratory. The data need to be extracted from a range of interfering factors, such as the geometrical spreading, multiple reflections, scattering, effects of the finite dimensions, and shapes of the specimens. The effects of some of these factors can be modeled and corrected for, but certain factors are difficult to control. This shows that the reliability and accuracy of the attenuation data depends on our understanding of the experimental techniques and the underlying theories.

There are two general categories of lab and field measurements leading to Q-type quantities (3.25) and (3.26): 1) using traveling waves and 2) using standing waves (vibrating systems). Let us briefly consider both of these categories, focusing on our main question, which is how well the resulting Q estimates can be related to some *in situ* property of the medium. In particular, we need to see whether some analogs of the "geometrical-spreading" correction may affect the determination of the frequency-dependent Q.

3.9.1 Methods using propagating waves

Several techniques for measuring Q in propagating waves are available (Chapter 4 in Bourbié *et al.*, 1987; Tonn, 1991). A broad variety of **transmission methods** are used both in the lab and in the field, with different sources and configurations of wave-propagation paths (Figure 3.11). These methods are based on the transient-signal approach (3.26) and use the following expression for the source- and receiver-effect corrected seismic amplitude:

$$A(t,f) = G(t,f)e^{-\chi t}, \tag{3.27}$$

corresponding to our definition of χ in eq. (1.6). Here, $G(t,f)$ is the factor accounting for all geometrical spreading, transmission, and reflection effects. According to the Q model, the attenuation coefficient here is usually taken in the form of $\chi = \pi f / Q$.

The success and accuracy of transmission methods is largely dependent on the ability to estimate $G(t,f)$ or eliminate this quantity from the amplitude relation (3.27). In **spectral-ratio methods**, $G(t,f)$ is regarded as frequency independent, which represents a good approximation in many cases. Such frequency-independent $G(t)$ can therefore be cancelled without knowledge of its exact form, by forming the logarithms of spectral ratios from pairs of different observation times,

$$\ln \frac{A(t_2,f)}{A(t_2,f)} = \ln \frac{G(t_2)}{G(t_2)} - \frac{\pi}{Q}(t_2 - t_1)f. \tag{3.28}$$

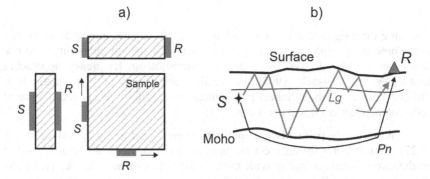

a)   b)

FIGURE 3.11

Principle of transmission Q measurements: a) three sample configurations and b) schematic field configurations (black – using P_n wave; gray – for L_g wave; S – the sources; and R – receivers.

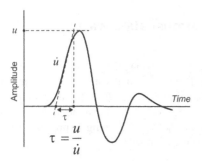

FIGURE 3.12
Signal rise-time definition
in attenuation
measurements.

This expression shows that Q^{-1} can be measured from the slopes of the resulting spectral-ratio amplitudes as functions of f. Like all other techniques, this method is most accurate for lower-Q rock. Because of its relating amplitude ratios at different frequencies, "spectral scalloping" and frequency-dependent coupling variations may also complicate identification of linear segments in the spectral-ratio responses. For low-amplitude signals, background noise typically causes the values of Q to increase. Overall, because of using a single, frequency-independent Q, the spectral ratio method is among the most reliable approaches.

Another notable approach to frequency-independent Q measurement is the **pulse rise-time method**. This approach is based on an empirical relation of the wave-rise time τ and Q^{-1},

$$\tau = \tau_0 + Ct^*, \tag{3.29}$$

where τ_0 and C are two empirical constants dependent on the source, and t^* is the accumulated attenuation parameter,

$$t^* = \int_0^T \frac{dt}{Q}. \tag{3.30}$$

Rise time for a signal can be measured as a ratio of its peak amplitude to the rate of the preceding amplitude build-up (Figure 3.12). Signal attributes similar to rise times can also be derived by using the **instantaneous-frequency** approaches (Matheney and Nowack, 1995). In practice, application of these methods is difficult because of the stringent requirements to low noise levels and the use of source-dependent empirical constants.

Absolute-Q transmission methods attempt measuring Q values in eq. (3.27) independently for different frequencies. Such methods are broadly used in earthquake seismology and provide most of the field evidence for the frequency-dependent Q of the Earth. All of the difficulties of the spectral-ratio techniques are present in these methods as well; however, the most important problem of the absolute-Q methods is their dependence on the models of background $G(t,f)$. As discussed in Chapter 4, these models are almost never sufficiently accurate, and this inaccuracy may often eliminate the frequency dependence of Q or even the entire absolute-Q signal. Absolute-Q values resulting from such assumed geometrical spreading are often 20 to 30 times lower than the corresponding Q_e values derived without such assumptions (Chapter 6).

   Coda Q methods represent a special category of the absolute-Q methods because they use a stochastic, scattered wavefield as the seismic source (Sato and Fehler, 1998). Seismic codas are measured at times exceeding twice the travel-times of source-receiver S-wave propagation, which practically ensures that scattered energy arrives at the receiver from all possible directions (Figure 3.13). Because of its averaging over large volumes of the lithosphere, coda-amplitude decays are relatively stable and can be measured with high consistency. The resulting time-domain equation for coda amplitude is identical to eq. (3.27), and in the traditional approach, it is similarly assumed that $G(t,f)$ can be "reasonably" accurately modeled. However, this assumption is also based on a major over-simplification of the scattering process and may be totally unreliable. The ambiguity of the resulting coda Q was discussed in Section 3.2.1, and in Section 5.6.4, we will give a different treatment of this problem.

   In **pulse-echo methods**, an ultrasonic signal is emitted by a piezoelectric crystal bonded to the rock sample (Figure 3.14a). The attenuation coefficient is determined from the amplitudes of two successive multiple reflections:

$$\alpha = \frac{1}{2L} \ln \left| \frac{A_1 R}{A_2} \right|, \qquad (3.31)$$

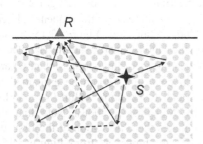

FIGURE 3.13
Principle of coda-Q measurement. A scattered wavefield is used, in which the information about the source is averaged. Singly (solid arrows) or multiply scattered (dashed arrows) waves can be used to model the scattering process. S –source; R – receiver.

where R is the reflection coefficient at the interface. The advantage of this method is in a direct measurement of α, which is a more meaningful quantity than Q. At the same time, this method may have significant difficulties in ensuring plane-wave propagation within the sample and accounting for scattering and coupling effects at the different boundaries, and also for reflections from the edges of the sample. All of these effects are frequency dependent, leading to a problem very close to the $Q(f)$–geometrical-spreading trade-off in transmission methods above. These effects can be difficult to model theoretically, but they can be partially compensated by

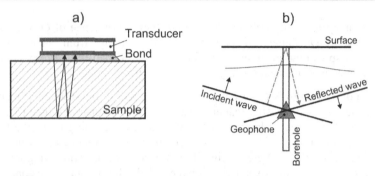

FIGURE 3.14
Principle of pulse-echo Q measurements: a) using lab sample and
b) in borehole recording of a local earthquake.

making measurements on reference samples.

As with all transmission methods, interpretation of pulse-echo
measurements in terms of material Q may be risky, and the attenuation-coefficient
approach is advisable. For example, Mashinski (2008) reported $Q(f)$ dependences
ranging from $Q \propto f^{0.8}$ to $Q \propto f^{1.2}$ in mono-crystallic quartz samples with various
inclusions. The data were presented in the conventional Q form (Figure 3.15a),
which was derived by following the standard amplitude corrections. However,
noting that these Q^{-1} values were actually calculated from the attenuation
coefficients similar to (3.31), we can transform these Q^{-1} back to χ by using
expression $\chi = \pi f Q^{-1}$. The result (Figure 3.15b) reveals that in all three cases, the
three general contributions discussed in Section 2.8 can be recognized: 1) a
positive frequency-independent $\chi(f) = \gamma$, 2) linear increase with frequency, likely
associated with Q_e^{-1}-type attenuation (dotted lines in Figure 3.15b); and
3) resonance peaks, particularly strong for the smoky quartz sample (labeled s)
near 1 MHz. The constant-$\chi(f)$ term dominates the recordings in the intact (i) and
fractured (f) samples. The existence of such constant χ could be explained by
reflectivity (back-scattering) on the defects and dislocations within the samples,
which were also extensively discussed by Mashinskii (2008). Note that the
presentation in the Q^{-1} form (Figure 3.15a) makes the observation of the first two
of these attenuation contributions most difficult; however, from Figure 3.15b,
these contributions also appear to be the most important.

A variant of the pulse-echo method is often applied to borehole recordings
of local earthquakes (*e.g.*, Abercrombie, 1998). In this case, the *P*- or *S*-wave
arrival amplitude is recorded twice by the same geophone, first in the direct-wave
arrival and the second after its reflection from the free surface (Figure 3.14b). This
case provides a clear illustration of the "geometrical-spreading" problem arising in
the pulse-echo method. The free-surface reflection coefficient may differ from the
theoretical value of $R = -1$, and the incident waves travel non-vertically, leading to

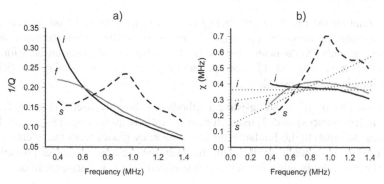

FIGURE 3.15
Interpretation of ultrasonic pulse-echo lab measurements from
Mashinskii (2008). a))Original interpretation in terms of Q^{-1} and
b) the same data in $\chi(f)$ form. Labels indicate quartz samples:
intact (i), fractured (f), and smoky (s). Dotted lines in plot b)
show linear $\chi(f)$ relations for the corresponding samples.

significant spurious frequency-dependent Q inverted from formula (3.31). This
example is considered in detail in Section 6.4.

3.9.2 Methods using vibrating systems

Apart from the whole-Earth's free oscillations, measurements using
vibrating systems have only been used in the laboratory. The use of standing
waves offers clear technical advantages, such as the ability to work with
monochromatic signals, high signal-to-noise ratios, and repeatability of
observations. At the same time, standing waves represent a major handicap in
comparing the results to seismological observations using traveling waves. The
only direct comparison of standing-wave lab data to seismology can be made in
the Earth's normal oscillations, which occupy the most remote part of the seismic
spectrum. In addition, as with most laboratory methods, these approaches are most
effective for low Q below about 20–30, whereas most materials of interest in
deep-crustal and mantle studies have much higher quality-factor values in the
range of $Q \approx 100$–2000.

From the viewpoint of our discussion, the most difficult aspect of
standing-wave methods is the definition of Q. The values of Q measured
according to indirect relations (3.25) represent the quality factors of the whole
measurement devices, and deriving the corresponding medium attributes from this
Q involves substantial theoretical idealizations and modeling. The *in situ* material
Q inferred from vibrating-system studies in fact always represents the Q
postulated in some visco-elastic model of the material and inverted by fitting its
predictions to the observations. If one questions the validity of such models, as we

do here, relating the resulting material-Q values to those derived from traveling-wave experiments becomes problematic. In most studies, measurements are performed at audio frequencies (1–10 kHz) or higher, which adds an additional problem of traversing ~8–15 octaves in frequency for correlating to field observations in global seismology.

Two groups of standing-wave methods are broadly used: 1) using free vibrations (resonances) and 2) using forced vibrations away from resonance. In both cases, to separate the fundamental low-frequency modes from the overtones, rods with diameters much smaller than their lengths are used whenever possible. For axial modes, the frequencies of mode numbers $n = 1, 2, 3...$ are equidistant,

$$f_n = \frac{nV}{2L},$$
(3.32)

where L is the length of the rod, and V is the wave velocity, equal $\sqrt{E/\rho}$ for longitudinal modes (E is the Young's modulus) and $\sqrt{\mu/\rho}$ for shear modes. For flexural modes in bending rods, the resonance frequencies are no longer equidistant (Schreiber *et al.*, 1973):

$$f_n = m_n^2 \frac{R_g \sqrt{E/\rho}}{2\pi L^2},$$
(3.33)

where R_g is the radius of gyration, and m_n is a constant depending on the order of resonance and the type of boundary condition used.

In the vicinity of a resonance f_n, vibrational devices behave as the linear mechanical oscillator discussed in Section 2.4.1. Similar to the oscillator, they also allow "visco-elastic" interpretations (Section 2.5.8); however, the complex-frequency interpretation (eq. 2.20) still represents a more robust view from first principles. According to this interpretation, attenuation causes each f_n to shift into the lower half of the complex-frequency plane,

$$f_n \rightarrow f_n \left(1 - \frac{i}{2Q_n} \right).$$
(3.34)

These shifts manifest themselves by time-domain logarithmic decrements of amplitudes (in the free-vibration approach), by broadening of spectral peaks (for measurements near resonance), or by strain- stress phase shifts (in the forced-oscillation approach).

Note that the measured quality factors Q_n in eq. (3.34) correspond to the individual oscillation modes. Strictly speaking, observation of a set of $\{Q_n\}$ values does not mean that a common Q can be assigned to the material of the sample or even to the whole measurement apparatus. Differences between Q_n values are principally due to the differences between the vibration modes. When plotted against the corresponding values of f_n, quality factors produce an *apparent Q* dependence, which simply means that Q_n varies concurrently with f_n. To invert this implicit dependence into an explicit $Q(f)$ to reveal the "material Q," a significant inversion effort is required. Unfortunately, this inversion is impregnated with applications of numerous simplifying assumptions, models, and data corrections. A number of such corrections were described by Bourbié *et al.* (1987, p. 166-169). This situation is quite analogous to the studies of whole-Earth oscillations.

Because rocks are not truly homogeneous, the velocities and Q's measured for the individual peaks may not necessarily represent the averages over the entire sample. At different frequencies, different parts of the sample may dominate the relaxation spectra. Once again, whole-Earth oscillations give an excellent illustration of this point. Sampling non-uniformity appears to be among the most significant problems of vibration methods. It is therefore important to use samples as homogeneous as possible and to verify the "purity" of the spectrum by checking the regularity of the peak-frequency distribution.

Phase-lag methods represent an important variation of vibrating-system measurements. With recent developments in laboratory techniques, phase-lag measurements have become practical for strain amplitudes below 10^{-5}, seismological periods of 1–1000 s, temperatures to ~1300°C, and pressures up to ~300 MPa (Jackson and Paterson, 1993). Faul *et al.* (2004) and Jackson *et al.* (2004) reported such measurements of Q on polycrystals simulating mantle olivine and discussed their seismological implications.

In phase-lag approaches, uniaxial or torsional, forced oscillations at frequencies far below the resonance ($\omega << \omega_0$) are induced, and both the stresses and strains are measured with very high accuracy. Torsional stresses are determined by measuring strains in standard elastic elements with known properties, connected in series and having similar shapes as the samples being tested (Figure 3.16). In particular, phase lags δ are measured between the harmonic strain and stress functions, and the resulting quality factor is determined as $Q = 1/\tan\delta$ (eq. 3.25). As before, thus-defined parameter Q is a property of the entire vibrating system. Its interpretation as a material Q once again relies on the assumptions of sample uniformity, negligible effects of the remaining parts of the apparatus, and validity of the visco-elastic model.

Fortunately, and also similar to the above, Q values resulting from phase-lag measurements can be transformed into a more tangible property possessing a direct physical meaning. Regardless of the visco-elastic model assumptions and of the shape of its amplitude spectrum, for the strain lagging the stress by phase angle δ during forced oscillations, the attenuation coefficient equals $\chi = \omega\sin\delta/2$, and therefore it can be derived from the reported Q^{-1},

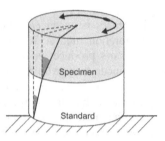

FIGURE 3.16

Principle of phase-lag Q measurements (Jackson and Paterson, 1993). From the two shaded angles, the ratio of strains in the sample and the standard element under the same torque is determined.

$$\chi = \frac{\omega \sin \delta}{2} = \frac{\pi f Q^{-1}}{\sqrt{1 + Q^{-2}}}. \quad (3.35)$$

This transformation leads to an interesting and somewhat different interpretation of phase-lag results. For example, Figure 3.17a shows the results of torsional forced-oscillations tests on an olivine sample under temperatures from 1000°–1200°C, as presented in $\lg(Q^{-1})$ form by Jackson et al. (2002). As expected, the data show a regular and monotonic increase of attenuation with temperature, but further inferences are difficult to make from this presentation.

When transformed into the attenuation coefficient by using the expression above, the data show a clear increase of $\chi(f)$ slopes with temperature, which are of course related to the increase of Q^{-1} (Figure 3.17b). At this point, we notice that the data may be viewed as a dominant linear trend in $\chi(f)$ at high frequencies, which rolls of to zero at low frequencies. This can be emphasized by picking the intercepts (γ) and slopes of these trends (κ, here shown by values Q_e from eq. $\kappa = \pi/Q_e$) (Figure 3.17c). Finally, by subtracting the linear dependences $\chi_0(f) = \kappa f$, the behavior of $\chi(f)$ at lower frequencies can also be emphasized (Figure 3.17d).

Notably, both the "effective attenuation" $1/Q_e$ and "geometrical attenuation" γ increase with temperature (Figure 3.17c). In particular, from ~1100° to 1200°C, $1/Q_e$ appears to increase especially quickly. As in other cases (Figure 3.15b and examples in Chapters 4 and 6), increased γ can be associated with increased heterogeneity and random (back-) scattering. The rapid decrease of the attenuation coefficient below ~0.2–0.4 Hz appears to be similar for all temperatures, although this cut-off may also be increasing with temperature. Logarithmic frequency sampling used by Jackson et al. (2004) (also inspired by the assumed visco-elastic model!) does not allow establishing this cut-off with confidence.

3.9.3 Frequency dependences of observed Q and χ

Unfortunately, observations of a frequency-dependent Q within a single experiment are never model free. The resulting $Q(f)$ can be viewed as arising from the particular models used (i.e., from expecting Q to be frequency dependent), or

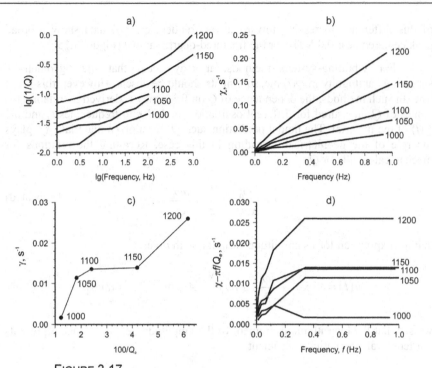

FIGURE 3.17

Representative results of torsional forced-oscillation phase-lag measurements in Fo$_{90}$ olivine (Jackson et al., 2002).

a) Interpretation in terms of Q^{-1} by Jackson et al. (2002); b) the sane data re-calculated into χ; c) slopes and intercepts of high-frequency trends in $\chi(f)$; and d) $\chi(f)$ with high-frequency trends removed. Labels indicate observation temperatures (in °C).

alternatively, the frequency dependence can be removed by using conservative models discouraging the quality-factor–based interpretation. Such conservative models use the general attenuation-coefficient form (2.128), as opposed to the $Q(f)$ form, which assumes that $\gamma \equiv \chi(f)|_{f\to0} = 0$.

For transmission methods, the above statement was illustrated in Figure 3.15, with several additional examples in Chapter 6. Non-zero γ should be expected from imperfectly corrected geometrical spreading and reflectivity. In the observations, γ is clearly positive, and the slopes of the linear $\chi(f)$ trends suggest frequency-independent values of $Q_e \approx 10$ for smoky, 37 for fractured, and infinite Q_e for intact quartz (Figure 3.15b). These values are much greater than the frequency-dependent Q (~3–15 for fractured quartz, 4–10 for smoky; Figure 3.15a) derived in the original interpretation by Mashinskii (2008). Also note that the difference between the fractured- (f) and intact-quartz (i) dependences between 0.4–0.6 MHz is exaggerated by the transformation into Q (Figure 3.15a). Instead of this difference increasing toward lower frequencies, $\chi(f)$ data show a broad peak centered near 0.8 MHz for the fractured-quartz sample (Figure 3.15b).

For vibrating-system methods, it may seem that the "geometrical spreading" ambiguity in $Q(f)$ measurements should be absent. However, this is not true. In such methods, the dependence of Q on frequency is derived by relating the values of Q_n measured for different oscillation modes' n to frequencies f_n and the $Q(f)$ of the material. The model predicting such Q_n for a given material $Q(f)$ plays the role of the geometrical spreading in this case. Recalling the relations for mechanical-oscillator Q,

$$Q = \frac{\omega_0}{\Delta\omega_{1/2}} = \frac{1}{\xi} = \frac{m\omega_0}{\zeta},$$

(3.1 repeated)

and the expression for its amplitude variation with time,

$$u(t) = A\exp\left(-\frac{\xi\omega_0}{2}t \pm i\omega_0 t\right) \equiv A\exp\left(-\frac{\zeta}{2m}t \pm i\omega_0 t\right),$$

(3.36)

we see that the logarithmic decrement of the amplitude near resonance represents our temporal attenuation coefficient:

$$\chi = \frac{\zeta}{2m}.$$

(3.37)

This quantity depends only on the friction coefficient (*i.e.*, on the ratio of the dissipation function D to the kinetic energy; see Section 2.4.1) and is *independent* of the resonance frequency ω_0. Therefore, one should expect a constant positive shift in $\chi(f)$ resulting from friction. In the case of a resonant bar or another rock specimen in a vibration-test apparatus, such friction should arise from sample heterogeneity, viscous coupling with the pore fluid, details of specimen suspension and mount (which affect the boundary conditions), and scattering into higher harmonics (related to the sample shape). As in other experiments, correcting for all of these factors by means of theoretical modeling may be an intractable task. Nevertheless, such corrections could be readily available from the analysis of the measured $\chi(f)$ data.

3.9.4 General character of Q(f) observations

Different types of measurements are used within the bands of the broader seismological frequency range (Figure 3.18). The general character of frequency-dependent Q observations is formed by the $Q(f)$ dependences revealed in some of these measurements, but mostly by comparing the different sub-bands.

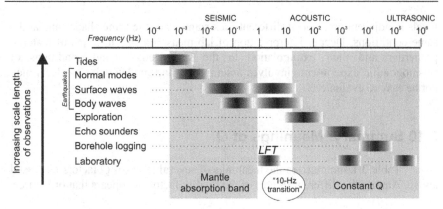

FIGURE 3.18
Schematic frequency spectra of seismic, acoustic, and ultrasonic measurements. Note that apart from low-frequency tosional lab measurements (labeled *LFT*), the frequency-depenent *Q* generally arises with increasing scale lengths of the observations.

Generally, the most reliable of the higher-frequency observations are above ~10 Hz and suggest a frequency-independent *Q* (Figure 3.18). *Q* values used in exploration seismology are, as a rule, frequency independent, as well as the models arising from higher-frequency ultrasonic measurements. At frequencies below ~1 Hz, an increase in the attenuation known as the absorption band occurs (Aki, 1980). Between the low- and high-frequency bands, a transition is located, referred to as the "10-Hz transition" by Abercrombie (1998) (Figure 3.18). However, as noted in Chapter 6, all observations below this frequency are of transient-wave, absolute-*Q* measurement type, and they may be affected by inaccuracies in the geometrical-spreading corrections. Both the absorption band and the 10-Hz transition disappear when these corrections are re-examined (see Chapter 7).

In summary, in correlating the *Q* values arising from different types of observations, it is important to keep in mind the types of quantities that are being measured. The meanings of *Q* are significantly different in the different types of experiments. While looking for a frequency-dependent attenuation within the mantle, laboratory evidence or theoretical conjectures can hardly outweigh rigorous and robust analysis of seismological data. Insufficient accuracy of background models may bias both field and laboratory *Q(f)* estimates, making them difficult to compare across broad ranges of frequencies and experiment scale lengths. In their Chapter 3, Bourbié *et al.* (1987) summarized a number of laboratory- and field-based attenuation measurements and noted that although most of them can be described by the corresponding and visco-elastic models, there is little agreement between the resulting values of *Q*. Experimental *Q* data are much more reliable when used "laterally," *i.e.*, for comparing the attenuation levels in adjacent areas or of different rock samples of the same shapes and at the same conditions. Deeper interpretation of lab measurements in terms of material properties and their extrapolation to the scale lengths and conditions of seismographic experiments involves assumptions and models whose effects may not be easy to evaluate.

3.10 Summary: Meanings of Q

Table 3.1 summarizes the meanings of several types of Q factors discussed above. Among the different usages of the quality factor, it appears that only three,

TABLE 3.1 Meanings of various theoretical and observed Q values.

	Attenuation parameter	Intended factors	Actual factors addressed
MATERIAL	Visco-elastic (Q_{VE})	Material property	Any phenomenology with logarithmic amplitude decay
	Intrinsic (Q_i)	Elastic-energy dissipation	Elastic-energy dissipation. May trade-off with GS[*] and Q_s if viewed as frequency-dependent
	Scattering (Q_s)[**]	Random heterogeneity	Structure, model; inseparable trade-off with GS and Q_i
	Total in situ (Q)	$Q^{-1} = Q_i^{-1} + Q_s^{-1}$	Equivalent to χ[***]; includes GS variations
OBSERVED	Apparent attenuation coefficient (χ)	Amplitude decay rate	Structure, attenuation, scattering
	Effective (Q_e)	Frequency-dependent part of χ	Q_i and frequency-dependent apparent GS
	Geometrical (γ)	Zero-frequency limit of χ	Frequency-independent apparent GS
	Apparent Q (Q_a)	Equivalent to χ	Same as χ
	Logarithmic decrement (coda, lab)	Q_a	Model, background GS, lapse-time window, χ
	Absolute-amplitude Q (field, lab)	Q_a	Background GS, χ
	Spectral ratio	Q_a	Q_e (only frequency-dependent χ)
	Rise time	Q_{VE}	Q_e, source and path parameters, noise level
	Spectral peak width (whole-Earth, lab)	Q_{VE}	Model, specimen shape and uniformity, energy distribution, Q_e, scattering
	Phase-lag	Q_{VE}	Same as above

[*]GS – geometrical spreading

[**] See Chapter 4

[***] See Chapter 5

indicated by rounded-corner boxes in this table, correspond to directly observable properties, and therefore the corresponding Q's are physically significant. Of these three, only for the apparent Q (which is equivalent to χ, dashed-line box), does the actual physical significance correspond to the meaning intended in the interpretation. For the other two cases related to vibrating systems (dotted-line box in Table 3.1), the Q factors are well-defined mechanically, but their visco-elastic explanations may be strongly doubted.

For all remaining types of Q, strong assumptions about the models and experimental environments are required to define the Q's and also to reconcile their actual and intended meanings. Such assumptions usually fail, making the observed Q values subjective, and difficult to compare and to relate to physical properties. Therefore, the use of the different Q's requires care in staying within the contexts, and preferably within the dashed-line boxes outlined in this table.

By contrast to the multiplicity of definitions of Q and their sensitivity to the modeling and observational contexts, the attenuation coefficient is a common property that can be found in all of these definitions. The reason for this is that, at least in principle, χ is an unambiguously observable quantity directly related to energy dissipation. Because of its origin in the fundamental concept of energy, χ also represents a better alternative than Q in describing the physics of attenuative elastic medium.

Chapter 4

Geometrical spreading

and scattering

Many things are incomprehensible to us not because
our comprehension is weak, but because they are
beyond the limits of our comprehension.

Kozma Prutkov, Fruits of Reflection (1853-54)

Geometrical spreading is critical for the understanding of many field
attenuation measurements. Geometrical spreading is either explicitly or implicitly
present in all descriptions of attenuation, and it receives the most emphasis in the
ongoing methodological debate[13]. The attenuation coefficient is determined by
correcting the observed seismic amplitudes for some estimated "background"
geometrical spreading, $G_0(t,f)$:

$$\chi(t,f) = -\frac{d\ln\left[A(t,f)\Big/G_0(t,f)\right]}{dt}. \quad \text{(1.1 and 1.6 combined)}$$

---

[13] See Morozov, 2009a, 2009b, 2009c, 2010a, 2010c, 2010d; Xie and Fehler, 2009;
Mitchell, 2010; Xie, 2010, and further comments in December 2010 issue of *Pure and
Applied Geophysics*.

Consequently, any error in G_0 becomes recorded as χ (and consequently Q^{-1}), and the typical error may absorb the entire frequency-dependent Q signal (see Section 3.7). Nevertheless, the properties, and even the very definition of geometrical spreading are among the most uncertain in attenuation studies.

In discussing geometrical spreading, it is important to stay clear of intuitive associations suggested by its name. If wave spreading is literally understood as "geometrical," *i.e.*, only related to a spreading wavefront of some simple shape, this term becomes impractical. Wavefronts do not exist in real wavefields and are never encountered in observations, which always use seismic amplitudes averaged over certain time windows. Therefore, the notion of geometrical spreading requires a careful definition. Our definition offered below can be more precisely named "non-dissipative spreading," yet we keep the conventional term "geometrical spreading" corresponding to the role of factor G_0 in the expression for χ above.

In view of the acute sensitivity of attenuation measurements to geometrical spreading, it is important to model, and particularly to *measure* the actual geometrical-spreading variations within the structures in which the attenuation is studied. As a first-order effect of the Earth's structure, geometrical spreading should be about as variable as its attenuation, and therefore assuming a constant geometrical spreading could be a greater fallacy than regarding Q as frequency independent. For example, Frankel *et al.* (1990) pointed out that the effective geometrical spreading of S waves in the northeastern United States was significantly faster than r^{-1}, where r was the hypocentral distance. From the numerical modeling of that paper, one can also see that the geometrical spreading does not follow any simple $r^{-\nu}$ dependence, but a similar spreading law can be used within the distance range before the near-critical *SmS* reflection. Note that for S waves at epicentral distances $\Delta < 100$ km, the values of ν were consistently higher (~1.5–1.9) than the expected theoretical $\nu = 1$. For L_g waves ($\Delta = 100$–400 km), ν equaled ~0.7, whereas it was postulated to equal 0.5 in the preceding frequency-domain studies (Frankel *et al.*, 1990).

As will be shown in Chapter 5, geometrical spreading can be measured as a part of the attenuation measurement, as the zero-frequency limit of the attenuation coefficient defined in Section 2.8, $\gamma = \chi|_{f \to 0}$. Importantly, observations carried out to date (Chapter 6) suggest that γ values are in most cases positive, stable within wave groups, and correlate with tectonic types of the lithosphere. For example, in short-period S-wave studies, the lithospheric geometrical spreading was found to be systematically under-compensated by the standard t^ν corrections, with γ taking positive values ranging from ~0.002 to ~0.2 s^{-1} (Morozov, 2008). For long-period body waves, $\gamma \approx 0.018$ s^{-1}. Values of γ are also positive for L_g and Rayleigh waves at up to ~100-s periods, beyond which a slight negative γ is found. For body, coda, and Rayleigh waves, the values of γ are consistently increased in active tectonic structures, which is probably related to the simplification and "homogenization" of the lithosphere with age. An increase of coda γ in the more complexly layered upper crust was suggested by numerical modeling in realistic lithospheric structures.

Such systematic values of γ and their relations to the crustal and lithospheric structures deserve a special analysis, which is performed in this chapter. Note that although the variations of parameters γ or ν are rarely analyzed in attenuation studies, the issue discussed here is already well-addressed indirectly, by mapping the attenuation parameters Q_0 and η in various geological structures. For example, it is generally accepted that tectonically active structures typically have lower Q_0 and increased η (Aki, 1980). However, because Q_0 and η trade-off with each other and with the background geometrical spreading, it is important to look for their combinations that are trade-off free and measurable. Correlating such characteristics with the structure and tectonic types would be of great importance for the analysis. As shown below, γ represents one such characteristic.

In this chapter, numerical modeling of geometrical spreading in several velocity structures is used to explain the systematically positive levels of γ observed in crustal and lithospheric attenuation studies. The main conclusion from this analysis is that the increased and variable geometrical spreading could be caused by the crustal and, particularly, upper-crustal reflectivity. Because the apparent frequency-dependence parameter η is approximately proportional to γ (Section 5.4), this observation could therefore also help explain the predominance of positive η values in lithospheric-scale attenuation measurements.

Numerical waveform modeling in realistic lithospheric structures shows that the upper-crustal structure and the position of the hypocenter within it determine the character of geometrical spreading. Reflectors above the earthquake hypocenter tend to increase γ, and reflectors immediately below the source – to decrease it. With strong reflectivity below the source, γ may become negative and lead to geometrically compensated amplitude peaks at 30–70-km hypocentral distances, which are also often attributed to the scattering Q. Such sensitivity of geometrical spreading to the upper-crustal structure may explain the inferred decrease of γ with tectonic age (Section 7.3).

The general conclusion about geometrical spreading is that the upper-crustal structure and position of the earthquake source within it are the most important factors controlling the geometrical-spreading patterns. The geometrical spreading determines the results of most frequency-dependent Q measurements.

In regards to scattering, we take the view that it is practically indistinguishable from geometrical spreading in all aspects, including the way it affects the wavefield, is modeled by the scattering theory, and measured from the data. The only difference between these attenuation processes could be the coherency of the scattered wavefield, which is, however, far beyond today's recording capabilities. For this reason, we discuss scattering in this chapter, and in close combination with geometrical spreading. Throughout the book, we treat scattering as a "hitherto not-modeled" part of the empirical geometrical spreading.

4.1 Definition of geometrical spreading

The principal difficulty in understanding the concept of geometrical spreading is its theoretical abstraction combined with the lack of a definition suitable for interpreting realistic datasets. Usually, the geometrical spreading is described as the effect of elastic wave energy spreading within an expanding wavefront. However, wavefronts exist in practically no real cases of interest, because they are destroyed by multi-pathing, triplications, reflections, mode conversions, and dispersion within heterogeneous structures. Only a few analytically tractable solutions exist, typically in uniform and isotropic models. "Practical" geometrical-spreading models usually represent empirical generalizations (*e.g.*, Frankel *et al.*, 1990; Zhu *et al.*, 1991; and below) and disagree with the theoretical spreading-wavefront models. Note that common empirical definitions, such as $G_0(t) = t^{-\nu}$ with arbitrary ν, already depart from the notion of a spreading geometrical wavefront.

Along with the disappearance of wavefronts in real seismic wavefields, the notions of "seismic phases," their "amplitudes," and "propagation paths" disappear as well. Strictly speaking, geometrical spreading cannot be associated with any particular phase, an inferred propagation mechanism, or with any particular recording time or distance. Therefore, the realistic geometrical spreading should be viewed as characterizing the entire wavefield.

On the other hand, the term "geometrical spreading" is typically used with respect to some "attenuation" property, which is expected to be a local property of the propagating medium sensitive to the wave frequency. Attenuation is something that can, in principle, be "turned off" at any point in the medium, but geometrical spreading cannot be removed and represents an imprint of the structure in which the wave propagation takes place.

Therefore, for a general, but practical definition of geometrical spreading, we can only use the following: *geometrical spreading is the effect of the background structure on seismic amplitudes in the absence of anelastic attenuation and small-scale scattering*. In terms of the conventional Q-factor terminology, such a geometrical limit is attained by setting $Q^{-1} = 0$ everywhere within the structure. The key to this definition is the "deterministic" nature of the structure, meaning that geometrical spreading is only one aspect of its more complete description. Geometrical spreading is therefore variable, unknown, and represents one of the most important goals of attenuation analysis. It can be described by a general space-time and frequency dependence of wavefield amplitudes, $G(\mathbf{r}, t, f)$, without regard for any specific propagation model (Figure 4.1).

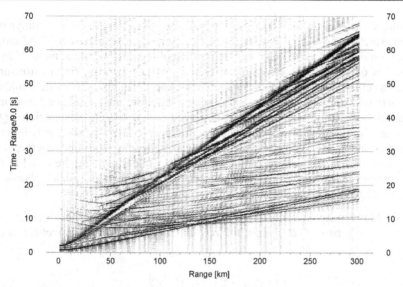

FIGURE 4.1

Numerical model of realistic vertical-component "geometrical spreading" in IASP91 lithospheric structure (Kennett and Engdahl, 1991). This section was obtained by attenuation-free waveform modeling by using the reflectivity method (Fuchs and Müller, 1973). Reduction velocity of 9 km/s was used for plotting.

Contrary to the conventional assumptions of geometrical spreading being sufficiently closely approximated by models such as $G_0(t) = t^\nu$, note that small errors of this approximation (estimated as ~10% by Morozov, 2008) can absorb the entire frequency-dependent Q signal in many cases. Because of the lack of information about the 3-D structure of the lithosphere, accurate modeling of realistic $G(\mathbf{r},t,f)$ appears practically impossible. Nevertheless, geometrical spreading can be easily *measured* and modeled by using the attenuation-coefficient technique described in this book.

At least conceptually, geometrical spreading is relatively easy to differentiate from anelastic dissipation. The distinctive characteristic of anelastic dissipation is in its frequency dependence. When using the Lagrangian approach, the distinctive characteristic of anelastic dissipation is in its being described by the dissipation function, as compared to the potential or kinetic energy for geometrical spreading. Dissipation should lead to effects vanishing at zero frequency, and therefore the zero-frequency limit contains only the effects of geometrical spreading and scattering.

Separation of geometrical-spreading and scattering effects is much more difficult, because we can no longer rely on simplified wavefront models. For a phenomenological definition of geometrical spreading that reflects its intuitive use in interpretation, we could use: "geometrical spreading is the effect of the large-scale, dissipation-free structure on seismic amplitudes." The meaning of "large-scale" is of course subjective and determined by our viewing certain structures (such as crustal gradients, boundaries, blocks, and topography) as model-building or "deterministic." By contrast, "small-scale" structures are considered "random," treated statistically, and may be described by scattering attenuation (Q_s). Fortunately, although the inversion for the full geometrical spreading may be very complex and uncertain, its small variations can be readily measured together with the variations of attenuation. On the other hand, while improving the detail of the background structural model, the scattering-attenuation model becomes assimilated by the geometrical spreading (Figure 4.1).

Mathematically, the above argument suggests that the geometrical spreading can be estimated from the zero-frequency limit of the wavefield in eq. (1.6),

$$G(t,f)\big|_{f \to 0} = A(t,f)\big|_{f \to 0}, \qquad (4.1)$$

provided that such a limit exists and can be determined from the data. This approach is quite practical and used for determining the empirical geometrical spreading in several data examples in Chapter 6. Unfortunately, the zero-frequency limit does not constrain the frequency dependence of $G(t,f)$, which has to be assessed from purely theoretical considerations. This uncertainty is a part of the fundamental ambiguity of distinguishing the small-scale structures causing random scattering from those responsible for the deterministic behavior of the wavefield.

4.1.1 Mathematical *versus* phenomenological models

Discussions about the analytically acceptable or the most accurate forms of geometrical-spreading models often take the central part in the debate of the validity of attenuation measurements (Xie and Fehler, 2009; Xie, 2010). Generally, the types of geometrical-spreading models preferred by different authors depend on the desired balance of theoretical simplicity with accuracy of fitting the seismic amplitude data. In seismic hazard studies, the requirements of accuracy are paramount, and the geometrical-spreading models are therefore path specific and contain many empirical parameters (*e.g.*, Atkinson, 2004; Pasyanos *et al.*, 2009). Such dependences typically change their characters between distances of ~70–140 km, where the near-critical Moho reflections disrupt the uniform-space approximations. Several complex geometrical-spreading models simulating the frequency-dependent effects of Moho refractions were also proposed for P_n waves (*e.g.*, Yang *et al.*, 2007). Note that in these studies, the estimation of Q^{-1}, including its frequency-dependent part, is done concurrently with the empirical geometrical-spreading estimation.

However, in studies focusing specifically on *in situ* Q measurements, fixed geometrical-spreading models are usually preferred. This preference is dictated by the need to have a well-defined background on top of which the Q can be measured. However, a risk of misrepresenting the errors in geometrical spreading as a frequency-dependent Q arises, which is not easy to handle in these cases. Three different points of view can be recognized with respect to this risk:

1) Simple geometrical-spreading corrections, such as multiplication by t^ν, are thought to be sufficiently accurate for practical attenuation measurements (*e.g.*, Aki, 1980). It is more important to maintain a common standard suitable for comparing results than to look for an ultimate, accurate solution.

2) Current geometrical-spreading models are inaccurate, but we have to live with them in order to know the Q. With 3-D numerical modeling, geometrical spreading would eventually be predicted with accuracy sufficient for attenuation measurements.

3) Geometrical-spreading and $Q(f)$ errors are inconsequential, and therefore any geometrical-spreading model that is somewhat close to the data, such as the one using uniform-space background, is adequate. This point of view is valid in the studies focusing on reproducing the empirical amplitude decay dependences rather than on revealing the true attenuation within the subsurface. Seismic magnitude calibration and nuclear-test monitoring are examples of such applied studies.

The first two of these approaches treat χ or Q as true characteristics of the subsurface and are therefore of interest to the present study. Both of these approaches rely on modeling for resolving the double geometrical-spreading/scattering/Q trade-off. Unfortunately, it appears that this trade-off is nevertheless an intractable problem, principally because geometrical spreading cannot be conceptually separated from attenuation. To understand such separation, or rather its absence, we need to compare the characters of the geometrical-spreading and attenuation models.

Considering the nature of the dependence of the observed attenuation on background geometrical-spreading models, it is important to differentiate two categories of these models: 1) "mathematical," defined on the basis of some theoretical considerations (*e.g.*, using some "wavefront" spreading with propagation time) and 2) phenomenological, characterizing the effects of geometrical spreading by empirical descriptions of the observations without attempting their detailed mathematical simulation.

Unfortunately, the existing paradigm of not only modeling, but also measuring $Q = Q_0 f^\eta$ leans heavily toward the "mathematical" models above. The reported Q_0 and η are always either explicitly or implicitly dependent on the assumed mathematical geometrical-spreading forms. However, such forms are *insufficiently accurate* for practically all cases of interest. The concept of a wavefront itself breaks down in all crustal models with velocity gradients and contrasts, in which triplications, reflections, and mode conversions are abundant. The concept of "rays" which could be followed in order to track the $G_0(t)$ dependences is also absent in a realistic medium. This also means that "multi-pathing" is common within the lithosphere. Even in the purely theoretical cases of pronounced structural layering, such as P_n, S_n (*e.g.*, Yang *et al.*, 2007), or P_L (Aki and Richards, 2002, p. 323), the t^ν dependence is violated and becomes frequency dependent. Therefore, wavefront-based models provide useful theoretical asymptotics but are not helpful for defining realistic geometrical spreading.

By using full numerical modeling in 3D, one could hope, in principle, to accurately solve for the wave amplitudes in the absence of energy dissipation. This may still be hampered by insurmountable limitations of the knowledge of the velocity/density structure and uncertainties in the source and receiver effects. Quality of numerical models will certainly continuously improve with further research; however, even the best model cannot be considered sufficiently accurate and accepted without verification.

By contrast, phenomenological models do not require detailed descriptions of the mechanisms of the wave processes but may be based on very general principles, such as the conservation of energy and time/spatial continuity. Such a model for geometrical spreading was proposed by Morozov (2008) and is employed here. In this model, the geometrical-spreading factor is allowed to weakly deviate from some best-known "theoretical" background $G_0(t,f)$, which may be generally time and frequency dependent,

$$G(t, f) = G_0(t, f)e^{-\gamma t}. \qquad (4.2)$$

Here, γ is our "geometrical-attenuation" parameter, which is the zero-frequency limit of the attenuation coefficient. This parameter can be adjusted to fit the amplitude data (see Section 2.8 and data examples in Chapter 6). The correction factor $e^{-\gamma t}$ is hereafter referred to as the "residual geometrical spreading".

Thus, apparently the only correct and feasible approach to measuring attenuation in the presence of variable Earth structure can be outlined as follows:

(i) Derive the best-possible model of the background geometrical spreading, $G_0(t,f)$;

(ii) By using the seismic amplitude data, measure the residual geometrical spreading, $e^{-\gamma t}$ and interpret the value of γ in respect to the background structure;

(iii) Evaluate the scattering-theory criterion $|\gamma t| \ll 1$ for the time ranges t used in the attenuation measurements. If the criterion is violated significantly[14], a reconsideration of the background model (*e.g.*, considering reflections or surface waves contributing to the coda together with body waves) and repetition of procedures (i) to (iii) may be advisable.

4.1.2 Functional forms

Given the uncertainty of the concept of geometrical spreading discussed above, the choice of its functional form should be inconsequential as long as it describes the wave amplitudes within the ranges of observations. Nevertheless, this choice also appears to be the key argument against the approach presented in this book (Xie and Fehler, 2009; Xie, 2010). The principal critique by Xie (2010) is that the exponential geometrical-spreading model of (4.2) lacks a physical basis. His argument is that $G_0(t,f)$ may be a valid approximation to reality, but introduction of an exponential factor $e^{-\gamma t}$ breaks this validity. This point is very instructive and deserves some consideration.

The pathos of our debate with Xie (2010) is about whether $G_0(t,f)$ can be considered adequate for attenuation measurements or does it need to be corrected before Q is measured in a specific area. If $G_0(t,f)$, in any of its functional or numerical forms, is already accurate, then one can safely set $\gamma = 0$, and approximation (4.2) becomes equivalent to the traditional model. However, it is always better to test this accuracy than to simply declare it to be accurate. Without looking for a physical basis of model (4.2), this expression can be viewed as a perturbation of the geometrical spreading in order to test whether $G_0(t,f)$ is indeed accurate. As shown in many examples in Chapter 6, this test fails in almost all cases, and values of $\gamma \neq 0$ are usually indicated by the data. In addition, these values also turn out to be *positive* in most cases, showing that there exists a systematic deviation from the commonly assumed background $G_0(t,f)$.

As shown above, the physical basis of the exponential geometrical-spreading correction $e^{-\gamma t}$ in eq. (4.2) lies in the perturbation theory. Consequently, this approximation is only rigorously valid for $|\gamma t| \ll 1$, and it could be in principle replaced by other similar forms, such as $(1 - \gamma t)$. To use any form of such geometrical-spreading corrections, $G_0(t,f)$ should be made sufficiently close to reality by using the appropriate model for the wavefield.

Among the possible forms of corrections applied to $G_0(t,f)$, the exponential form used in eq. (4.2) is advantageous because of its affinity to the descriptions of

---

[14] In practical measurements, this criterion does not appear very critical. Refining the background model may become quite costly for the analysis, althogh most useful for interpretation.

scattering and attenuation. This form also offers insightful analogies with other areas of theoretical physics and can be supported by theoretical arguments and models of short-scale reflectivity and wavefront-curvature variations (Sections 5.5 and 5.6). From these examples and affinity to the scattering theory, it is likely that the $|\gamma t| \ll 1$ criterion might be relatively relaxed in the $e^{-\gamma t}$ form for the residual geometrical spreading.

However, the same affinity to scattering theory is the reason for model (4.2) being disliked by some researchers studying attenuation in terms of Q. Because the attenuation-coefficient approach emphasizes the similarity between the residual geometrical spreading, $e^{-\gamma t}$, and attenuation, $e^{-\varkappa f t}$, and measures them concurrently, it directly demonstrates that errors in the geometrical spreading could often be responsible for the observed frequency-dependent Q.

Looking for a physical basis for selecting the functional forms of $G(t,f)$, note that the commonly used $G(t) \propto t^{\nu}$ law has in fact little basis beyond convention. The t^{ν} spreading law is only justified in unrealistic end-member cases of uniform and isotropic media, and only for integer values of $\nu = 0$ for plane, $\nu = \frac{1}{2}$ for cylindrical, and $\nu = 1$ for spherical waves. The causes for such discrete values are the discrete wavefront topologies and assumptions of their shapes being invariant while expanding in sizes during propagation. With a ν not equal 0, $\frac{1}{2}$, or 1, the approximation $G(t) \propto t^{\nu}$ is completely empirical, even more so than our approximation (4.2) derived from the perturbation theory. With such ν, the "fractal" wavefront shapes leading to their areas expanding as t^{ν} are not easy to visualize, and such spreading is definitely not dictated by the physics of wave propagation.

Further complication of the parametric geometrical-spreading laws comes from their frequency dependence and effects of dispersion. Because of dispersion, wave packets spread out with propagation time, and the peak amplitudes decay faster than those of their harmonic components. As a result, in the time-domain, geometrical spreading is usually faster than in the frequency domain. However, the laws of dispersion again do not lead to spreading rates in the form of t^{ν}.

In reality, geometrical spreading never follows the t^{ν} law, and all geometrical-spreading functions used in practical data analysis represent only convenient *ad hoc* approximations to the data. Such approximations generally need to be constructed differently for different wave types and within different distance ranges. In particular, body waves at near ranges (usually 0–50 or 0–70 km) decay faster than t^{1} (such as with $\nu \approx 1.3$; Atkinson, 2004). At greater ranges, beyond ~100–140 km, the wavefield becomes dominated by the waves trapped within the crust and spread slower, with $\nu \approx 0.5$. Between these ranges, there is an intermediate range in which the direct waves are joined by post-critical reflections from the Moho, and spectral amplitudes increase with distance as about $r^{0.2}$ (Atkinson, 2004). The anelastic-attenuation (Q^{-1}) effects are typically easier to measure only at the far ranges, where the geometrical-spreading effects are weaker and simpler.

Empirically, without considering the theoretical differences between geometrical spreading and scattering, the amplitudes could be expected to behave asymptotically as $r^{-V_{near}}$ at $r \to 0$ and $r^{-V_{far}}$ at $r \to \infty$. Both of these asymptotics can be combined by using piecewise–power-law dependences, as in Atkinson (2004). Alternately, interpolation between such dependences can be performed, for example, by using the following, smooth geometrical-spreading function (for simplicity, considered frequency independent here):

$$G(t) = t^{-v(t)} e^{-\gamma t}, \tag{4.3}$$

where $v(t)$ varies from $v(0) = v_{near}$ to $v(\infty) = v_{far}$, and we also added factor $e^{-\gamma t}$ to characterize the potential deviations of geometrical spreading at far distances. Note that another advantage of this exponential correction is that it does not affect the near-range asymptotic. This form of empirical $G(t)$ will be illustrated in Section 6.9 . Certainly, other parametric forms can be constructed, and therefore it is important to measure the effects of their variability on the resulting estimates of attenuation.

4.2 Numerical models in realistic lithospheric structures

Numerical waveform modeling can be used to study the variability of geometrical effects within the lithosphere. The following examples use the 1-D reflectivity method (Fuchs and Müller, 1971), which allows producing waveform synthetics in finely layered crustal and lithospheric structures. All reflections, mode conversions and all orders of multiples are included, reproducing the realistic wavefield (Aki and Richards, 2002, p. 157). To model the effects of attenuation, this program uses the visco-elastic approach, and consequently Q to characterize the attenuation within model layers. At present, however, we set these Q values very high in order to simulate the elastic wave propagation.

In this modeling, we produced 800-s long, three-component synthetic records sampled at 200-ms intervals and output at 1-km spacings from near-zero to 600-km distances from the epicenters. This allowed examining the wavefield to large offsets and avoiding any numerical wrap-around effects. The modeled frequency band was 0.2–2.4 Hz by using a "spike" source function suitable for spectral measurements. Sufficiently dense phase velocity spectrum from 1 to 120 km/s was selected in order to avoid frequency aliasing during numerical mode summations.

For each three-component record produced by the modeling, a sample-by-sample root-mean square (r.m.s.) trace was formed, and its peak vector amplitude and the total-trace energy were measured. The peak amplitudes were further squared, and both quantities multiplied by r^2 to correct for the reference r^{-1}

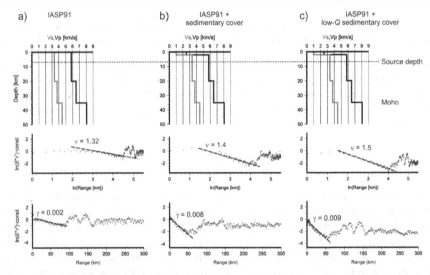

FIGURE 4.2
Results of numerical simulations of geometrical spreading in:
a) IASP91 model, b) IASP91 with a low-velocity sedimentary
layer, and c) the same as b) but with $Q_P = 20$ and $Q_S = 10$ within
the sedimentary layer. *Top row* – the V_P and V_S velocity models;
Middle row – geometrical spreading within near-offset ranges, in
logarithmic distance scale; and *Bottom row* – the complete
distance range in linear scale. Small black crosses show the peak
energy in two (radial and vertical) components combined. Both
amplitudes are geometrically compensated by using the
theoretical $(range)^2$ factor. Dashed lines labeled with v values
indicate the approximations of geometrical spreading using the $t^{\,v}$ law at near offsets, and lines with labels γ show the same
ranges approximated by $e^{\gamma t}$ dependences.

geometrical spreading of body waves. The amplitudes were finally scaled and
presented together in Figures 4.2 to 4.4.

Several velocity models were tested, most of them based on the global
IASP91 model (Kennett and Engdahl, 1991) consisting of a simple three-layer
crust and mantle without strong gradients and low-velocity zones (Figures 4.2 and
4.3). The densities were set equal to 2.8 g/cm³ within the crust and 3.2 g/cm³
within the mantle. In addition to the standard IASP91 model (Figure 4.2a), two of
its modifications were also considered: one containing a 2-km–thick low-velocity
sedimentary layer with the same high Q's (Figure 4.2b), and another one with
strongly attenuating sediments: $Q_P = 20$ and $Q_S = 10$ (Figure 4.2c). Point sources
were located at 7-km depths in all models.

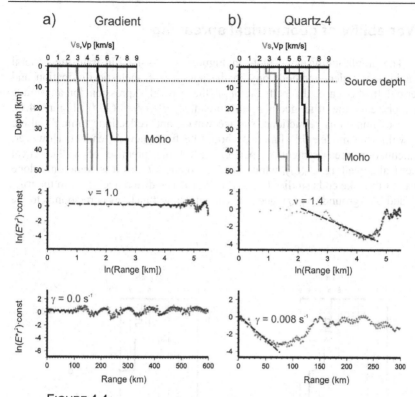

FIGURE 4.4

Geometrical-spreading modeling results for: a) a hypothetical gradient model of the crust and b) realistic, detailed structure from inversion of the travel-time data from nuclear explosion Quartz-4 in Russia (Morozova *et al.*, 1999). Geometrical compensation of the amplitudes, lines, and labels as in Figure 4.2.

Because of the crustal and mantle structure, the resulting wavefields are complex (Figure 4.1). Clearly, the "geometrical spreading" represented by these sections is far from any of the theoretical "spreading-wavefront" models and contains pervasive "multi-pathing" represented by multiple reflections and mode conversions. Nevertheless, this amplitude distribution is close to those commonly observed from shallow sources. This modeling illustrates our point in Section 4.1 that geometrical spreading cannot be associated with any particular seismic phase, but only with the time or distance dependence of the wavefield. Note that although the free-surface, Moho, and intra-crustal reflections cause rapid and persistent variations of the amplitudes recorded at the surface, distinct trends can be recognized in the pre-critical Moho reflection range (0–100 km) and beyond it (Figures 4.2 to 4.4).

4.3 Variability of geometrical spreading

The amplitude-distance plots in Figures 4.2 to 4.4 illustrate the general attenuation-coefficient methodology in the absence of anelastic dissipation and scattering. In this case, the path factor in the general expression for amplitude (1.8) represents the true geometrical spreading: $P_G(t,f) = G(t,f)$ (Figure 4.1). Because of numerous refractions, surface waves, and reflection "tuning" taking place within the model, this function should be frequency dependent. However, the attenuation measurement is performed on the premise of a theoretical geometrical spreading, $G_0(t,f)$, which is taken to equal t^{-1}, as it is commonly done in local-earthquake coda studies (e.g., Aki, 1980). The difference between the true, $G(t,f)$, and background, $G_0(t,f)$, geometrical-spreading functions corresponds to the

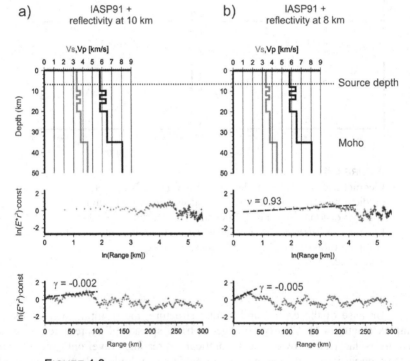

FIGURE 4.3
Geometrical-spreading modeling results with a reflectivity zone below the source at 7 km: a) reflectivity starting at 10-km depth and b) starting at 8-km depth. Geometrical compensation, lines, and labels are as in Figure 4.2. Note the slower geometrical spreading at near offsets (ν and γ) dropping below the theoretical level of $\nu = 1$ and $\gamma = 0$ in case b).

zero-frequency limit of χ, which we denote by γ, as in Section 2.8. Therefore, the corrected estimate of the true geometrical spreading is:

$$G(t,f) = P_G(t,f) = G_0(t,f) e^{-\gamma t}. \tag{4.4}$$

Note that in this formula, the residual geometrical spreading is approximated as frequency independent. This is an important assumption without which the separation of the effect of geometrical spreading from Q^{-1} would hardly be possible. However, note that by dropping the use of Q^{-1} and undertaking a consistent $\chi(f)$ approach, such separation may also become unnecessary. The attenuation coefficient (2.119) should be best treated as a single quantity, and the geometrical spreading and a Q^{-1} not differentiated until final interpretation.

Values of γ determined from eq. (4.4) are shown in bottom plots in Figures 4.2 to 4.4. Alternately, the amplitude-decay corrections can be applied by adjusting the power-law parameters, $\nu \rightarrow \nu + \delta\nu$, such that:

$$G(t,f) = P_G(t,f) = G_0(t,f) t^{-\delta\nu}. \tag{4.5}$$

The corresponding values of $\nu = 1 + \delta\nu$ are shown in the middle-row plots in Figures 4.2 to 4.4. Note that such dependence can only be fit beyond ~5-km offsets, apparently where the reflections from the free surface become significant (middle-row plots in Figures 4.2 to 4.4). This shows that the t^{ν} law represents a poorer approximation for the actual geometrical spreading whereas dependence eq. (4.6) covers the modeled geometrical spreading well from 0 to ~50–100-km distances (Figures 4.2 to 4.4). Power-law (t^{ν}-type) geometrical-spreading functions, with the appropriate values of ν, should be more important near the source (for $t \rightarrow 0$), where the accuracy of theoretical predictions improve and deviations from the true structure are less significant. The perturbation-type (such as our $t^{\nu}e^{-\gamma t}$) geometrical-spreading law should be preferable at all distances at which $\gamma t \ll 1$, because it does not distort the t^{ν}-type spreading in the immediate vicinity of the source and allows accumulation of empirical geometrical-spreading corrections during wave propagation.

In all cases in Figures 4.2 to 4.4, the peak amplitudes (tiny crosses in Figure 4.2 middle and bottom) show approximately r^{-1} (or t^{-1}) behaviors (near-horizontal slopes in Figure 4.2 middle and bottom), but only when averaged and considered beyond ~100-km distance ranges. Closer than ~60 to 70 km from the source, the amplitudes drop off quickly, corresponding to $\nu \approx 1.32$ for the IASP91 model and $\nu \approx 1.4$–1.5 for models with sedimentary layers (Figure 4.2). At ~100-km hypocentral distances, near-critical Moho reflections arrive, whose geometrical-spreading–corrected amplitudes may exceed those near the epicenter (in particular, for the IASP91 model, Figure 4.2a). At greater offsets, the P- and S-wave Moho reflections are followed by numerous multiples, which develop a more uniform geometrical spreading at large distances (Figure 4.2, bottom).

The introduction of a sedimentary layer above the source increases the near-source geometrical-spreading exponent from $\nu \approx 1.3$ to 1.4 (Figure 4.2b). This trend of increasing pre-critical ν was also observed in other models with heterogeneities located above the source. This effect was also noted by Frankel *et al.* (1990), who explained it by waves reflecting downward from the base of the sedimentary layer. The increased attenuation within the sediments appears to somewhat increase the spreading exponent to $\nu \approx 1.5$ (Figure 4.2c).

Notably, when additional reflectivity is placed below the source region, ν decreases and may drop below 1, and γ becomes negative (Figure 4.3). When the reflectors are located close beneath the source, the geometrically compensated amplitudes rise monotonically from the source to about ~50 km (with $\nu \approx 0.93$), followed by a decay at larger offsets (Figure 4.3b). This behavior resembles the one observed in total-energy measurements (Figure 6.48), in which such increased amplitudes were typically attributed to backscattering (*e.g.*, Wu, 1985). This similarity is not surprising, as the reflectors below the source can indeed be viewed as "scatterers" returning the energy to the surface. However, "backscattering" is still a very loose term for such reflections, because it creates a connotation with random scattering in an otherwise uniform crust (*e.g.*, Wu, 1985) whereas in reality we have predominantly upward reflections within a pronounced layered crustal structure. The distance dependence of these amplitudes should also be largely caused by wide-angle reflection coefficients varying with distance, and not by a $r^{-\nu}$-type geometrical spreading. Also note that the position of the peak at ~50 km corresponds to the fixed source depth of 7 km in this modeling, and with deeper sources and reflective zone depths, the peak should accordingly move to longer offsets.

Two additional numerical tests show a simple crustal model with a constant velocity gradient (Figure 4.4a) and a realistic platform model named "Quartz-4" and derived from detailed studies of Peaceful Nuclear Explosions (PNE's) in Russia (Morozova *et al.*, 1999) (Figure 4.4b). The model in Figure 4.4b was also used in our PNE coda studies (Morozov and Smithson, 2000). As one would expect, in the gradient-crust model, the amplitude decay curves are the simplest and show the best agreement with the theoretical r^{-1} dependence (Figure 4.4a). This is the only numerical example of a good agreement with the assumed theoretical geometrical spreading we have found so far. By contrast, because of its greater crustal thickness, the Quartz-4 model shows a range of amplitudes decaying faster than in any of the IASP91-based models, with $\nu \approx 1.7$, followed by strong *PmP* and *SmS* onsets at ~150 km. Note that this large geometrical-spreading exponent is still within the range observed by Frankel *et al.* (1990).

The relatively small values of γ (Figures 4.2 to 4.3, bottom) show that the uniform-space, body-wave background approximation ($G_0(t, f) \propto t^{-\nu} \propto r^{-1}$, with $\nu = 1$) is acceptable within the ~0 to 100-km distance ranges. For other types of waves, ν should be different (such as ½ or 0.83 for L_g; Campillo, 1990), and γ would vary accordingly. Therefore, γ may also vary with frequency bands, offset ranges, and observations using different types of $G_0(t, f)$. Also, for refracted body waves (P_n or S_n; Yang, 2007) and in fact for any realistic "colored" reflection sequences, $G(t, f)$ is inherently frequency dependent. However, all this does not alter the role of γ as the measure of geometrical-spreading deviations from its best-known reference level.

Finally, note that in the synthetic data (Figures 4.2 to 4.4), parameter γ was measured from the distance-corrected amplitudes shown in bottom plots in by using the following relation:

$$\left\langle \ln\left[E(t,f)t^2 \right] \right\rangle_f = \left\langle 2\ln\left[\frac{P_G(t,f)}{G_0(t,f)} \right] \right\rangle_f = const - 2\gamma t \cdot \qquad (4.6)$$

Angular brackets $<...>_f$ here denote the averaging of the frequency dependence in our time-domain measurements. This frequency dependence is dominated by the source spectrum produced by the numerical simulator combined with the "colored" response of the structure mentioned above. Therefore, the γ values measured here represent averages of the frequency-dependent geometrical spreading,

$$\gamma_{measured} = \left\langle \chi(f) \big|_{Q^{-1}=0} \right\rangle_f, \qquad (4.7)$$

and not exactly the values of $\chi|_{f\to 0}$ defined in eq. (2.119). However, we ignore this frequency dependence of the geometrical spreading for now.

4.4 Effects of velocity dispersion

Another common cause of positive γ is velocity dispersion and pulse broadening. Such effects are always present whenever energy dissipation, reflectivity, or scattering is present. Note that time-domain geometrical spreading is typically measured for wave arrivals, which represent wave packets localized in time. In the presence of dispersion, wave packets spread out with propagation distance, causing the amplitudes to decrease in addition to their reduction due to energy loss.

Xie and Nuttli (1988) included pulse broadening in geometrical spreading and proposed a method for its estimation, which is in broad use today (*e.g.*, Li *et al.*, 2009). These authors suggested that because of pulse broadening, energy density in a propagating wave reduces by factor $1/U$, where:

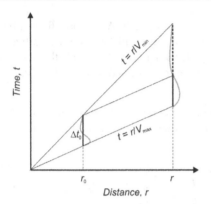

FIGURE 4.5
Effect of velocity
dispersion on geometrical
spreading. Because of pulse
broadening, its amplitude
additionally decreases.

$$U(r) = \frac{r}{\Delta t_0}\left(\frac{1}{V_{min}} - \frac{1}{V_{max}}\right), \quad (4.8)$$

r is the travel distance, and V_{min} and V_{max} are the minimum and maximum group velocities within the frequency band of interest. Parameter Δt_0 here is a constant used to normalize this factor so that $U(r_0) = 1$ at some reference distance, r_0. However, eq. (4.8) further simplifies to $U(r) = r/r_0$ and actually does not depend on the velocity dispersion parameters. This is not surprising, because geometrical spreading, as well as other wave-amplitude factors, is defined up to an arbitrary scaling which has to be removed by normalization. The factor containing velocities in eq. (4.8) is simply a part of such scaling.

The reason for missing the dispersion effect in expression (4.8) is in assuming the wave-pulse width to be zero at $r = 0$. To correct this problem, let us consider a pulse of finite duration Δt_0 at distance r_0 and linearly expanding from this point (Figure 4.5). Equation (4.8) then modifies to:

$$U(r) = 1 + \frac{r - r_0}{\Delta t_0}\left(\frac{1}{V_{min}} - \frac{1}{V_{max}}\right). \quad (4.9)$$

For weak dispersion, r_0 can be selected so that the second term in (4.9) is small within the range of observation distances. Therefore, $U(r)$ can be rendered in the attenuation-coefficient form,

$$U(r) \approx e^{2a_d(r-r_0)}, \quad (4.10)$$

where α_d is the spatial attenuation coefficient caused by dispersive pulse broadening,

$$\alpha_d = \frac{1}{2\Delta t_0}\left(\frac{1}{V_{min}} - \frac{1}{V_{max}}\right) \approx \frac{\delta \ln V}{2V \Delta t_0}, \quad (4.11)$$

and $\delta \ln V$ is the relative group velocity variation across the observation frequency band. For example, for fundamental-mode Rayleigh waves at frequencies $f < 0.3$ Hz, $(\delta \ln V)/V \approx 0.07$ s/km (see surface-wave examples in Section 6.1). By taking $\Delta t_0 \approx 500$ s, we obtain $\alpha_d \approx 0.7 \cdot 10^{-4}$ km^{-1}, which is comparable to the observed values of α for surface waves. Note that higher-frequency wavelets with smaller Δt_0 broaden stronger; however, within the same wave, the effect of α_d is frequency independent, *i.e.,* "geometrical".

Dispersion also affects the relation between the spatial and temporal attenuation coefficients. Without taking pulse broadening into account, $\chi = V\alpha$, where V is the group velocity (Aki and Richards, 2002, p. 293). This relation was obtained by equating the harmonic-wave amplitude at the dominant frequency, $\exp(-\chi t)$, with the amplitude of the wavelet maximum, $\exp(-\alpha r)$, which is located at travel distance $r = Vt$ at time t. However, because of pulse broadening, the relation between these amplitudes should actually be:

$$\frac{1}{\sqrt{U}} \exp(-\chi t) = \exp(-\alpha r), \qquad (4.12)$$

leading to,

$$\chi = V\alpha - \frac{\ln U}{2t} \approx V(\alpha - \alpha_d). \qquad (4.13)$$

This expression shows that in the presence of dispersion, α should always exceed α_d. This appears reasonable, because the limit of $\alpha = \alpha_d$ corresponds to a wave attenuated by pure pulse broadening without energy loss.

4.5 Scattering

The term "scattering" refers to the theoretical description of a wave process rather than some specific process. In wave mechanics, all types of wave propagation are described by solutions to the corresponding wave equations and, in this sense, there exists no special "scattering" processes. From another point of view, many useful wave-equation solutions were obtained by using the scattering approximation. Deviations of geometrical spreading from the background level or other inaccuracies in the forward model can be handled by the general scattering theory. Note that our attenuation-coefficient approach also represents a scattering-theory type approximation, in which the wave states are characterized by their amplitudes (see eqs. (1.2) and (1.3)).

Scattering is a type of solution to the wave equations in which the wavefield is represented by a superposition of some asymptotic states denoted as 'in' and 'out'. The 'in' states are approximated as freely traveling in some "distant

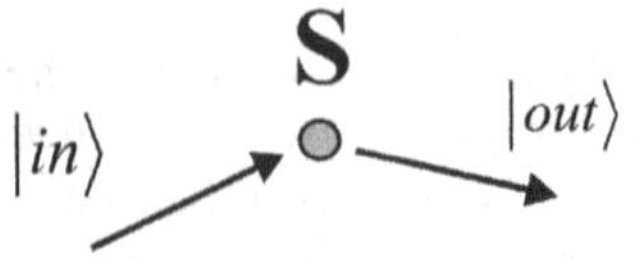

$$S$$

$$|in\rangle \quad\quad |out\rangle$$

$$|out\rangle = |in\rangle + S|in\rangle$$

FIGURE 4.6

Scattering-matrix approximation. The output state of the field, $|out>$, is a combination of the unperturbed field, $|in>$, and a linear perturbation caused by the scattering operator, S, acting on the input state.

past," whereas the 'out' states are considered similarly free from interaction in some "distant future". The case with only two states involved is called "single scattering" and, by using series of such states, the "multiple-scattering" approximation is obtained.

The general relation between the 'in' and 'out' states is given by the "scattering matrix" (Figure 4.6). As the equation in this figure shows, the scattering-matrix approximation only affirms that any 'out' state consists of the incident one plus a perturbation which is linear in terms of the incident state.

In principle, multiply scattered solutions can represent as close approximation to reality as needed and can be useful in solving very complex theoretical problems. For example, in the 1960's, the "scattering-matrix" formulation of the quantum field theory was viewed as a fundamental physical principle striving to explain most processes in the Universe. However, this trend later gave in to favoring the more general field theory based on the Lagrangian formulation.

The choice between the single-, multiple-scattering, or other models strongly depends on the accuracy of the background solution. Scattering theory is used when the background solution is close to the reality, but still needs to be improved. For example, to model wave propagation through a stack of horizontal elastic layers, iterative multiple scattering could be used in order to account for multiple reflections and conversions. However, with the use of the propagator method (Aki and Richards, 2002, p.393–406), an exact solution accounting for all multiples can be found, and therefore scattering-type solutions would not be needed at all.

4.5.1 Lippmann-Schwinger equation

A general relation between single and multiple scattering is illustrated by the Lippmann-Schwinger equation, which is very important in scattering theory. Using the notation from quantum mechanics, some stationary field $|\phi>$ (e.g., a plane wave) represents an eigenstate of some unperturbed operator, H_0 (in this example, the Hamiltonian):

$$H_0|\phi\rangle = E|\phi\rangle.\tag{4.14}$$

In the presence of some perturbation of the model, δH, the eigenstate changes to $|\psi\rangle$, which should satisfy, for the same eigenvalue E,

$$\left(H_0 + \delta H\right)|\psi\rangle = E|\psi\rangle.\tag{4.15}$$

By subtracting these two equations, we can write:

$$|\psi\rangle = |\phi\rangle + \left(E - H_0\right)^{-1}\delta H|\psi\rangle.\tag{4.16}$$

However, operator E-H_0 is singular, and it should be regularized by adding infinitesimal imaginary shifts to it:

$$|\psi^{\pm}\rangle = |\phi\rangle + \left(E - H_0 \pm i\varepsilon\right)^{-1}\delta H|\psi^{\pm}\rangle.\tag{4.17}$$

This is the Lippmann-Schwinger equation.

The interpretation of the $|\phi\rangle$ and $|\psi^{\pm}\rangle$ states requires some care (Weinberg, 1995). Nevertheless, for our purposes, it is sufficient to intuitively view $|\psi^-\rangle$ and $|\psi^+\rangle$ as the 'in' and 'out' states, respectively. Note the recursive nature of this equation. Starting from $|\psi^{\pm}\rangle = |\phi\rangle$ in the interaction-free case, we have for single scattering:

$$|\psi_1^{\pm}\rangle = |\phi\rangle + \left(E - H_0 \pm i\varepsilon\right)^{-1}\delta H|\phi\rangle.\tag{4.18}$$

This is known as the Born approximation. Note that it does not preserve the energy because the 'in' state $|\phi\rangle$ is left intact in the 'out' solution. Plugging this solution into the right-hand side of (4.17), we obtain double scattering:

$$\begin{aligned}|\psi_2^{\pm}\rangle = |\phi\rangle + \left(E - H_0 \pm i\varepsilon\right)^{-1}\delta H|\phi\rangle + \\ + \left(E - H_0 \pm i\varepsilon\right)^{-1}\delta H\left(E - H_0 \pm i\varepsilon\right)^{-1}\delta H|\phi\rangle,\end{aligned}\tag{4.19}$$

and so on.

Thus, the main point to be learnt from the above derivation is that if the perturbation to the eigenproblem operator δH is known, the corresponding perturbations to the eigenstates can be obtained, to any degree of accuracy, by using recursive eq. (4.17). A similar method is used to derive the surface-wave attenuation in Section 6.1.

4.5.2 Types of scattering

Depending on the ways the wave states are defined, special cases of "forward-" and "back-scattering," or "small-angle scattering" are often identified, in cases when the wave states are characterized by propagation directions. Other instances of the general scattering approach include solving for interactions between the Earth's normal modes, splitting, and other specific relations between the states of the wavefield. Here, we only mention several cases of scattering that are important for subsequent discussions.

Single scattering

In seismic S-wave coda-envelope studies, scattering also has a more specific meaning related to the type of model of the medium. The medium is represented by some background structure (usually, uniform space or half-space) with known wave propagation. On this background structure, random perturbations are superimposed, and the amplitude of the input-to-output state coupling can be determined by using some kind of perturbation-theory approximation, such as those by Born or Rytov. At the microscopic scale of a single scatterer, scattering is described by the differential cross section,

$$\frac{d\sigma}{d\Omega} = \frac{\left(J_{out} r^2\right)\big|_{r \to 0}}{J_{in}}, \tag{4.20}$$

where J_{in} is the incident-wave energy-flux density, J_{out} is the scattered energy-flux density at distance r from the scatterer, and $d\Omega$ is the solid angle in which the scattering is measured (Figure 4.7). Because J_{out} is proportional to r^{-2} at $r \to 0$, this expression is constant. If n scatterers are present in a unit volume of the medium, then the corresponding differential scattering coefficient from this unit volume is defined as:

$$g = 4\pi n \frac{d\sigma}{d\Omega}. \tag{4.21}$$

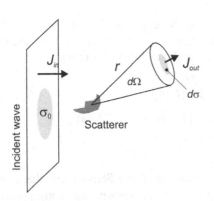

FIGURE 4.7

Differential ($d\sigma$) and total (σ_0) scattering cross sections for a single scatterer.

By averaging over-all scattering directions, the total scattering coefficient is obtained:

$$g_0 = \frac{1}{4\pi} \int g\,d\Omega = n\int d\sigma = n\sigma_0 \,, \qquad (4.22)$$

where g_0 is called the turbidity, or "haziness" of the medium, and σ_0 is the total scattering cross section. The latter quantity can be visualized as the area on the incident wavefront from which the energy is removed by a single scatterer (Figure 4.7). The dimension of g_0 is reciprocal distance and therefore its inverse is another useful characteristic of scattering, the mean free path: $\ell = 1/g_0$. The meaning of ℓ is the distance over which the energy flux in the incident wave decreases by e times due to scattering. Finally, the decrease of the incident-wave amplitude caused by such scattering is described by the spatial attenuation coefficient α as defined in eq. (1.5). In the above case of pure scattering, $\alpha = g_0/2$.

Multiple scattering

Single-scattering solutions are limited to weak scattering and short propagation distances, $d \ll \ell$. For longer paths, the scattering approximation can be improved by including multiple scattering events. Within the general scattering framework, virtually any solutions to the wave equations can be presented as infinite multiple scattering. For example, wavefront bending within a smoothly varying medium can be represented by an infinite series of refraction events. However, such solutions may become cumbersome and difficult for interpretation if the problem is tractable by other approaches.

In coda seismology, a specific type of multiple-scattering solution is used with great success. Because of the averaged character of the wavefield, it only depends on the time, distance from the source, and sometimes the anisotropy. By averaging scattering occurring at different angles near time t (Figure 4.8), the Radiative Transfer Equation governing the distribution of the wave-energy flux from the source can be obtained. For example, in 2D, this equation reads (Ishimaru, 1978):

$$\left(\frac{\partial}{\partial t} + V\hat{s}\cdot\nabla + V\ell^{-1}\right)J\left(\mathbf{r},t\,|\,\hat{s}\right) = \frac{V\ell^{-1}}{2\pi}\int f\left(\hat{s}\,|\,\hat{s}'\right)J\left(\mathbf{r},t\,|\,\hat{s}'\right)d\hat{s}' + S\left(\mathbf{r},t\,|\,\hat{s}\right), \qquad (4.23)$$

where V is the velocity of energy propagation, J is the energy flux per unit angle, $\hat{s}$ is the unit vector in the direction of energy propagation, S is the source function, and f is called the phase function. This function describes the energy transfer in a single scattering event (Figure 4.6) and is proportional to turbidity. A medium with parameters slowly varying at the scale of ℓ is assumed. Note that multiple scattering acts as an additional, distributed source (first term in the right-hand side of eq. (4.23)). At large distances, this term dominates over S, which is described as the "diffusive regime." In this regime, the rate of amplitude decay with distance is reduced, for example, from r^{-1} to $r^{-3/4}$ in a uniform 3-D space.

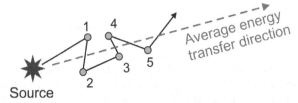

FIGURE 4.8

Multiple-scattering model. Multiple single-scattering events (gray circles) are combined and averaged to produce an averaged distribution of wave intensity.

In using the radiative transfer equation, there also appears to exist a caveat that is not always appreciated. Note that if coda intensity satisfies eq. (4.23) and exhibits different decay rates at the near and far distances, this still does not prove that random heterogeneities are indeed present within the medium. "Multiple scattering" indicated by this equation can occur by bending rays, as mentioned above, or by reflections underneath or above the zone of interest. For example, the mentioned change of the amplitude decay from r^{-1} to $r^{-\nu}$, with $\nu < 1$, could also be caused by rays refracting upward. Such refractions are common within the crust and mantle with downward velocity gradients.

4.5.3 Scattering in random media

As mentioned above, in coda studies, wave scattering is often understood in a more specific sense as interaction with random heterogeneities within the medium. In such cases, two characteristic cases need to be considered, differentiated by the scale-lengths of medium heterogeneities, L, compared to the incident wavelengths, λ.

Scattering on larger structures

If larger size heterogeneities dominate scattering, $L \gg \lambda$, then g_0 is approximately frequency independent (Warren, 1972). For a constant g_0, the total path-attenuation factor is frequency independent:

$$P(x) = e^{\frac{-g_0 x}{2}},$$ (4.24)

which is exactly the same as for geometrical spreading. Consequently, Q_s is proportional to the frequency, which is often observed and interpreted as scattering on large-scale structures within the lithosphere (Dainty, 1981; Sato and Fehler, 1998). However, such scatterers are in fact represented by the first-order, "deterministic" crustal and mantle structure, which contains much more information than implied by the single parameter Q_s. In association with larger scatterers, Q_s bears a confusing connotation of random heterogeneity in a uniform-background, isotropic-scattering model, which is clearly inappropriate for describing the layered lithospheric structures with pronounced reflectivity and numerous other properties.

In the presence of large-scale heterogeneity with $L \gg \lambda$, observations of scattered waves lose their statistical character. In such cases, one never measures the averaged wavefield properties, but only their local variations near the point of observation. Therefore, Q_s also loses its statistical meaning and only describes the average amplitude decay of the wavefield. However, this amplitude decay is more precisely described by the term "geometrical attenuation," which is an empirical measure of the residual, variable geometrical spreading closely related to the structure. Thus, we can definitely recommend using γ instead of Q_s to describe large-scale scattering.

Small-scale scattering

Unlike scattering on large structures, scattering on small-scale heterogeneities ($L \ll \lambda$) *can* lead to a behavior with $\chi|_{f \to 0} = 0$, which suggests a meaningful Q_s-type interpretation. This behavior follows from destructive interference of closely spaced reflectors and leads to scattering amplitudes increasing with frequency. For a simple illustration, consider 1-D propagation in a medium with bimodal distribution of reflectivities, which take on only two alternating values equal $r_i = r$ for odd and $r_i = -r$ for even i. Because the mean of a reflectivity time series equals zero, a stack of N such reflectors acts on a long wave roughly as only zero or one reflector per half wavelength (Figure 4.9):

$$\sum_{i=1}^{N} r_i \approx \begin{cases} r_N & \text{for } N = 2k+1, \\ 0 & \text{for } N = 2k. \end{cases} \tag{4.25}$$

Therefore, the finely layered scattering medium becomes progressively more "transparent" for longer waves, and the attenuation coefficients are proportional to the number of wavelengths per unit travel path, $\alpha \propto \lambda^{-1}$ or $\chi \propto f$. In such cases, the "scattering-quality" factor Q_s can be defined in a meaningful way:

$$Q_s^{-1} = \frac{1}{\pi} \frac{d\chi}{df}, \tag{4.26}$$

which should be approximately constant with frequency. However, the meaning of this quantity is still different from the one suggested by its name of a "Q". Q_s^{-1}, as defined by formula (4.26), represents the average reflection amplitude rather than a friction or spectral-width constant in some oscillatory process. Also, definition

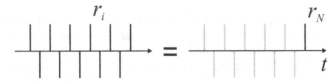

FIGURE 4.9
Time-average of a zero-mean, alternating reflection sequence is equivalent to a single reflection.

(4.26) does not exclude the possibility of $\gamma = \chi|_{\neq 0} \neq 0$ for the scattered field, which should arise when the reflectivity sequence is no longer not purely alternating, as in our example. Non-zero γ should also arise from bending rays, reflections, or other perturbations of geometrical spreading.

4.5.4 Relation to geometrical spreading

Accuracy of the background model is important for using the scattering approximation, because it only works when the corresponding scattered term S|*in*> is relatively small (Figure 4.6). When using the specific picture of scattering in a heterogeneous medium and attempting isolation of scattering effects, an accurate background model is particularly critical. Unfortunately, in real seismic measurements, such an accurate background model is practically never achieved, and it may likely be unfeasible with the accuracy required for Q modeling. In consequence, the inverted Q_s often represents a proxy for the incompletely corrected effects of the structure, *i.e.*, for the residual geometrical spreading. The ambiguity of scattering Q leads to attributing structural effects (such as diving waves, reflections) to scattering in total-energy and coda studies. The "scattering Q" terminology appears to be unfortunate, as it associates systematic refractions and reflections caused by lithospheric structures with scattering, which is usually considered random and described by using the path-lengths only (and not, for example, by reflection angles and amplitudes). As a result, the "scattering Q" as well as η in the $Q(f) = Q_0 f^\eta$ law may principally be distorted representations of the velocity gradients and depths to major reflectors within the lithosphere.

In attenuation measurements, the background model is represented by the corresponding geometrical-spreading correction. The effect of this correction is particularly difficult to differentiate from scattering on medium heterogeneities. The reason is that scattering is also caused by the structure and does exactly what the generalized geometrical spreading does — redistributes the elastic energy among different observation points and times. Its only difference is in its "random" character, which is a subjective characteristic imposed by an observer. Scattering can only be defined in relation to some theoretical background model and therefore scattering also absorbs any inaccuracies in such a model. For example, reflections from major, spatially separated contrasts, such as the crustal basement or Moho, are nearly frequency independent, and consequently the associated Q_s is nearly proportional to the frequency. However, treating major crustal features as random scatterers in a uniform space and describing them by scattering attributes such as Q_s would miss most points of seismic interpretation. Lithospheric features, such as the topography, Moho, and crustal discontinuities are significantly richer in information and much more valuable for interpretation than the average "Q_s". Such features should be included in the background model and accounted for in geometrical spreading, and consequently they should disappear from Q_s.

4.5.5 Observable scattering

There are two contrasting approaches to interpreting seismic scattering from attenuation observations. First, in the common practice, models for the background and scattering types are formulated, and based on them, scattering is recognized in the data. For example, from single-scattering coda models, only the total $Q^{-1} = Q_i^{-1} + Q_s^{-1}$ can be measured (Aki and Chouet, 1975), whereas multiple-scattering models are capable of separating these two Q^{-1}'s (Wu, 1985), but still be relative to a uniform background. Although uniform background models are still in common use, they are being continuously refined and now include smooth variations and source anisotropy (e.g., Carcolé and Ugalde, 2008).

Regardless of how good the models may be, they can hardly guarantee the <10% accuracy required for attenuation measurements. The uncertainty and variability of the subsurface structure makes this task impractical in the foreseeable future. Also, the need for scattering methods reduces as the detail of the structure is obtained and incorporated in the models. In a hypothetical, perfectly known structure, the entire wavefield is predictable and non-random, and consequently there is no room for scattering at all. In terms of Figure 4.6, the |in> state becomes an "eigenstate" of the structure, leading to $\mathbf{S} = 0$. Therefore, at any stage in the background/scattering model refinement, Q_s^{-1} represents the model error, i.e., the difference between the true and modeled structures. However, we attributed the same meaning to the residual geometrical spreading. Attenuation data contain no evidence of such errors due to random heterogeneities and, as shown in Chapter 5, they may arise from a variety of deterministic as well as random factors.

The second approach suggested in Morozov (2008) and taken in this book is to bypass the above complications and to use empirical geometrical-spreading models. The measured residual geometrical spreading then absorbs any model errors that are currently attributed to Q_s. For these reasons, Morozov (2009a, 2009c) suggested that the notion of Q_s is redundant and should be abandoned in favor of the more general geometrical spreading and Q_e.

Thus, when considering the phenomenological geometrical spreading, scattering becomes a part of it, and therefore unobservable by itself. The above trade-off problem reduces to separating the geometrical spreading from Q_e^{-1}. As shown in Chapter 5, both of these quantities may contain contributions from small-scale heterogeneities. These quantities are also not entirely separate, but can be viewed as parts of a single entity, which is the intrinsic temporal attenuation coefficient, χ_i. Note that the attenuation coefficient may, in fact, *not need* to be split into parts for consistent data interpretation. The total attenuation coefficient contains no trade-off, and interpretation based on it can be unambiguous and independent of geometrical-spreading models. However, if separation of dissipation from geometrical spreading is still desired, it can be based on the frequency dependence of χ, as described in Chapter 5. If small-scale heterogeneity of the medium is of interest, then it can only be extracted from χ_i by using the appropriate models; however, this is where we run the risk of errors and artifacts caused by the uncertainties of background models.

Beyond the measurements and modeling of averaged wave-amplitude envelopes, scattering could, in principle, be identified by the lack of coherence in the recorded events. However, this is still far beyond the current measurement capabilities, and also beyond the level of detail included in modeling. Furthermore, with improving sampling in both recording and modeling, scattering should retain its subjective, relative meaning. Larger portions of the wavefield would be marked as "coherent" (*i.e.*, understood and predicted) and more small-scale detail should be found in the model; as a result, scattering should again diminish in its significance.

Chapter 5

Attenuation Coefficient

Do not seek unity in the whole, but rather in the
uniformity of distinction.

Kozma Prutkov, Fruits of Reflection (1853-54)

As argued in the preceding chapters, the quality factor, Q, generally does
not represent a consistent property of the Earth's medium. By contrast to Q, the
attenuation coefficient, χ in eq. (1.6) (or α in (1.5)), can be viewed as at least a
useful approximation for such a property. According to the different groups of
attenuation mechanisms, χ can be subdivided into the geometrical spreading,
anelastic attenuation, and elastic scattering. However, when considering a realistic
(variable and measurable) geometrical spreading, scattering becomes
indistinguishable from the other two attenuation mechanisms in practical
observations.

The attenuation coefficients obtained from measurements are "apparent"
quantities, and they can be represented in the forms of the frequency-dependent Q
or t^*. In this chapter, we discuss the relations between these apparent quantities,
and develop a nomenclature of attenuation factors. We further propose a general
decomposition of the apparent attenuation coefficient χ in terms of the
corresponding local property of the medium. This local property is called the
"intrinsic attenuation coefficient" and denoted by χ_i. Similar to the apparent
attenuation coefficient, χ_i combines the variations of geometrical spreading and

anelastic attenuation, which become associated with the local properties of the medium. For traveling waves, a fundamental relation presenting the path factor as an exponential integral over the propagation path is obtained:

$$P(t,f) = G_0(t,f)\exp\left(-\int_0^t \chi_i d\tau\right).$$ (5.1)

This corresponds to eq. (1.8), in which χ now becomes χ_i. Such similarity between the apparent properties accumulated with observation time and the corresponding intrinsic properties accumulated over propagation paths or volumes is very general and also applies to other properties related to χ: the geometrical attenuation (γ) and effective attenuation (Q_e). In addition, according to another way to subdivide χ, this path-integral form applies to the frequency-independent and frequency-dependent attenuation coefficients.

To demonstrate that χ_i can indeed be attributed to small-scale structures and local material properties, two theoretical examples are considered in this chapter. Analytical solutions for χ_i are derived for three end-member cases: 1) short-scale incoherent, 2) coherent reflectivity at normal incidence in a layered medium, and 3) ray bending in a smoothly varying velocity structure. These solutions show that the intrinsic attenuation coefficient is related to the squared average reflectivity and the residual wavefront curvature. Both of these quantities are taken relative to the background model, so that when this model becomes more detailed, parts of χ_i transform into the corresponding contributions to the model's geometrical spreading.

With respect to the observations and inversion, both χ and χ_i can also be subdivided into their zero-frequency and frequency-dependent parts. As shown in this chapter, under reasonable assumptions, the frequency-independent part can be interpreted as an estimate of the residual geometrical spreading, and the frequency-dependent part can be interpreted as an estimate of the effective Q_e (which is apparent when χ is considered, or "intrinsic" when derived from χ_i). The non-geometrical part of the attenuation coefficient is directly related to the elastic-energy dissipation, including short-scale scattering within Earth's materials.

These models form the basis for quantitative interpretation of attenuation measurements and for their modeling in realistic Earth structures.

5.1 Apparent attenuation coefficient, Q, and t^*

As discussed in Chapter 2, the apparent χ in most cases is non-zero when extrapolated to frequency $f = 0$, from which it generally increases. The cumulative attenuation coefficient χ^* therefore has a similar behavior. Denoting the frequency-independent contribution in χ by $\gamma = \chi|_{f\to 0}$, we can explicitly separate it from the frequency-dependent part:

$$\chi^* = \gamma^* + \kappa^* f \text{, and } \chi = \gamma + \kappa f \text{,} \qquad (5.2)$$

where parameters κ can also be expressed through the "effective attenuation" quality factors Q_e and Q_e^*, suggesting some parallels with the Q picture:

$$\chi^* = \gamma^* + \frac{\pi}{Q_e^*} f \text{, and } \chi = \gamma + \frac{\pi}{Q_e} f \text{.} \qquad (5.3)$$

Both γ and Q_e are apparent quantities. Note that γ^* and Q_e^* are measured in frequency units.

In the conventional approach, the attenuation coefficients are further replaced with new parameters t^* and Q (Der and Lees, 1985; Aki and Chouet, 1975, respectively):

$$t^* = \frac{\chi^*}{\pi f} \text{, and } Q^{-1} = \frac{\chi}{\pi f} \text{,} \qquad (1.11 \text{ and } 1.9 \text{ again})$$

which are used in subsequent interpretations. From the general expressions in eq. (5.2), we therefore have:

$$t^* = \left(Q_e^* \right)^{-1} + \frac{\gamma^*}{\pi f} \text{ and } Q^{-1} = Q_e^{-1} + \frac{\gamma}{\pi f} \text{.} \qquad (5.4)$$

With Q_e = const, the second of these formulas was used by Dainty (1981) to describe the S-wave frequency-dependent Q^{-1} of S waves at 1–30 Hz. Dainty (1981) attributed the $\gamma/\pi f$ term in it to scattering Q^{-1} and emphasized its characteristic f^{-1} dependence. As discussed above, however, scattering Q can hardly be separated from the residual geometrical spreading, and an explanation in terms of γ, i.e., of the lithospheric structure affecting the geometrical spreading, appears more reliable.

In addition to t^*, which is measured from the geometrical-spreading–compensated amplitudes, Der and Lees (1985) also introduced the "apparent t^*," here denoted by $\bar{t}^*$,

$$\bar{t}^* = -\frac{1}{\pi} \frac{\partial}{\partial f} (\ln \delta P) = \frac{1}{Q_e^*} \left(1 - \frac{\partial \ln Q_e^*}{\partial \ln f} \right) \text{.} \qquad (5.5)$$

This quantity is determined from spectral-ratio measurements and is independent of geometrical spreading, and consequently from γ. Comparing eqs. (5.4) and

(5.5), note that the relative difference of t^* and $\bar{t}^*$ is also caused by the geometrical-spreading factor and is inherently frequency dependent:

$$\frac{t^* - \bar{t}^*}{\bar{t}^*} = \frac{\dfrac{\gamma^* Q_e^*}{\pi f} + \dfrac{\partial \ln Q_e^*}{\partial \ln f}}{1 - \dfrac{\partial \ln Q_e^*}{\partial \ln f}} \approx \frac{\gamma^* Q_e^*}{\pi f}. \tag{5.6}$$

This ratio is close to the "t^*-bias function" by Der and Lees (1985).

Another useful way to understanding the effects of the residual geometrical spreading on t^* measurements is by relating $\bar{t}^*$ to t^* (Der and Lees, 1985):

$$t^* + f\frac{dt^*}{df} = \bar{t}^*. \tag{5.7}$$

If the apparent $\bar{t}^*$ is known, this differential equation can be integrated to obtain the "true" $t^*(f)$. However, this integration is non-unique, and its uncertainty is given by the solution to the homogeneous counterpart of eq. (5.7). This homogenous solution reads $t^* = af^1$, where a is an arbitrary constant. By taking $a = \gamma^*/\pi$, we see from eq. (5.4) that this homogenous solution is again nothing more than the residual geometrical spreading.

5.1.1 Single-station measurement using spectral ratios

The temporal form of the attenuation coefficient is particularly useful in single-station attenuation measurements. With our definition of χ, measurement of attenuation parameters typically reduced to inverting some observed amplitude $A(t, f)$ corrected for geometrical spreading:

$$A_{GS}(t,f) = \frac{A(t,f)}{G_0(t,f)} = S(f)R(t,f)e^{-\chi t}, \tag{5.8}$$

or in logarithmic form,

$$\ln A_{GS}(t,f) = \ln S(f) + \ln R(t,f) - \chi t, \tag{5.9}$$

where t is the observation (travel) time, f is the frequency, $G_0(t,f)$ is the approximate geometrical-spreading factor determined from theoretical considerations or modeling, $S(f)$ is the common (typically, the source) spectrum, and $R(t, f)$ is the time-frequency response of the receiver. In multiple-station, traveling-wave studies, t in this expression could also be viewed as a proxy for different station locations, and for coda studies, t corresponds to the different time windows extracted within the coda. It is usually assumed that factor $R(t, f)$ can be removed from the amplitude (e.g., by deconvolution), and therefore the remaining inversion of eq. (5.9) implies simultaneously inverting for $S(f)$ and parameters γ and κ in $\chi = \gamma + \kappa f$.

When viewed as frequency dependent, parameter $\kappa = \pi/Q_e$ trades off with both $S(f)$ and γ, making the inversion poorly constrained. In most practical cases, it is sufficient to consider only a frequency-independent κ. The inverse problem above then becomes stable and reduces to detecting the dependence of the amplitudes on the time-frequency product, ft. This dependence can be isolated by taking "spectral ratios" of log-amplitudes at different observation times,

$$\frac{-1}{(t_2 - t_1)} \ln \frac{A_{GS}(t_2, f)}{A_{GS}(t_1, f)} = \chi = \gamma + \kappa f \cdot \tag{5.10}$$

Therefore, measurement of parameters γ and κ reduces to fitting a linear frequency dependence to the spectral ratios in eq. (5.10). If only the value of Q_e^{-1} is of interest, γ can be removed by forming the following double spectral ratios:

$$\kappa = -\frac{\ln A_{GS}(t_1, f_1) + \ln A_{GS}(t_2, f_2) - \ln A_{GS}(t_1, f_2) - \ln A_{GS}(t_2, f_1)}{t_1 f_1 + t_2 f_2 - t_1 f_2 - t_2 f_1}, \tag{5.11}$$

where $t_{1,2}$ and $f_{1,2}$ are pairs of observation times (or receiver stations) and frequencies, respectively. Note that in this expression, raw amplitudes $A(t, f)$ can be used instead of $A_{GS}(t, f)$, provided that the geometrical-spreading correction $e^{-\gamma t}$ can be approximated as frequency independent.

5.1.2 Stacked spectral ratios

With the use of the attenuation coefficient, the concept of "stacked spectral ratios" developed by Xie and Nuttli (1988), and used in many single-station studies can be explained very naturally. Similar to expression (5.10), the stacked spectral ratios simply represent estimators of χ from log-amplitude data. To see this, consider the expression for geometrical-spreading amplitude (5.8) again. In this expression, Xie and Nuttli (1988) considered only the traditional parameterization $\chi = \pi f/Q$, with $Q = Q_0 f^n$. However, this is unnecessary, and we will re-formulate the stacked spectral ratios by directly using χ.

In the stacked spectral-ratio technique, the observation time range is subdivided into several time windows centered at times t_i, $i = 1 \ldots N$, from which M mutual spectral ratios (5.10) are formed,

$$SR_m(f) = \frac{-1}{(t_j - t_i)} \frac{\ln A_{GS}(t_j, f)}{\ln A_{GS}(t_i, f)}, \tag{5.12}$$

where $m = 1 \ldots M$ is the spectral-ratio number, i and j are time-window numbers, and a different pair of (i,j) values is selected for each m. For amplitudes obeying model (5.8), both the source and receiver factors cancel in these ratios, and each of these values should equal χ. Therefore, the observed values of SR_m represent unbiased estimates for χ at the corresponding frequencies. The final estimate $\bar{\chi}$ is obtained by averaging ("stacking") these ratios at each frequency:

$$\bar{\chi}(f) = SSR(f) = \frac{1}{M} \sum_{m=1}^{M} SR_m(f). \tag{5.13}$$

Thus, the stacked spectral ratios also represent estimates for χ. The advantage of this estimator is in reduced uncertainty of χ, which is achieved by averaging.

Up to this point, the analysis by Xie and Nuttli (1988) and by many of their followers is equivalent to the attenuation-coefficient method of this book; however, these approaches strongly diverge afterwards. The above authors assume the frequency dependence of $\chi(f)$ to occur by means of a power-law $Q(f)$, and consequently they seek the corresponding power-law for $\chi(f)$,

$$\chi(f) = \pi Q_0^{-1} f^{1-\eta}. \tag{5.14}$$

The two parameters of this power law are then obtained by using a linear regression in the $(\ln f, \ln \bar{\chi})$ plane:

$$\ln \bar{\chi}(f) = \ln \frac{\pi}{Q_0} + (1-\eta)\ln f. \tag{5.15}$$

By contrast, in this book, we allow a non-zero zero-frequency limit of χ, and therefore look for a straight line in the $(f, \bar{\chi})$ plane:

$$\bar{\chi}(f) = \gamma + \kappa f. \tag{5.16}$$

Several data examples comparing these two parameterizations are given in Chapter 6. As discussed in Chapter 1, the available field data do not warrant preferring one of these approaches over the other. However, as shown throughout this book, our final selection in favor of (5.2) is based on: the physics of attenuation; the character of its frequency dependence; the simplicity, generality, and consistency of interpretation; and correlation with other geophysical and geological data.

5.2 Intrinsic attenuation coefficient

The relation of χ^* to t^* suggests a useful general relation between the apparent and *in situ* attenuation properties. For traveling waves for which "ray paths" can be traced, t^* is given by path integrals of the *in situ* Q^{-1} (Der and Lees, 1985):

$$t^* = \int_0^t \frac{d\tau}{Q} .$$ (part of 1.11)

Although we do not agree that the *in situ* Q^{-1} actually exists, the generalized χ^* should indeed be similarly accumulated along the travel paths. Loosely speaking, such accumulation means that the total scattering can be represented by a superposition of "single-scattering" events occurring within intervals $d\tau$ (Figure 4.8). Let us call the corresponding local property the "intrinsic attenuation coefficient," and denote it by χ_i:

$$\chi^* = \int_{Path} \chi_i d\tau .$$ (5.17)

Note that, although the integral form of this equation is similar to eq. (1.11), the meaning of integration of the apparent (χ) and intrinsic (χ_i) coefficients are different. Integration of χ is performed at a fixed receiver, whereas χ_i values are accumulated along the actual propagation path (Figure 5.1).

The new quantity, χ_i, combines the local variations of geometrical spreading, scattering, and anelastic attenuation. Among these three factors, the anelastic attenuation is the one which definitely requires a frequency-dependent χ_i (compare to eq. (1.10)). As argued in Section 3.7, the other two factors can be separated only by making additional simplifications, such as the frequency independence of geometrical spreading. The

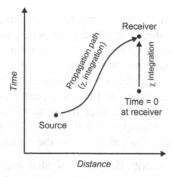

FIGURE 5.1
Integration paths of the apparent (eq. (1.11)) and intrinsic (eq. (5.17)) attenuation coefficients.

difficulty of their separation is related to the fundamental ambiguity in the definitions of geometrical spreading and scattering. In many practical cases, however, separation of these quantities is unnecessary, and χ_i can be treated as a single-medium property.

A most important consequence of eq. (5.17) is that δP represents a path integral which can be rendered in either temporal or spatial form,

$$\delta P = \exp\left(-\int_{\text{Path}} \chi_i d\tau \right) = \exp\left(-\int_{\text{Path}} \alpha_i ds \right), \qquad (5.18)$$

where s is the ray-path length, and α_i and χ_i are the corresponding spatial and temporal intrinsic attenuation coefficients. This shows that variations of geometrical spreading, scattering, and attenuation have similar characters and are accumulated over wave-propagation paths. The exponential form possesses important general properties and similarities to ray-, wave-, and quantum-field mechanics, which will be discussed below.

For body waves, if the ray paths can be considered frequency independent, eq. (5.17) can be decomposed into the frequency-independent and dependent parts,

$$\gamma^* = \int_{\text{Path}} \gamma_i d\tau, \qquad (5.19)$$

and

$$\kappa^* = \int_{\text{Path}} \kappa_i d\tau, \text{ or } Q_e^{-1} = \frac{1}{t} \int_{\text{Path}} Q_i^{-1} d\tau. \qquad (5.20)$$

For low-frequency waves for which the ray approximation is inappropriate (such as surface waves or normal modes), integral (5.17) can be generalized to integration over the volume of the field. To perform such generalization, let us denote the part of the energy density dissipated from the elastic field by E_D. The determination of the character of E_D presents the greatest difficulty in the approach. As argued in Chapter 2, the choice made in the visco-elastic theory, which equates E_D to the elastic energy, is physically inadequate. As also shown in Chapter 2, the kinetic energy, E_k, could be the appropriate choice for E_D, given the lack of data and a more precise theory of energy dissipation within the mantle.

Regardless of the choice for E_D, the energy density dissipated over time interval δt equals $E_D \chi_i \delta t$. This means that areas with higher $E_D \chi_i$ make the strongest contributions to the dissipated energy. On the other hand, considering a low dissipation rate, we can expect that the energy within the wave is continuously redistributed, and the relative distribution of energy (including E_D) is approximately preserved. This approximation is required in order for the general wave amplitude distribution to remain corresponding to the same mode, such as to the selected fundamental mode of the field. After this equilibration, energy densities at all points in the wavefield should decay at a common rate, which equals the observed decay rate χ. Therefore, the total dissipated energy can be written in two equivalent forms:

$$\int E_D \chi_i \delta t dV = \tilde{E} \chi \delta t, \qquad (5.21)$$

where the total energy of the field is $\tilde{E} = \int E_D dV$. Consequently, the observed attenuation coefficient is a weighted average of the intrinsic attenuation coefficients,

$$\chi = \frac{1}{\tilde{E}} \int E_D \chi_i dV. \qquad (5.22)$$

Recalling that in the conventional notation, both the apparent and intrinsic χ are proportional to the corresponding Q^{-1}, we see that integral expression (5.22) corresponds to the conventional forward model for Q^{-1}. However, the Fréchet kernel in (5.22) equals $E_D / \tilde{E}$, which differs from the traditional kernels derived from treating $\tan(Q^{-1})$ as phase shifts of the bulk and shear elastic moduli (e.g., Dahlen and Tromp, 1998, p.347). Velocity kernels are generally the largest in areas of greatest strains (e.g., near the crust-mantle boundary) and low near the free surface. By contrast, considering $E_D = E_k$, the attenuation kernels should be the largest in areas of high velocities, which would typically occur near the surface.

5.2.1 Linearity with frequency

As shown on data examples in Section 3.9 and Chapter 6, linearity of $\chi(f)$ with respect to frequency appears to be a common observation from many datasets at periods shorter than ~100 s. Apparently, there could exist some general reason for such linearity. The exponential path form (5.1) suggests that such reason could be the corresponding linearity of the intrinsic attenuation coefficient, $\chi_i(f)$. Such linearity within the available frequency bands may represent the most significant observation from the above theory, suggesting that the *in situ* attenuation parameters γ_i and κ_i are in fact frequency independent. This frequency independence would greatly simplify the interpretation and increase the constraining power of the resulting models.

Several theoretical examples in Section 5.6 show that $\chi_i(f)$ is indeed linear in f in several important cases. Interestingly, this linearity (or more generally,

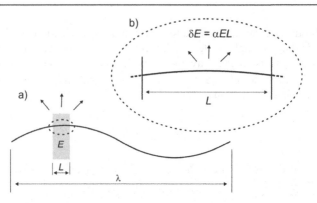

FIGURE 5.2

Attenuation in a long seismic wave: a)in the definition of Q (see Figure 1.1, p. 5), dissipation from a small volume (gray) is related to the entire wavelength λ; b)zoom in on the dissipation area, showing that for $\lambda >> L$, wave field within the scattering area is approximately uniform, and the dissipation rate should not depend on the amplitude gradient, i.e., on λ. The attenuation coefficient, α, can then be treated as a unique parameter of the medium.

increase with f) can be derived from a very general principle of localization of $\chi_i(f)$ and from the "off mass shell" behavior of the wave-mechanical system.

5.2.2 Localization and "off mass shell" dissipation

The greatest conceptual difficulty in modeling attenuation is what can be called *localization*, *i.e.*, justifying the existence of a χ_i as a true medium parameter. Strictly speaking, χ_i does not exist as an independent quantity, but results from a superposition of various attenuation mechanisms, such as the effects of fluids or fractures. At present, we have no models of the specific attenuation within the mantle, and therefore consider the attenuation coefficient χ_i itself as a phenomenological property responsible for seismic wave dissipation by the medium. This approximation means that the energy dissipation rate at any point in the medium depends on the local wave amplitudes at that point, but not on their gradient. Such an approximation should likely be reasonable for waves which are significantly longer than the medium's heterogeneities. In this limit, local energy dissipation should occur as if caused by a spatially uniform, oscillating wavefield (Figure 5.2).

Our phenomenological model of attenuation thus considers energy dissipation as occurring entirely due to the oscillations at a given point within the medium, and independently of the gradient of the wavefield amplitude. This approximation appears to be natural, because the processes causing dissipation, such as pore- or fracture-fluid flows are local and do not "see" other parts of the wavefield (Figure 5.2).

From the concept of localization, an interesting general property of attenuation can be suggested. Consider a plane wave with $\omega \neq Vk$, *i.e.*, not satisfying the wave equations. Such waves are called "off mass shell" in quantum field theory. Even though this wave does not propagate by itself, what attenuation rate would it have? According to the local character of the process, χ_i is a function of only frequency ω but not wavenumber k (Figure 5.2), and therefore χ_i *should increase with ω, but not k*. In the Lagrangian description, such behavior is also expected, because both the kinetic energy and dissipation function are local and proportional to ω^2 even "off mass shell". In Section 6.1.2, this principle will allow us to solve the Love-wave dissipation problem.

Note that in the visco-elastic theory, the above "off mass shell" test cannot be conducted. Because this theory starts not from the Lagrangian, but from the differential equations of motion, all its waves are always "on mass shell." Similar to the response to a variable ω, the response to the gradient of the field (*i.e.*, k) is therefore always encoded in the visco-elastic Q. Recall that for the same reasons, the kinetic and potential energies, and also their dissipation rates are not clearly differentiated in visco-elasticity.

5.3 Nomenclature of attenuation factors

The eqs. of (5.18) are very general and principally only state that the energy decay δE during a short propagation time δt is proportional to δt and also to the current wave energy E. This is the perturbation- (scattering-) theory approximation, which is valid for all processes involved in seismic-wave attenuation. In the same perturbation-theory sense (*i.e.*, for $\delta E/E \ll 1$), contributions of all these processes in α_i and χ_i should be additive, and thus we can talk about the attenuation coefficients associated with each of them individually. However, separation of these contributions in the observed quantities may be subtle and difficult. To approach such separation, we need to clarify the relevant terminology first. Several types of attenuation factors can be differentiated, as described below.

Geometrical spreading: The geometrical spreading, G, is a fundamental property which is either explicitly or implicitly present in all descriptions of attenuation. The definition of this quantity is not trivial, as discussed in Chapter 4. To summarize our conclusions in that chapter, geometrical spreading is variable and can be modeled only approximately, with accuracy that is typically insufficient for measuring Q^{-1}. Nevertheless, geometrical spreading can be easily *measured* and modeled empirically by means of the attenuation-coefficient technique described here.

Background geometrical spreading: By contrast to the true geometrical spreading, background geometrical spreading represents the best-known theoretical approximation to the attenuation- and scattering-free amplitudes, denoted $G_0(\mathbf{r},t,f)$. It can be derived by theoretical or numerical modeling, or by empirical fitting of the observed amplitudes. The only requirement is that the G/G_0 ratio should be close to one for the distance and time ranges of interest. This ratio is referred to as the **residual geometrical spreading**. Background geometrical spreading is corrected for when calculating the path factor with eqs. (1.2) and (1.3), and consequently the corresponding attenuation coefficient, $\chi_{BGS} \equiv 0$.

Apparent attenuation coefficients (χ and χ^*): These coefficients represent the measured attenuation and need to be differentiated from the *in situ*, or "intrinsic" coefficients below. The general problem is inverting the frequency-dependent apparent attenuation coefficients for the various intrinsic properties below.

Geometrical attenuation: This represents the deviation of the true geometrical spreading from its background model. It represents an estimate of the residual geometrical spreading in the attenuation-coefficient form: $\exp(-\chi_{GA}t) \approx G/G_0$. The corresponding parameter χ_{GA} therefore gives the correction applied to the background model in order to obtain at approximation for the true geometrical spreading. We can also talk about the *apparent* (as part of χ) and *intrinsic* (as part of χ_i) geometrical attenuation. The first of these quantities represents the logarithmic decrement of attenuation-free amplitudes, and the second describes ray curvatures and short-scale reflectivity, as shown in Section 5.6.

Anelastic (also called dissipation, χ_d) and elastic (also scattering, χ_s) attenuation coefficients: We use the term "anelastic" dissipation, χ_d, to denote the energy leaving the kinetic- and potential-energy parts of the Lagrangian of the field. By contrast, χ_s corresponds to the "elastic" scattering energy which is redistributed, but remains in the elastic field. Both of these coefficients have the "intrinsic" character of χ_i. Scattering is the most difficult to define rigorously, because it is associated with "random" structural variations. In the existing treatments (Sato and Fehler, 1998), scattering is only considered with respect to featureless, uniform-space, isotropic backgrounds. With the increase of detail recognized in the background structure, scattering effects are eliminated and become absorbed by the empirical geometrical spreading. On the other hand, for any approximation of the structure, the residual short-scale scattering effects become indistinguishable from those of χ_d if no special assumptions are made about the latter. For these reasons, it appears that χ_s is observationally intractable, similarly to scattering Q_s (Section 4.5.5). Nevertheless, scattering attenuation in uniform isotropic media is also the best-studied theoretically (*e.g.*, Sato and Fehler, 1998) and provides many insights into the mechanisms of elastic energy dissipation.

Intrinsic attenuation coefficient (χ_i): The use of term "intrinsic" here is only related to the local and macroscopic character of this quantity, as a property of the propagating medium. This use is different from the conventional association of "intrinsic" with pure absorption. Parameter χ_i represents the total effect of the residual geometrical spreading, anelastic dissipation, and elastic scattering taken at a point within the propagating medium: $\chi_i = \chi_{GA} + \chi_d + \chi_s$. Because χ_d and χ_s are observationally inseparable, χ_s can be incorporated into χ_d, and the problem of decomposing the intrinsic attenuation coefficient then simplifies to: $\chi_i = \chi_{GA} + \chi_d$.

Zero-frequency attenuation coefficients (γ_i and γ): Assuming that it can be inverted from the data, the frequency dependence of χ_i provides the best clues for separating χ_{GA} and χ_d in this quantity. The zero-frequency attenuation coefficient, γ_i, provides a good approximation for χ_{GA}. This is due to the approximate frequency-independence of χ_{GA}, as follows from the theoretical examples below in this chapter. Similarly to χ_i, γ_i is related to the corresponding observable (apparent) γ by eq. (5.17).

Parameters of frequency dependence (κ_i and κ): By their definitions, terms containing κ in the expression for χ (5.2) vanish at zero frequency. From the comparison with mechanics in Chapter 2, energy dissipation χ_d should approach 0 at $f \to 0$, which is also suggested by its traditional definition as $\chi_d = \pi f Q^{-1}$ (which is not really fulfilled because of the frequency dependence of Q). The character of $\chi_d = O(f)$ follows from dissipation occurring from the kinetic-energy part of the Lagrangian, which is itself proportional to f^2. Thus, the significance of the *in situ* κ_i and its apparent counterpart κ is in their containing the entire contributions of dissipation and small-scale scattering.

Further, if $\chi_d|_{f=0} = 0$, then $\kappa_i = \chi_d/f$ becomes a useful **non-geometrical**, or **energy-dissipation factor** of the medium. If geometrical spreading is frequency independent, then $\kappa_i = (\chi_i - \gamma_i)/f$ combines the effects of anelastic material absorption and scattering at scale-lengths not accounted for by the deterministic structural model. For an analog to the conventional terminology, parameter κ_i can be transformed into an **intrinsic quality factor, Q_i,** by $Q_i = \pi/\kappa_i$. Note once again, that term "intrinsic" here means related to the local, small-scale properties of the material, which may contain contributions from scattering on small random heterogeneities. By its meaning, Q_i represents the *in situ* material property corresponding to the observable (effective) Q_e.

To complete our nomenclature, note that the ratio:

$$f_c = \frac{|\gamma_i|}{\kappa_i} \tag{5.23}$$

represents the **intrinsic cross-over frequency** of the medium, at which the attenuation levels from the geometrical and dissipation/scattering mechanisms are equal. Similarly to other quantities discussed above, f_c also has its apparent counterpart, which is also denoted f_c. For wave frequencies above f_c, dissipation mechanisms dominate attenuation, and the apparent Q usually shows weaker frequency dependences. For $f < f_c$, geometrical spreading effects dominate, and Q is strongly frequency dependent. By taking the ratio of the geometrical and total attenuation coefficients, we obtain another useful relative attribute:

$$B_0 = \frac{\gamma_i}{\chi_i} = \frac{1}{1 + f/f_c} \cdot$$ (5.24)

Because γ_i corresponds to the "scattering Q^{-1}" in scattering-attenuation studies, this ratio equals the "**seismic albedo**," which is discussed below. However, as the above relation shows, seismic albedo contains no additional information compared to f_c. Similar to Q, the use of this quantity may be complicated by its embedded frequency dependence, and f_c should still be preferable for characterizing the relative contributions of the geometrical and dissipation factors.

5.4 Relation of (γ, Q_e) to (Q_0, η)

Although the (γ, Q_e) description appears to be preferable physically, numerous attenuation measurements and interpretations were reported in the (Q_0, η) form. Fortunately, the two parameterizations can be approximately related to each other in many practical cases of not very high η. This relation provides a way to estimate γ and Q_e from many published observations. In addition, it explains the general physical meaning of parameter η.

To infer a mapping between these two parameterizations, consider an attenuation coefficient governed by eq. (5.2) but parameterized as $\pi f Q^{-1}$, where Q is given by the power-law equation, $Q = Q_0 f^\eta$. Therefore, we have:

$$\gamma + \frac{\pi f}{Q_e} \approx \frac{\pi f}{Q_0 \left(f/f_0 \right)^\eta} = q f^{1-\eta},$$ (5.25)

where $q = \pi f_0^\eta / Q_0$ is a constant. By taking the logarithm of both sides, an approximation for η is obtained:

$$\eta = 1 - \frac{\Delta \ln \left(1 + \dfrac{f}{f_c} \right)}{\Delta \ln f},$$ (5.26)

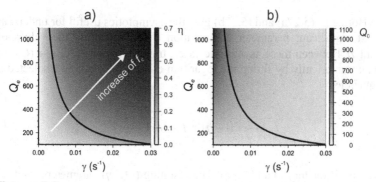

FIGURE 5.3

Relation of (γ, Q_e) parameters to (Q_0, η) approximated within the frequency range of 1–10 Hz: a) η and b) Q_0. The hyperbolas drawn in both plots correspond to $f_c = Q_e \gamma / \pi = 1$ Hz.

where ΔX denotes the variation of quantity X across the frequency band of interest. In this expression, f becomes the characteristic frequency of the measured band, and f_c is defined in eq. (5.23). For two asymptotic cases of high and low observation frequencies, we have, respectively:

1) For $f >> f_c$, the logarithm above is dominated by the second term:

$$\eta \approx 1 - \frac{\Delta \ln \frac{f}{f_c} + \Delta \left(\frac{f_c}{f} \right)}{\Delta \ln f} \approx \frac{f_c}{f^2} \frac{\Delta f}{\Delta \ln f} \approx \frac{f_c}{f}, \qquad (5.27a)$$

and therefore $\eta << 1$. This case corresponds to weak geometrical attenuation and falls far below the hyperbola $\gamma Q_e = \pi f$ in Figure 5.3a.

2) For a strong geometrical contribution to attenuation, $f << f_c$ (upper-right corners of the diagrams in Figure 5.3) and the second term under the logarithm is small. Therefore η approaches one:

$$\eta \approx 1 - \frac{1}{f_c} \frac{\Delta f}{\Delta \ln f} \approx 1 - \frac{f}{f_c}. \qquad (5.27b)$$

In summary, η is generally determined by the ratio of the central measurement frequency to the cross-over frequency, f_c. High $\eta \approx 1$ should be expected in areas of low attenuation and high geometrical attenuation (where f_c is high), or also for measurements performed at lower frequencies.

Both eqs. (5.27a) and (5.27b) give the asymptotics useful for understanding the most significant parameter combination (f/f_c) controlling the values of the apparent η. Between these asymptotic cases, transformation (γ, Q_e) → (Q_0, η) can be found numerically by fitting the logarithm of eq. (5.25) within a selected range of frequencies:

$$\ln\frac{q}{\gamma}+(1-\eta)\ln f \approx \ln\left(1+\frac{f}{f_c}\right), \qquad (5.28)$$

and solving it for $\ln(q/\gamma)$ and $(1-\eta)$. The resulting (Q_0, η) parameters for 1–10 Hz frequency range are shown in Figure 5.3. Note that q (and Q_0^{-1}) is proportional to γ, and η is principally controlled by the values of f_c. For $\gamma = 0$, $Q_0 = Q_e$, and Q_0 quickly decreases with increasing γ (Figure 5.3).

5.5 Path-integral form of attenuation

Exponential path integrals similar to those in eq. (5.18) are broadly used for describing wave propagation in quantum mechanics and field theory, where they are known, for example, as "Feynman integrals". Such an integral form is characteristic for "propagators" and arises from their fundamental factorization property requiring that some type of a "product" of two propagators also represents a propagator. For example, Green's functions $G(\mathbf{x}_1,\mathbf{x}_2)$ of some field possess this propagator property, which in this case reads:

$$G(\mathbf{x}_1,\mathbf{x}_2) = \int G(\mathbf{x}_1,\mathbf{x})G(\mathbf{x},\mathbf{x}_2)d^3\mathbf{x}. \qquad (5.29)$$

The dynamic ray theory (Červený, 2001) also illustrates the origins of exponential form (5.18). In this theory, factorization (5.29) arises from the multiplicative property of ray propagator Π (eq. (4.4.86) in Červený, 2001), which in logarithmic form can be written as:

$$\ln\Pi(R,S) = \ln\Pi\left(R,\tilde{Q}_N\right)+\sum_{i=N}^{1}\Upsilon\left(\tilde{Q}_i,\tilde{Q}_{i-1}\right). \qquad (5.30)$$

In this expression, S is the source, R is the receiver, and Q_i and $\tilde{Q}_i$ are the incidence and emergence points at the i-th interface, respectively (Figure 5.4), and

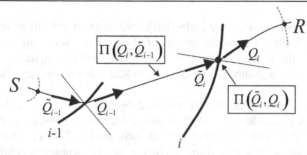

FIGURE 5.4
Notation in ray propagator in a layered medium (eq. (5.30)). Geometrical
spreading is related to the ratio of wavefront curvatures (gray dashed lines)
at the receiver (R) and source (S).

$$\Upsilon\left(\tilde{Q}_i,\tilde{Q}_{i-1}\right) = \ln\left[\Pi\left(\tilde{Q}_i,Q_i\right)\Pi\left(Q_i,\tilde{Q}_{i-1}\right)\right]. \tag{5.31}$$

In our notation (eq. (5.18)), δP corresponds to $\Pi(R,S)$, and
$\Upsilon\left(\tilde{Q}_i,\tilde{Q}_{i-1}\right)$ corresponds to $\int \chi_i d\tau$, where the integral is taken from point Q_{i-1} to Q_i
along the ray.

Partitioning of $\Upsilon\left(\tilde{Q}_i,\tilde{Q}_{i-1}\right)$ into the interface- and path-related factors in eq.
(5.31) shows that both reflection/conversion and ray-bending effects influence the
values of γ_i. In Section 5.6, we derive exact expressions for such effects in several
end-member examples.

5.6 Theoretical models

In this section, we discuss several theoretical examples showing what
mechanisms could create the linear frequency dependences of χ in the amplitude
path factors. As we will see, the residual geometrical spreading γ_i and the
corresponding frequency-independent κ_i occur in several end-member cases:

1) **Refraction in a medium with smoothly varying velocities:** This
    example shows that γ_i is also related to the variations of wavefront
    curvature (*i.e.*, to perturbations of the traditional geometrical spreading).

2) **Incoherent normal-incidence reflectivity**, corresponding to large
    numbers of sparse reflections occurring during long propagation paths. In
    this case, case χ_i is proportional to the gradient of the acoustic
    impedance.

3) **Coherent small-scale reflectivity**, with random reflectivity scale lengths much shorter than the length of the incident wave. In this case, the reflectivity becomes "coherent" at $f \rightarrow 0$, and consequently $\gamma_i = 0$. This example is studied numerically, as in Richards and Menke (1983).

4) **Coda:** This case is characterized by the absence of a predominant propagation direction, random omnidirectional short-scale scattering, and averaging over multiple scattered-wave modes. The apparent frequency-dependent χ arises from averaging of the scattering amplitudes and attenuation properties over large areas surrounding the receivers. The models for local- and long-range coda sampling developed here will also be used in the next chapter.

All these cases relate to elastic processes of refraction or reflectivity, which fall under the category of "scattering," or more generally of the "geometrical" attenuation processes described in this book. Our specific goal here is to illustrate the origins of the geometrical parameter γ_i in theoretically tractable cases. The physics of anelastic attenuation is not discussed in these examples, and its effects are simply incorporated by factor $\exp(-i\kappa_i ft)$.

5.6.1 Refraction in a medium with smoothly varying velocities

In the absence of interfaces and caustics, geometrical spreading is caused by variations in the wavefront curvature (Figure 5.4). In dynamic ray theory, this curvature is measured by the trace of wavefront curvature matrix, $H = \frac{1}{2} \operatorname{tr} \mathbf{K}$, which is obtained from second derivatives of the travel-time field T with respect to the wavefront-orthonormal coordinates y_k (eq. 4.6.15 in Červený, 2001):

$$K_{ij} = V \frac{\partial^2 T}{\partial y_i \partial y_j}, \tag{5.32}$$

where V is the wave velocity. Curvature H is related to the ray-theoretical geometrical spreading by the following differential equation (eqs. (4.10.28) and (4.10.29) in Červený, 2001):

$$H = L^{-1} \frac{dL}{ds}, \tag{5.33}$$

where L is the geometrical-spreading denominator and s is the ray arc length. The solution to this equation relating $L(R)$ at the receiver to $L(S)$ at the source is:

$$L(R) = L(S) \exp\left(\int_S^R H ds\right), \tag{5.34}$$

which again has the exponential path-integral form of the path factor in eq. (5.18). Ratio $G = L(S)/ L(R)$ represents the desired geometrical-spreading factor, which equals $G_0 \delta P = G_0 e^{-\alpha s}$, where G_0 is the background approximation for geometrical spreading. In the presence of anelastic attenuation κ_i, the path factor becomes:

$$\delta P = \frac{1}{G_0} \frac{L(S)}{L(R)} \exp\left(-f\int_S^R \frac{\kappa_i ds}{V}\right),$$ (5.35)

and consequently the spatial attenuation coefficient is,

$$\alpha_i = H - \ln G_0 + \frac{\kappa_i}{V_i} f,$$ (5.36)

with the corresponding relation for χ_i. This expression shows that for smoothly refracting waves, α_i contains a frequency-independent "geometrical" part $(H - \ln G_0)$, which equals the difference of the actual wavefront curvature from the one predicted by the geometrical-spreading law selected as the background reference.

5.6.2 Incoherent normal-incidence reflectivity

To understand the relation of the *in situ* attenuation coefficient to the properties of the medium, it is instructive to analyze its properties in a simple 1-D medium. For plane-wave propagation, the theoretical geometrical-spreading factor equals one; however, reflections in a heterogeneous medium cause deviations from this value. Because transmission coefficients can be completely described by the reflection-coefficient series, the geometrical part of the attenuation coefficient should also be expressed through reflectivity. In fact, as shown below, the geometrical attenuation coefficient equals half the average of squared reflection coefficient.

To begin, consider a boundary between two layers of acoustic impedances Z_{j-1} and Z_j (Figure 5.5). The specific expression for impedance depends on the local properties of the medium, wave type, and the angle of its incidence on the boundary. From Section 2.7, we know that the complex-valued acoustic impedance for a P wave or S wave in an attenuative medium at normal incidence is:

$$Z = \rho V\left(1 + \frac{i}{2Q_i}\right),$$ (5.37; repeated (2.117)

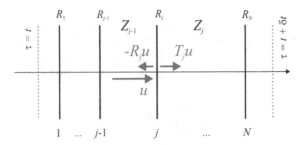

FIGURE 5.5

One-dimensional plane-wave reflection-transmission problem. Solid lines are reflectors; dashed lines are incident-wave wavefronts at times t and $t + \delta t$, respectively. Multiple reflections are ignored.

where ρ, V, and Q_i^{-1} are the mass density, wave velocity, and parameter of anelastic attenuation, respectively. Note that although we generally argue that Q cannot be considered as a medium property, this parameter is retained here for convenience of comparison to the current terminology. Such use of Q is possible, and $Q = Q_i$ because we are considering an otherwise uniform background in which the geometrical spreading is accurately known, and a single wave type is used.

Considering, for simplicity, the normal-incidence case and denoting the displacement in the incident wave by u, the displacements in the reflected and transmitted waves become ($-R_j u$) and $T_j u$, respectively (Figure 5.5), where, R_i is the reflection coefficient,

$$R_j = \frac{Z_j - Z_{j-1}}{Z_j + Z_{j-1}}, \tag{5.38}$$

and $T_j = 1 - R_j$ is the transmission coefficient,

$$T_j = \frac{2Z_{j-1}}{Z_j + Z_{j-1}}. \tag{5.39}$$

The corresponding energy transmission coefficient is:

$$T_{E,j} = \frac{Z_j}{Z_{j-1}} T_j^2 = \frac{4Z_{j-1}Z_j}{\left(Z_j + Z_{j-1}\right)^2}, \tag{5.40}$$

and the energy reflection coefficient equals $R_{E,j} = 1 - T_{E,j}$.

For small impedance contrasts, the above coefficients are:

$$R_j = \frac{1}{2}\delta_j(\ln Z),$$ (5.41)

$$T_j = 1 - \frac{1}{2}\delta_j(\ln Z), \text{ and}$$ (5.42)

$$T_{E,j} = 1 - \frac{1}{4}\left|\delta_j(\ln Z)\right|^2 = 1 - \left|R_j\right|^2,$$ (5.43)

where $\delta_j(X)$ denotes the contrast in parameter X across the j-th boundary. Switching to a continuous $Z(t)$ description, the impedance contrasts over an infinitesimal propagation time interval $[t, t+\delta t]$ can be considered small, and therefore from eq. (5.43) we get,

$$\ln T_E \approx -\sum_{j=1}^{N}\left|R_j\right|^2 = -\int_t^{t+\delta t}\left|r\right|^2 d\tau,$$ (5.44)

where $r(t)$ is the root-mean square density of reflectivity.

Equation (5.44) only gives the transmission loss caused by reflections on the boundaries passed by the wave between propagation times t and $t + \delta t$. The anelastic medium attenuation over the same time interval leads to an additional energy decay:

$$\ln T_E \approx -\int_t^{t+\delta t}\left|r\right|^2 d\tau - 2f\int_t^{t+\delta t}\kappa_i d\tau,$$ (5.45)

where κ_i is the non-geometrical attenuation factor.

If the transmitted waves interfere incoherently, the energy transmission coefficients combine multiplicatively with propagation time, and their logarithms are therefore additive. For a wave traversing N boundaries in a finite propagation time t, the energy density $E(t)$ is (Figure 5.5):

$$E_N = E_0\prod_{j=1}^{N}T_{E,j} = E_0\exp\left[\sum_{j=1}^{N}\ln T_{E,j}\right],$$ (5.46)

or, in terms of the continuous reflectivity function, $r(t)$,

$$E(t) = E(0)\exp\left\{-\int_{\tau=0}^{t}\left[\left|r\right|^2 + 2\kappa_i f\right]d\tau\right\}.$$ (5.47)

This expression shows that the logarithm of the transmitted energy loss is a path integral:

$$\ln E(t) - \ln E(0) = -\int_0^t \left[|r|^2 + 2\kappa_i f \right] d\tau,$$
(5.48)

and consequently the temporal attenuation coefficient equals:

$$\chi_i = -\frac{1}{2} \frac{d \ln E(t)}{dt} = \frac{|r|^2}{2} + \kappa_i f.$$
(5.49)

The corresponding spatial attenuation coefficient, $\alpha = \chi/V$, equals:

$$\alpha_i = \frac{|r|^2 + \kappa_i}{2V} = \frac{|r_{\text{spatial}}|^2}{2} + \frac{\kappa_i}{V} f.$$
(5.50)

Note the difference between the temporally and spatially averaged root mean square reflectivities denoted by r and r_{spatial}, respectively.

Thus, in the case of incoherent 1-D acoustic-wave propagation, the geometrical attenuation coefficient equals half the corresponding path-averaged squared reflectivity. As path-averaged properties, α and χ can be evaluated over finite propagation-time intervals, and therefore they can be time-dependent.

Note that when $\kappa_i = 0$, the resulting α_i or χ_i are associated with geometrical attenuation, which is geometrical forward scattering in this case. In the approximation considered here (normal incidence and absence of multiple reflections), these geometrical α or χ are independent of the frequency and length of the incident wave.

If multiple reflections are present, frequency-dependent effects (tuning) should arise even in the geometrical limit. Such effects should likely have the form of resonance peaks rather than a continuous trend with frequency. In Chapter 3, we identified such peaks in ultrasonic laboratory attenuation measurements (*e.g.*, Figure 3.15), and similar phenomena can be identified in local-earthquake coda χ data (Section 6.8.1).

5.6.3 Coherent small-scale reflectivity

The example above assumed incoherent interference of scattered arrivals, which occurs when the scatterers are large and spaced at large distances compared to the incident wavelength. In this section, we consider the opposite limit of scatterers that are small and relatively closely spaced. In this case, destructive

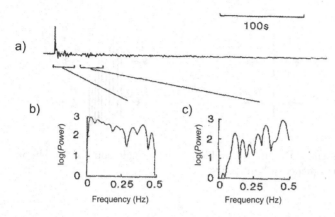

FIGURE 5.6
Transmission responses of a simulated 100-boundary sequence (modified after Richards and Menke, 1983): a) the transmitted record resulting from a single initial pulse; b) the power spectrum of its initial part (main pulse with early forward scattering; and c) spectrum of the later forward-scattered waves.

interference of scattered waves occurs, and the attenuation coefficient exhibits strong frequency dependence. For simplicity, we consider the 1-D case, in which scattering reduces to normal-incidence reflectivity. Originally, this example was analyzed by Richards and Menke (1983), who demonstrated the frequency-dependent effects of scattering (Figure 5.6) and presented them in terms of the "scattering Q". Let us briefly review this important example from a somewhat different angle, and in particular look closely at the decay of spectral amplitudes with time.

During 1-D propagation, the wavefronts remain perfectly planar, and consequently the theoretical geometrical spreading equals exactly one. Therefore, all perturbations of the wavefield are due to elastic scattering on the boundaries and anelastic attenuation between them. In particular, scattering causes a part of the wave energy to be reflected backward (this is referred to as "back-scattering"), and part of it continues propagating forward while being delayed relative to the primary wave. This delayed part of the propagating wavefield is called "forward-scattered" (Figure 5.7). From numerical simulations and real data, the initial wave pulse and both the forward- and back-scattered waves exhibit linear spectral variations, which increase with propagation time. Note that the senses of these variations are opposite for the initial pulse and back- and forward-scattered waves, whose high frequencies are progressively depleted and enhanced, respectively (Figure 5.6).

Conventionally (*e.g.*, Richards and Menke, 1983), the relative changes in the spectra (Figure 5.6) are assumed to be proportional to the number of

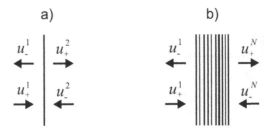

FIGURE 5.7
One-dimentional scattering problem: a) on a single boundary and b) on a random sequence of boundaries.

wavelengths traveled by the incident wave. The resultant spectra are fit by using approximations of the form $\exp(-\pi f t/Q_s)$, and scattering is automatically viewed as analogous to the anelastic attenuation, for which the corresponding spectral amplitude decay is $\exp(-\pi f t/Q_i)$. However, as shown below, the proportionality to the number of wavelengths is only valid for coherent superposition of scattered waves, and for incoherent scattering, the attenuation becomes independent of the incident wavelength. We therefore take a conservative view (Section 3.6) and use Q-type parameters only for the frequency-dependent part of the attenuation coefficient.

To derive the attenuation coefficient for the general case including both weak and strong, back and forward scattering, note that the scattering problems for both the single- and multi-boundary cases (Figure 5.7) can be described equivalently by using the general scattering-matrix formulation,

$$\begin{pmatrix} u_+^N \\ u_-^N \end{pmatrix} = \mathbf{T}_{N,1} \begin{pmatrix} u_+^1 \\ u_-^1 \end{pmatrix},$$ (5.51)

where $N = 2$ for the single-boundary case, and $\mathbf{T}_{N,1}$ is the transmission matrix relating the states on the right to those on the left in Figure 5.7. Waves with subscripts '+' travel to the right, and those with '-' to the left. For a single interface, transmission matrix $\mathbf{T}_{2,1}$ combines coefficients (5.38) and (5.39) for forward- and backward-wave propagation,

$$\mathbf{T}_{2,1} = \frac{1}{2Z_2} \begin{pmatrix} Z_1 + Z_2 & Z_1 - Z_2 \\ Z_1 - Z_2 & Z_1 + Z_2 \end{pmatrix} \approx \mathbf{I} - \begin{pmatrix} 0 & r_2 \\ r_2 & 0 \end{pmatrix},$$ (5.52)

where the second equation corresponds to the small-reflectivity approximation, r_2, the reflectivity at the boundary, and $\mathbf{I}$ is the identity matrix. Alternately, the amplitudes of the waves traveling away from the boundary can be related to those incident on it from both sides:

$$\begin{pmatrix} u_-^1 \\ u_+^N \end{pmatrix} = \mathbf{S}_{N,1} \begin{pmatrix} u_+^1 \\ u_-^N \end{pmatrix}. \tag{5.53}$$

In this expression, $\mathbf{S}_{N,1}$ is called the scattering matrix. For $N = 2$, this matrix combines the reflection and transmission coefficients in both propagation directions:

$$\mathbf{S}_{2,1} = \frac{1}{Z_1 + Z_2} \begin{pmatrix} Z_2 - Z_1 & 2Z_2 \\ 2Z_1 & Z_1 - Z_2 \end{pmatrix}. \tag{5.54}$$

In the absence of anelastic attenuation, the elastic energy is preserved in the outgoing states:

$$Z_1 \left[\left| u_-^1 \right|^2 + \left| u_+^N \right|^2 \right] = Z_2 \left[\left| u_+^1 \right|^2 + \left| u_-^N \right|^2 \right], \tag{5.55}$$

for any N, and consequently the sum of powers of back- and forward-traveling waves is constant at any frequency.

For $N-1$ interfaces, matrix $\mathbf{T}_{N,1}$ is a product of wave-mode transformations at all boundaries:

$$\mathbf{T}_{N,1} = \prod_{i=2}^N \mathbf{T}_{i,i-1} \begin{pmatrix} e^{i\Delta\varphi_i} \\ & e^{-i\Delta\varphi_i} \end{pmatrix} \approx \exp\left[-\sum_{i=2}^N \begin{pmatrix} 0 & r_i e^{i\Delta\varphi_i} \\ r_i e^{-i\Delta\varphi_i} & 0 \end{pmatrix} \right], \tag{5.56}$$

where $\Delta\varphi_i$ is the phase shift of the forward-traveling wave during its propagation in layer i. Let us denote the elements of this "propagator" matrix across the stack of all $N-1$ boundaries (Figure 5.7b) by:

$$\mathbf{T}_{N,1} \equiv \begin{pmatrix} G^{++} & G^{+-} \\ G^{-+} & G^{--} \end{pmatrix}. \tag{5.57}$$

The total reflection amplitude, u_-^1, can be found from the requirement that in the right-hand side of Figure 5.7b, there should be no incoming wave traveling to the left:

$$u_-^2 = G^{-+} u_+^1 + G^{--} u_-^1 = 0, \tag{5.58}$$

and consequently,

$$u_-^1 = -\frac{G^{-+}}{G^{--}}u_+^1.$$ (5.59a)

This gives the total back-scattered amplitude. The total transmitted amplitude is therefore:

$$u_+^N = G^{++}u_+^1 + G^{+-}u_-^1 = \left(G^{++} - \frac{G^{+-}G^{-+}}{G^{--}}\right)u_+^1.$$ (5.59b)

Equations (5.59a) and (5.59b) can be used to numerically model propagation of a long seismic wave through a stack of thin random layers. We use an example similar to that by Richards and Menke (1983), with 1000 layers of uncorrelated random velocities drawn from Gaussian distribution with a mean 3.0 km/s and standard deviation of 0.25 km/s. The density is assumed constant. Using its scale-invariance, the impedance was normalized to yield a mean value of $Z = 1$, and the same value of impedance was placed at both ends of the random sequence (Figure 5.8). The travel time within each layer is taken equal to 1 s, which also gives the characteristic Nyquist frequency of $f_N = 0.5$ Hz, relative to which all frequencies in the propagation process can be measured.

To investigate the time "history" of scattering, the impedance time series (Figure 5.8) is truncated at boundaries $N = 2 \dots 1000$, and the remainders of the series are closed with a layer having $Z = 1$. The resulting variations of the reflected and transmitted wave intensities show great fluctuations with respect to the different statistical realizations of the impedance time series (Figure 5.9). However, after averaging over multiple realizations, the transmitted and reflected powers exhibit clear, mutually complementary exponential decays (Figure 5.10). By measuring the logarithmic decrements of these decays, temporal attenuation coefficients χ are measured for selected normalized frequencies f/f_N (Figure 5.11).

The above procedure was performed for impedance contrasts spaced at regular time intervals $\Delta t_i = 1$ s (Figure 5.8) and also repeated for another set of random impedance variations in which Δt_i were randomly distributed. A log-normal distribution of Δt_i was constructed so that the average $\langle \Delta t_i \rangle$ also equaled 1 s. As expected, the resulting attenuation coefficients are similar for lower frequencies $f < 0.3 f_N$. At $f > 0.3 f_N$, the attenuation in the random-Δt_i sequence saturates at a constant level (black line in Figure 5.11), but the attenuation in the regularly spaced sequence continues to increase to $f \approx 0.5 f_N$, after which it decreases to near zero at $f \approx f_N$. This pattern resembles the well-known "frequency folding" effect, which is characteristic for aliasing. As one can see, near $f \approx f_N$, phases of all reflections superimpose equivalently to the case of $f \approx 0$, and the elastic attenuation drops to zero. Interestingly, the regularly spaced impedance series exhibits a very narrow "notch" at $f \approx 0.5 f_N$, at which the attenuation drops sharply because of tuning of the incident wave with the reflectivity sequence

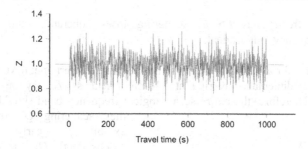

FIGURE 5.8

Random Gaussian distribution of impedance corresponding to mean velocity 3.0 km/s and standard deviation 0.25 km/s. The impedance is normalized to a mean value of 1.0.

(Figure 5.11). However, neither aliasing nor tuning are present in the more realistic random-Δt_i impedance series.

In summary, the general behavior of the attenuation coefficient in 1-D random media can be described as follows:

1) At near-zero frequencies, the attenuation is low ($\chi \approx 0$) because of the destructive interference of the impedance contrasts.

2) Up to certain frequency f_0, χ depends almost linearly on f. In this range, "scattering Q_s" can be meaningfully defined. However, Q_s is not a "quality factor," but a measure of the slope of $\chi(f)$ dependence, which is proportional to the mean stochastic reflection amplitude. The value of f_0 may generally depend on the statistics of the distribution of layer thicknesses and was $f_0 \approx 0.3 f_N$ in our example.

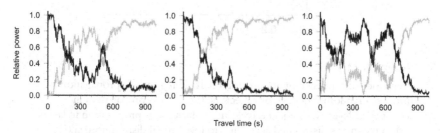

FIGURE 5.9

Wave attenuation in three statistical realizations of impedance time series (Figure 5.8) for frequency $f = 0.2 f_N$. Black line shows the transmitted power, gray shows the reflected power.

3) At frequencies $f > f_0$, scattering loses coherence, and χ becomes frequency independent.

The value of f_0 for a particular area may not be easy to determine; nevertheless, for an average sedimentary layering at ~10-cm thickness, f_0 may be quite high (~40 kHz). Therefore, the entire seismological frequency band should lie within the "scattering Q" regime and exhibit a nearly frequency-independent Q_s. However, at significantly lower frequencies and longer scale lengths, the 1-D approximation considered here breaks down because of the effects of structure (*i.e.*, geometrical attenuation), the attenuation coefficient saturates (Figure 5.11), and an apparent frequency-dependent Q_s is observed. Thus, once again, separation of the deterministic and stochastic wave-propagation regimes is critical when considering scattering.

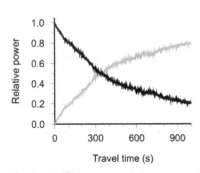

FIGURE 5.10

Transmitted (black) and reflected (gray) power averaged over 100 statistical realizations as in Figure 5.9.

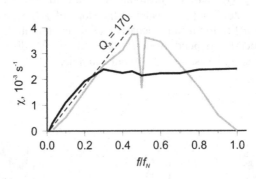

FIGURE 5.11

Frequency dependence of attenuation coefficient χ in 1-D propagation through a random sequence or reflectors. Gray line – propagation in a sequence of layers with equal travel times; black line – propagation in layers with travel-times distributed according to log-normal distribution; and dashed line – slope corresponding to "scattering Q" = 170.

5.6.4 Coda

Coda waves represent another instructive and practically important case of geometrical attenuation. The peculiarity of the coda problem is that, in this case, there exist no wave fronts and even no single propagation direction that can be followed in order to measure the geometrical spreading, and the time and spatial dependences of the wavefield are decoupled. Because of this property, coda represents an example in which the temporal attenuation coefficient, χ, can be directly measured in single-station observations, by contrast to the spatial attenuation coefficient in the traveling-wave examples above. This problem was first considered for 3-D cases by Aki and Chouet (1975), Sato (1977) and modified for a 2-D case by Xie and Nuttli (1988). Let us consider the 2-D case for simplicity, and also because it is related to L_g coda Q modeling discussed later in this book.

The relation of the seismic coda to scattering is well established (*e.g.*, Aki, 1969). This relation is evidenced by the stability of coda characteristics (in particular, its mean log-amplitude) in respect to the source parameters and time, which shows that coda represents a spatially and temporally averaged process. Broadening of S-wave seismogram envelopes with observation distance (Sato and Fehler, 1998, p.39–40) also clearly shows that scattering affects the propagation of seismic waves. The general effect of scattering on seismic arrivals makes them stochastic in character, which suggests statistical, rather than deterministic, methods of analysis.

However, while the *fact* of scattering is reliably established, its *origin* and the *spatial distribution* of scatterers still remain debatable. Unfortunately, in today's research literature, the transition from the general principle of scattering to detailed mathematical models is often motivated by the simplicity of analytical or numerical modeling. In consequence, the resulting interpretations may be affected by unrealistic simplifying assumptions. The resulting distributions of heterogeneities often appear to be biased toward uniform and large volumes of the lithosphere instead of being concentrated near the surface, where all attenuation mechanisms should likely be the strongest. Here, we will consider such assumptions on the examples of the local and L_g codas.

Earthquake codas provide spatially and temporally averaged sampling of the subsurface, and they are known for the stability of the resulting estimates. However, this sampling is not uniform. If scattering and not anelastic attenuation is the main object of analysis, term "coda sampling" refers to the locations and amplitudes of scatterers and not to the areas traversed by wave paths. Determining this sampling is of a non-trivial task, and it can hardly be uniquely established without making assumptions about the structure. However, the popular intuitive assumption of the lithosphere being "uniformly sampled" by coda waves is likely grossly incorrect. For example, local-earthquake codas are often inverted for variations of scattering properties of the lithosphere with depths, and L_g codas are often inverted for regional variations of crustal Q. In reality, coda attenuation should be determined by the distribution of scatterers near the surface and ray shapes and anelastic attenuation deeper within the lithosphere. Although assumptions about straight rays and uniform and isotropic scattering may often lead to "reasonable" values for Q, they still may be inaccurate with respect to describing its causes and spatial variations.

Despite the general "attenuation without Q" theme of the book, in this section, as well is in other places involving extensive comparisons to the conventional models and results, we utilize the Q-factor terminology in parallel with the attenuation coefficient. This provides a convenient language of presentation for readers used to attenuation levels given in terms of Q. Nevertheless, this still does not mean that Q represents a meaningful physical quantity in these cases. For consistent interpretation, values of Q^{-1} of various kinds below should always be understood as the corresponding attenuation coefficient divided by πf.

Local-earthquake case

Two types of coda models usually employed in local-earthquake studies are schematically illustrated in Figure 5.12a. Scattering heterogeneities are assumed to be distributed uniformly within the lithosphere, and both the direct and scattered waves are approximated as traveling along straight paths. Longer parts of the coda are assumed to "sample" larger volumes, and in particular, greater depths within the lithosphere. Single-scattering approximation is often used (e.g., Aki and Chouet, 1975), and in cases where deviations from the predictions of such a model are observed, they are typically attributed to multiple scattering (Figure 5.12a; Wu, 1985) or to the depth dependence of the distribution of scatterers. However, with all this increase in detail, the validity of the straight-ray and isotropic-scattering (i.e., structureless, uniform-velocity) model is practically never questioned and accepted as a "practical assumption".

Nevertheless, scattering within the lithosphere should occur differently, and likely as illustrated in Figure 5.12b. The upper crust, which is brittle and seismogenic, contains most of its heterogeneity and therefore should scatter the seismic waves most intensively. This heterogeneity is also enhanced by the effects of the free and near surface, which contain the strongest variations of the reflection coefficient anywhere within the Earth. Further, the upper part of the lithosphere is strongly layered, as evidenced by numerous controlled-source seismic investigations. The strongest velocity/density contrasts are located again within the uppermost crust and near the crust-mantle boundary. This layered reflectivity is strongly anisotropic and is dominated by upward- and downward reflections, P/S mode conversions, and multiple reflections. Finally, the background velocity structure is non-uniform and shows a general downward velocity increase, leading to predominantly upward-bending rays. Additionally, as discussed in Chapter 4, the upper-crustal reflectivity may increase the flux of

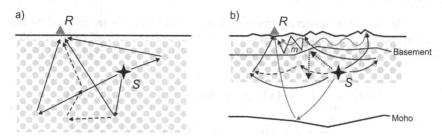

FIGURE 5.12
Schematic coda scattering models: a) Aki's (1969) model and
b) alternate model allowing a more realistic background
structure. Label S indicates the earthquake sources and label R
denotes the receivers. Gray pattern – scattering heterogeneities;
solid arrows – single-scattering ray paths; and dashed lines –
multiple scattering paths. In plot b) rays may be curved due to
velocity gradients, wavy line indicates a surface wave, and 'm' is
an upper-crustal multiple. Dotted arrows show waves downward
reflected within the upper crust, effectively increasing the
geometrical spreading.

downward-scattered waves, further altering the geometrical spreading for the
direct and scattered waves arriving from the deeper crust (Figure 5.12b).

For a direct wave propagating within an attenuative medium, the time-
averaged energy density from a source at point x_S at time $t = 0$ can be expressed
as:

$$E_d\left(\mathbf{x},t\middle|\mathbf{x}_S,\omega\right)=G_0\left(\mathbf{x}\middle|\mathbf{x}_S,\omega\right)e^{-2\chi^*(\omega)}L\left(t-\tau\left(\mathbf{x},\mathbf{x}_S\right),\omega\right),\qquad(5.60)$$

where $\mathbf{x}$ is the location of the receiver, t is the time, $\chi^*(\omega)$ is the total attenuation
coefficient accumulated along the path, and $L(t,\omega)$ is the mean source power per
unit angular frequency. As above, G_0 is the geometrical-spreading function, which
in this case may be frequency dependent because of refractions and reflections
within the lithosphere, but is approximated as directly dependent only on the
source and receiver coordinates and not on travel time. This approximation
corresponds to the physical picture of waves propagating in the form of band-
limited impulsive packets which may be dispersive, but generally retain their
compact characters. For such waves, $L(t,\omega)$ can also be approximated as:

$$L\left(t,\omega\right)\approx\delta\left(t\right)W\left(\omega\right),\qquad(5.61)$$

where $W(\omega)$ is the spectral density of the signal. Under this approximation,
ω serves as an external parameter not coupled with t. Below, parameter ω will be
omitted for brevity of notation.

In the single-scattering approximation, coda energy is represented by the sum of energies resulting from scattering at all points on some surface Se, at which the sum of the forward- and scattered-wave travel times equals t (Figure 5.13):

$$E_c(R,t) = \frac{W}{4\pi^2} \int d^3 x \lambda(\mathbf{x}) G_0(\mathbf{x}|\mathbf{x}_S) G_0(\mathbf{x}_R|\mathbf{x}) e^{-2\chi_1^* - 2\chi_2^*} \delta(t - t_1 - t_2)$$

$$= \frac{W}{4\pi^2} \oint_{Se} ds \lambda(\mathbf{x}) G_0(\mathbf{x}|\mathbf{x}_S) G_0(\mathbf{x}_R|\mathbf{x}) e^{-2\chi_1^* - 2\chi_2^*}. \tag{5.62}$$

For a uniform background velocity model, surface Se represents an ellipsoid with foci located at the source and receiver. In the more general case, the scattering surface should be additionally elongated downward due to the velocities increasing with depth, and its shape should also reflect local variations in the velocity structure. Here, $\lambda(\mathbf{x}, \omega)$ is the scattering-amplitude factor measuring the relative amplitude of scattered waves at this point. Note that for isotropic scattering in a uniform background, $\lambda = 2\alpha$, which in this case equals the turbidity of the medium (see Sato, 1977). In general, however, λ represents predominantly back scattering, whereas α represents forward scattering, residual geometrical spreading, and dispersion. These quantities are different, and they also correspond to spatially distinct locations within the model (Figure 5.13). Thus, generally, scattering amplitudes, λ, and attenuation coefficients, χ, should probably be treated separately.

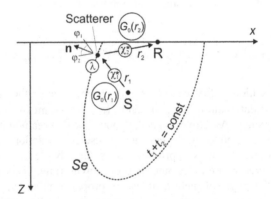

FIGURE 5.13
Single-scattering model for local-earthquake coda recording. S – the source; R – the receiver near the surface. Symbols indicate the background geometrical spreading [$G_0(r_{1,2})$], attenuation coefficients ($\chi^*_{1,2}$), and scattering amplitude (λ) in eq. (5.62). Curve labeled Se is the "scattering ellipsoid" from which a constant coda lag time is observed.

Expression (5.62) gives the coda energy envelope at time t. To obtain the apparent (*i.e.*, predicted for an observation) temporal attenuation coefficient, we need to differentiate $\ln E_c$ with respect to coda lag time:

$$\chi_c = \frac{-1}{2E_c} \frac{\partial E_c}{\partial t},$$ (5.63)

which gives an integral of the gradient of the product of G_0, λ and the attenuation factors in (5.68),

$$\frac{\partial E_c}{\partial t} = \frac{W}{4\pi^2} \int d^3 x V \frac{\partial}{\partial n} \left[\lambda(\mathbf{x}) G_0(\mathbf{x}|\mathbf{x}_S) G_0(\mathbf{x}_R|\mathbf{x}) e^{-2\chi_1^* - 2\chi_2^*} \right] \delta(t - t_1 - t_2)$$

$$= \frac{W}{4\pi^2} \oint_{Se} d^2 s V \frac{\partial}{\partial n} \left[\lambda(\mathbf{x}) G_0(\mathbf{x}|\mathbf{x}_S) G_0(\mathbf{x}_R|\mathbf{x}) e^{-2\chi_1^* - 2\chi_2^*} \right],$$ (5.64)

where operator $\dfrac{\partial f}{\partial n} \equiv \mathbf{n} \dfrac{\partial f}{\partial \mathbf{x}}$ denotes the projection of the gradient on the outer normal to the scattering curve, $\mathbf{n}$ (Figure 5.14). Therefore,

$$\frac{\partial E_c}{\partial t} = \frac{W}{2\pi} \oint_{Se} d^2 x e^{-2\chi_1^* - 2\chi_2^*} V \Psi,$$ (5.65)

where

$$\Psi = \frac{\partial}{\partial n} \left[\lambda(\mathbf{x}) G_0(\mathbf{x}|\mathbf{x}_S) G_0(\mathbf{x}_R|\mathbf{x}) \right] - 2\lambda(\mathbf{x}) G_0(\mathbf{x}|\mathbf{x}_S) G_0(\mathbf{x}_R|\mathbf{x}) \alpha_n(\mathbf{x}),$$ (5.66)

and the directional single-scattering attenuation coefficient α_n at point $\mathbf{x}$ is (Figure 5.14):

$$\alpha_n = \frac{\partial \chi^*}{\partial n} = \alpha_i (\cos \varphi_1 + \cos \varphi_2).$$ (5.67)

Thus, the single-trace, apparent attenuation coefficient for coda (5.63) is given by the ratio of two contour integrals (5.62) and (5.65). As one can see, spatial variations in G_0, λ, and α_i contribute to the coda energy.

The above derivation shows that coda $Q_c = \pi f / \chi_c$ physically describes the properties of the scattering amplitude, λ, combined with attenuation along the ray paths, χ_i, and averaged over a complex system of wave paths. If one disregards all of the factors related to the realistic lithospheric structure and also assumes $\lambda = $ const, then the resulting Q_c becomes equal to Q_i within the model. Alternately, if one assumes Q_i to be known and λ to be frequency dependent, then λ maps into the resulting Q_c (this seems to be the commonly used approach). However, if we realize that the above assumptions are far too unrealistic, and scattering occurs mostly within the upper crust and involves multiple modes shown in Figure 5.12b, then a much more rigorous model accounting for the actual subsurface structure is required before any properties of lithospheric scattering can be inferred from the observed Q_c.

L_g case

In long-range L_g coda Q studies, a different sampling pattern is observed. This problem is usually treated as 2D, with wave propagation approximated as occurring along the surface of the Earth, and attenuation parameters also being averaged in the vertical direction and varying horizontally. The basic forward model for this type of coda-wave propagation was studied in Section 5.6.4. The principal observation is that for coda lapse time t, the amplitude and the apparent χ correspond to scattering occurring at the ellipse and having the source and receiver as its foci (Figure 5.14). As coda lapse times increase, progressively larger elliptical areas are covered by coda rays.

Forward problem

For L_g coda lapse times, a complex mixture of seismic waves contributes to the coda, in which, however, crustal-guided S waves are believed to be the most important. From time- and spatially averaged coda observations, it is impossible to

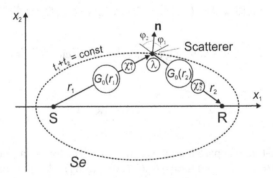

FIGURE 5.14
Schematic scattering ellipse (labeled Se) contributing to formation of the coda from source (S) at receiver point (R). Circled symbols indicate the background geometrical-spreading $[G_0(r_{1,2})]$, attenuation coefficients ($\chi_{1,2}$) and scattering amplitude (λ) in eq. (5.68).

discriminate between these guided waves, and therefore they are also considered in a depth-averaged sense only considering the horizontal variations in seismic properties. In this approximation, expression (5.60) for the direct-wave energy can still be used, with $\mathbf{x}$ and $\mathbf{x}_S$ becoming 2-D vectors on the Earth's surface. Separation of the time-and frequency-dependent direct-wave energies (5.61) can also be used under the same assumptions of incoherent wave-packet averaging as above. Further, within the accuracy of the above approximations and considering the long and weakly dispersive characters of short-period L_g waves, it appears sufficient to use a constant-group velocity background model, $V = \text{const}$. Under this simplification, rays become straight, and G_0 becomes isotropic and dependent on the source-receiver distance alone.

Similar to the single-scattering approximation in the local-coda case above, coda energy is represented by an expression similar to (5.62), in which the integration becomes 2D, and the scattering surface Se becomes an ellipse $r_1 + r_2 = Vt$ on the surface of the model (Figure 5.14),

$$
\begin{aligned}
E_c\left(R,t\right) &= \frac{W}{2\pi}\int d^2\mathbf{x}\,\lambda(\mathbf{x})\,G_0\left(r_1\right)G_0\left(r_2\right)e^{-2\chi_1^\cdot-2\chi_2^\cdot}\delta\left(t-\frac{r_1+r_2}{V}\right) \\
&= \frac{W}{2\pi}\oint_{Se} dl\,\lambda(\mathbf{x})\,G_0\left(r_1\right)G_0\left(r_2\right)e^{-2\chi_1^\cdot-2\chi_2^\cdot}.
\end{aligned}
\tag{5.68}
$$

and its time derivative,

$$
\begin{aligned}
\frac{\partial E_c}{\partial t} &= \frac{WV}{2\pi}\int d^2\mathbf{x}\,\frac{\partial}{\partial n}\left[\lambda(\mathbf{x})\,G_0\left(r_1\right)G_0\left(r_2\right)e^{-2\chi_1^\cdot-2\chi_2^\cdot}\right]\delta\left(t-\frac{r_1+r_2}{V}\right) \\
&= \frac{WV}{2\pi}\oint_{Se} d^2\mathbf{x}\,\frac{\partial}{\partial n}\left[\lambda(\mathbf{x})\,G_0\left(r_1\right)G_0\left(r_2\right)e^{-2\chi_1^\cdot-2\chi_2^\cdot}\right] \\
&= \frac{WV}{2\pi}\oint_{Se} d^2\mathbf{x}\,e^{-2\chi_1^\cdot-2\chi_2^\cdot}\Psi,
\end{aligned}
\tag{5.69}
$$

where:

$$
\Psi = \frac{\partial}{\partial n}\left[\lambda(\mathbf{x})\,G_0\left(r_1\right)G_0\left(r_2\right)\right] - 2\lambda(\mathbf{x})\,G_0\left(r_1\right)G_0\left(r_2\right)\alpha_n(\mathbf{x}).
\tag{5.70}
$$

As in the local-coda case, the principal contributions to χ_c come from the spatial variations in G_0, λ, and α_i. In addition, from eq. (5.67), the areas closest to the source-receiver line, in which $\cos\varphi_{1,2} \approx 1$, also contribute the strongest back scattering.

So far, we have considered a relatively general case. Now, let us assume uniform distribution of scatterers, uniform directivity of the source and receiver, isotropic scattering, absence of dispersion in both direct and scattered-wave paths, and ignore the Earth's curvature. With these approximations, we can use $G_0(r) = 1/(2\pi Vr)$ for the geometrical-spreading functions, and the integrals above can be evaluated analytically, giving (Xie and Nuttli, 1988):

$$E_c(R,t) = \frac{\lambda W e^{-2\chi_i t}}{2\pi U \sqrt{V^2 t^2 - R^2}} \theta\left(\frac{Vt}{R}\right), \qquad (5.71)$$

where relation $\chi_i = \alpha_i V$ was used, and θ is the Heavyside step function. This function equals 1 within the coda time range, in which $Vt > R$. In this solution, factor U (see Section 4.4) was re-introduced to correct for the effect of dispersion omitted from the geometrical-spreading functions.

Now, how can we define the geometrical spreading for a seismic coda? This is not obvious, because strictly speaking, coda does not exist in the absence of attenuation, and our definition of geometrical spreading as the attenuation-free limit does not work for it. Nevertheless, we can "turn off" the anelastic energy dissipation only by setting $\alpha = 0$ while keeping some notional (even infinitesimal) scattering $\lambda \neq 0$ in the above equation, leading to the "geometrical" limit of coda energy:

$$G_c(t) = A(R,t) \propto \sqrt{E_c(R,t)\big|_{\alpha=0}} \propto \frac{1}{\sqrt{U}\left(V^2 t^2 - R^2\right)^{1/4}}. \qquad (5.72)$$

This is the geometrical coda spreading by Xie and Nuttli (1988), in which the dispersion factor $U^{1/2}$ is also included. Thus, coda decays somewhat slower than at the surface-wave rate of $(tU)^{-1/2}$. In the local-coda limit, $R \ll Vt$, and the coda spreading law becomes $A(t) \propto (tU)^{-1/2}$ (Aki and Chouet, 1975).

With coda geometrical spreading defined by eq. (5.72), the corrected attenuation path factor (5.71) becomes:

$$P(t) = \frac{\sqrt{E_c(R,t)}}{G_c(t)} = e^{-\chi_i t}. \qquad (5.73)$$

This factor is purely exponential in t, and its logarithmic decrement, χ, equals χ_i. Thus, the apparent temporal attenuation coefficient of accurately corrected coda equals the anelastic attenuation of the medium. This relation is used in numerous inversions of local- and long-range coda observations. However, while using this model, it is important to keep in mind that several approximations were made for simplifying relations (5.68) and (5.70) to (5.71). These approximations may not automatically be valid for large areas covered by the codas. One way to avoid the pitfalls of relying on elaborate but inaccurate models is to derive empirical corrections for geometrical coda attenuation (5.71) from the frequency dependence of χ.

Inverse problem

Based on the general forward coda model of Section 5.6.4, we can now pose the corresponding inverse problem. The quantity measured in coda observations is the time- and frequency-dependent apparent attenuation coefficient, $\chi_c(t,f)$, which removes the source- and instrument-related amplitude scaling from the coda energy density:

$$\chi_c = \frac{-\dot{E}_c}{2E_c},$$
(5.63 again)

where dot denotes the partial derivative in time. The corresponding quantity can be predicted by ray-based modeling of $E_c(t,f)$ using on the *in situ* values of α_i and λ within the model (Section 5.6.4). Let us assume that coda attenuation values $\chi(t,f)$ are observed at stations $n = 1...N_S$ for the corresponding ranges of coda times $t \in [T_1^n, T_2^n]$. Therefore, the standard least-squares optimization consists in minimizing the quadratic form:

$$\Phi = \frac{1}{2}\sum_{n=1}^{N_S}\int df \int_{T_1^n}^{T_2^n} dt \left(\chi^{observed} + \frac{\dot{E}_c}{2E_c}\right)^2 \Bigg|_{\text{station } n},$$
(5.74)

in terms of variables α_i and λ. Furthermore, parameter α_i usually contains a significant frequency dependence, $\alpha_i = \gamma_i + \kappa_i f$, and therefore the functional minimization problem is:

$$\frac{\delta\Phi}{\delta\gamma_i} = 0, \ \frac{\delta\Phi}{\delta\kappa_i} = 0, \text{ and } \frac{\delta\Phi}{\delta\lambda} = 0.$$
(5.75)

Let us cumulatively denote all of parameters γ_i, κ_i, and λ by p_k, where $k \in [1, N_k]$.

Solving the equation of (5.75) for a realistic dataset requires a numerical approximation of the objective function. This can be done by using various parameterizations for the spatial distributions of model parameters. In the most general finite-element approach, spatial variations of each parameter p_k can be decomposed in terms of a set of N_e basis functions $\psi_{k,i}(\mathbf{x})$:

$$p_k(\mathbf{x}) = \sum_{i=1}^{N_e} p_k^i \psi_{k,i}(\mathbf{x}). \tag{5.76}$$

The regions in which these basis functions are non-zero are called finite elements. In the commonly used 2-D tomographic scheme, finite elements represent rectangular cells in model volume, and each basis function is a constant equal to $1/(cell\ volume)$ within the corresponding cell.

With perturbation δp_k^i in the discrete parameter p_k^i, the predicted contribution to the objective function (5.74) would change to:

$$\left. \frac{\dot{E}_c}{2E_c} \right|_{\mathbf{p}^0 + \delta\mathbf{p}} \approx \left. \frac{\dot{E}_c}{2E_c} \right|_{\mathbf{p}^0} + \sum_{k=1}^{N_e} A_k \delta p_k, \tag{5.77}$$

where the sensitivity kernels A_k are:

$$A_k = \frac{\partial}{\partial p_k}\left(\frac{\dot{E}_c}{2E_c}\right) = \frac{\dot{E}_c}{2E_c}\left(\frac{1}{\dot{E}_c}\frac{\partial \dot{E}_c}{\partial p_k} - \frac{1}{E_c}\frac{\partial E_c}{\partial p_k}\right). \tag{5.78}$$

Using these kernels in the vicinity of the current estimate $\mathbf{p}^0$, the objective function becomes quadratic in p_k:

$$\Phi = \frac{1}{2}\sum_{n=1}^{N_S}\int df \int_{T_1^n}^{T_2^n} dt \left(\chi^{observed} - \chi^0 + \sum_{k=1}^{N_e} A_k \delta p_k\right)^2 \bigg|_{\text{station } n}$$

$$= \frac{1}{2}\sum_{k,l=1}^{N_e} B_{kl}\delta p_k \delta p_l + \sum_{k=1}^{N_e} C_k \delta p_k + const, \tag{5.79}$$

where χ^0 is the value of the attenuation coefficient predicted by using parameters $\mathbf{p}^0$,

$$B_{kl} = \sum_{n=1}^{N_S}\int df \int_{T_1^n}^{T_2^n} dt A_k A_l \bigg|_{\text{station } n}, \tag{5.80}$$

and

$$C_k = \sum_{n=1}^{N_S}\int df \int_{T_1^n}^{T_2^n} dt \left(\chi^{observed} - \chi^0\right) A_k \bigg|_{\text{station } n}. \tag{5.81}$$

With this discrete matrix form (5.79) of the objective function, the equation of (5.75) can be solved numerically by using iterative techniques, such as the gradient, conjugate-gradient, or SIRT methods.

To evaluate the contributions from partial derivatives in A_k (5.78), one needs to perform ray tracing in the appropriate velocity model while replacing fields of $(\gamma_i, \kappa_i, \lambda)$ with the corresponding basis function φ_k. However, it is important to keep in mind that for scattered waves, multiple paths contribute to the value of coda energy from the same source-receiver pair and at any given time. Therefore, this summation occurs over multiple ray paths, and it is important to ensure the correct weighting of the contributions from different paths. This weighting depends on the knowledge of the source and receiver properties, such as the uniformity of their directivities. For example, in the single-scattering coda model (5.70), the integration over ray paths reduces to contour integral over the scattering ellipse at each time t (Section 5.6.4), and the derivatives of $\dot{E}_c$, with respect to the discrete parameter perturbations, are:

$$\frac{\partial \dot{E}_c}{\partial \lambda_k} = \frac{W}{4\pi^2} \oint_{Se} d^2\mathbf{x}\, e^{-2\chi_1^* - 2\chi_2^*} \left\{ \frac{\partial}{\partial n}\left[\psi_{\lambda_k}(\mathbf{x}) G_0(r_1) G_0(r_2) \right] - 2\psi_{\lambda_k}(\mathbf{x}) G_0(r_1) G_0(r_2) \alpha_n(\mathbf{x}) \right\},$$

$$\frac{\partial \dot{E}_c}{\partial \gamma_k} = \frac{-W}{2\pi^2} \oint_{Se} d^2\mathbf{x}\, e^{-2\chi_1^* - 2\chi_2^*} \lambda(\mathbf{x}) G_0(r_1) G_0(r_2) \psi_{\gamma_k}(\mathbf{x})(\cos\varphi_1 + \cos\varphi_2), \text{ and}$$

$$\frac{\partial \dot{E}_c}{\partial \kappa_k} = \frac{-fW}{2\pi^2} \oint_{Se} d^2\mathbf{x}\, e^{-2\chi_1^* - 2\chi_2^*} \lambda(\mathbf{x}) G_0(r_1) G_0(r_2) \psi_{\kappa_k}(\mathbf{x})(\cos\varphi_1 + \cos\varphi_2).$$

$$(5.82)$$

The style of this derivation and the subsequent inversion are very similar to the traditional velocity tomography. However, note that the expressions for the attenuation-sensitivity kernels above have practically no relation to the velocity-sensitivity kernels used in travel-time tomography.

Subsurface sampling

To invert for the *in situ* L_g Q, Xie and Mitchell (1990) proposed a heuristic back-projection scheme, and it would be interesting to relate it to the more rigorous derivation of subsurface coda sampling above. These authors subdivided the study area into 2-D cells and attributed a quality-factor value, denoted Q_m, to the m-th cell. They further assumed that the observed apparent attenuation factor for a given source-receiver pair is given by the average of the corresponding Q_m^{-1} of all cells enclosed within the scattering ellipse:

$$Q^{-1} = \frac{1}{S} \sum_{m=1}^{N_c} s_m Q_m^{-1} + \varepsilon \cdot \qquad (5.83)$$

In this expression, s_m is the area of the m-th cell lying within the scattering ellipse, S is the total area of the ellipse, N_c is the number of cells, and ε is the measurement and modeling error, which is considered as random. Recalling that both the apparent and *in situ* Q^{-1} can be viewed as proxies for the corresponding χ_c, this expression can be re-written for the attenuation coefficients:

$$\chi_c = \frac{1}{S} \sum_{m=1}^{N_c} s_m \chi_{i,m} + \varepsilon \cdot \qquad (5.84)$$

Note that on the left-hand side of this expression, we have the apparent χ_c, whereas on the right-hand side we have the intrinsic χ_i. Similar relations should hold for its zero-frequency, γ, and frequency-dependent, κ, parts.

Expression (5.84), in combination with our derivation of χ in Section 5.6.4, reveals the physical reasons for averaging the values of Q^{-1} given in eq. (5.83). Indeed, ray integrals in eqs. (5.68) and (5.70) are roughly proportional to the areas covered by the rays. Unlike in the local-coda case considered in the previous section, the uniformity of ray coverage in the horizontal–2-D L_g coda Q problem is likely to be good, and therefore models (5.83) and (5.84) could represent reasonable approximations.

However, being "reasonable" in the absence of additional information still does not guarantee that these approximations are sufficient for producing accurate results. Note that L_g coda inversions usually result in significant frequency-dependent Q estimates (*e.g.*, Mitchell *et al.*, 1997). In our notation, this frequency dependence suggests that the geometrical attenuation. γ, should be significant for L_g coda. At the same time, γ should be accumulated in the near-surface areas, where ray curvatures are the strongest. Note that the sensitivity kernels (5.82) are also concentrated near the sources and receivers, because of the $G_0(r_1)G_0(r_2)$ factors in them. Therefore, an alternate model could attribute γ_i values to the vicinities of the receivers and seek a slowly varying model for κ_i within the whole of the study area.

With the use of the more accurate forward model (5.74), L_g coda Q tomography could certainly be improved. In particular, model (5.84) (to say nothing of 5.83) ignores contributions from the scattering amplitude λ, whereas such contributions could be most significant. For example, spatially variable and frequency-dependent λ, such as related to surface topography, could cause apparent attenuation effects similar to those of χ_i. However, addressing these problems require revisiting the original datasets and take us away from the main goals of this book.

5.7 Separability of attenuation factors

In Section 3.6, we discussed the difficulty of separating the "scattering Q" from the variations of geometrical spreading and anelastic attenuation. Let us continue this discussion and pose the following question: can we separate the contributions from geometrical spreading, elastic scattering, and anelastic energy dissipation in the attenuation coefficient?

Generally, the residual geometrical spreading and attenuation only represent parts of a single attenuation coefficient, and therefore they should generally be modeled and interpreted together. However, if separate geometrical spreading and Q_i are of interest, they can be differentiated, but only by imposing additional constraints. As noted above, the frequency-dependent attenuation coefficient, χ, represents practically all the data available to interpretation in most attenuation measurements. As discussed in Section 2.8, by the characters of this frequency dependence, the observed χ or α data can be decomposed into three parts: 1) frequency independent, 2) monotonously increasing with frequency, and 3) oscillatory. The oscillatory part is typically related to the deterministic structure in which the attenuation is measured (such as tuning and mode interference), and therefore we only have two observable attributes to work with: the frequency-independent part of χ (denoted γ), and its rate of variation with frequency, κ.

To resolve the three desired quantities from only two measured parameters, one needs additional constraints. Three types of *a priori* constraints[15] were proposed for interpreting the $\chi(f)$ dependences in different studies. The first type used in the traditional approach consists in postulating that the true geometrical spreading is known sufficiently accurately and equals $G(t,f) = G_0(t,f)$. Consequently, the entire measured $\chi(f) = \ln[P(t,f)/G_0(t,f)]$ is attributed to attenuation according to formula (1.10). However, as shown in Chapter 3, this assumption is often so inaccurate that it does not allow measuring even the first order of the frequency dependence of Q.

The second approach, which is taken in this book, approximates the residual geometrical spreading, $\delta G(t,f) = G(t,f)/G_0(t,f)$, as a function weakly deviating from one and frequency independent, *i.e.* sets $\delta G(t,f) = \exp(-\gamma t)$. Compared to the first approach, this is a strongly relaxed assumption which allows measuring the residual geometrical spreading and removes the corresponding artifact from Q. This constraint appears natural for all cases of frequency-independent background $G_0(t,f)$, such as the commonly used $G_0(t) = t^\nu$. Note that under this approximation, the total $G(t,f)$ can still be frequency dependent (*e.g.*, such as suggested for P_n by Yang *et al.* (2007) or modeled using a "colored" reflection sequence). However, if some perturbation of the structure leads to a

---

[15] Another example of often-used terminology which should be used sparingly. There exists no *a priori* information, data, or constraints. In simple words, this term means "unsupported assumptions," or "postulates".

frequency-dependent $\delta G(t,f)$, the resulting values of Q may still become biased. The only solution to correct this bias would consist in finding a more accurate model for $G_0(t,f)$.

The third interesting alternative to these constraints could consist in setting $\delta G(t,f)$ so that the resulting *in situ* Q becomes frequency independent. This solution is the most difficult to implement, and its validity depends on whether we consider frequency-independent rheological Q as a significant reality. However, this approach could help define a frequency-dependent component in geometrical spreading that could be associated with "scattering." For special types of models (*e.g.*, self-affine or stochastic), other types of constraints can apparently be devised as well.

Considering the inherent uncertainty of frequency-dependent geometrical spreading and Q separation, isolation of the additional effects of scattering in either κ or γ appears problematic. The existing approaches to such separation (*e.g.*, Wu, 1985; Jin *et al.*, 1994) rely on comparing the values of χ at different frequencies and distances from the source, which again implies a perfectly known $G(t,f)$. The commonly used uniform half-space models for $G_0(t,f)$ are insufficient for this purpose, and common observations of values $\eta \approx 1$ for Q_s may be the indicators of such inaccuracy. Ideally, if perfect $G(t,f)$ models were available, scattered-wave amplitudes could be "migrated" to invert for zones of increased scattering; yet even then the resulting models would likely be better described in terms of structure, and not Q_s.

The theoretical models described above show that the anelastic part of attenuation can be isolated on the basis of its characteristic increase with frequency. For both sparse and multiple-free reflectivity, smoothly bending rays, and coda, the observed geometrical attenuation ($\chi|_{Q_i^{-1}=0}$) is frequency independent, and therefore we can expect that the attenuation-free limit equals the zero-frequency one: $\chi|_{Q_i^{-1}=0} = \chi|_{f=0} = \gamma$, with a similar relation for the anelastic quantities. Thus, it appears that in many practical cases, γ_i can be interpreted as related to the residual geometrical spreading.

The above interpretation of the anelastic dissipation and residual geometrical spreading exhausts the available observed parameters, and therefore scattering cannot be unambiguously isolated from them. Based on the frequency dependence of its attenuation coefficient, scattering can be mixed with the anelastic dissipation. Based on the time- or distance-dependence of the scattered amplitudes, scattering also trades off with realistic geometrical spreading and can be absorbed by it.

5.8 Seismic albedo

The "seismic albedo" describes the relative effect of scattering compared to the anelastic attenuation of seismic waves. Similar to Q, this concept was introduced from an analogy to optics, and defined as the ratio of the scattering and total dissipation strengths (Wu, 1985):

$$B_0 = \frac{Q_s^{-1}}{Q_s^{-1} + Q_i^{-1}} . \qquad (5.85)$$

This quantity in fact does not rely in the Q-factor paradigm of attenuation and can be expressed in a simpler form by using the attenuation coefficients:

$$B_0 = \frac{\chi_s}{\chi_i} , \qquad (5.86)$$

where χ_i is our total intrinsic attenuation coefficient defined above, and χ_s is its part caused by elastic scattering.

Seismic albedo is a well-defined theoretical quantity; however, in observations, this definition has a significant problem related to the uncertainty of χ_s. As argued above (Sections 3.6 and 5.7) and also in Section 7.4, apparently the only productive view to this uncertainty is to incorporate the scattering effects in empirical geometrical spreading, γ_i, and to abandon the notions of χ_s and Q_s. The expression for seismic albedo therefore becomes:

$$B_0 = \frac{\gamma_i}{\chi_i} = \frac{1}{1 + f/f_c} . \qquad (5.24 \text{ again})$$

This shows that similar to f_c, seismic albedo may in fact represent the relative significance of structural effects (such as reflections from the sediment-basement contacts and the Moho) to those of attenuation.

Chapter 6

Applications

Throwing pebbles into the water, look at the ripples they form on the surface; otherwise, such occupation becomes an idle pastime.

Kozma Prutkov, Fruits of Reflection (1853-54)

In this chapter, the attenuation-coefficient method is illustrated on several data examples. We selected several examples of different wave types (body and surface waves, normal modes, refractions, and coda) within several frequency bands ranging from ~500 s to ~100 Hz. Our main objective is to establish that in most datasets, the dependence of the attenuation coefficient on frequency can be satisfactorily represented by the linear law discussed in Chapters 1, 2 and 5. Also, in each of these examples, we attempt isolating the effects of the residual geometrical spreading discussed in Chapter 4 and give it a specific interpretation.

The general conclusion of this chapter is that residual geometrical spreading takes different forms in different experiments; however, in all cases, it represents the most significant characteristic of the wave-attenuation process. Geometrical spreading is regionally variable and correlates with geological structures and crustal tectonic types and ages. In one particularly interesting example (Section 6.8.6), we are able to detect temporal variations in the geometrical-spreading parameters and relate them to the seasonal and volcanic cycle variations.

Along with the purely empirical results, we also consider numerical modeling of surface-wave attenuation within the mantle (Section 6.1.2) and short-period coda-wave attenuation from nuclear explosions (Section 6.8.2). These numerical examples show that: 1) once again, the attenuation coefficients are approximately linear in frequency and 2) the "geometrical" part of this attenuation coefficient is predictable from purely structural information. The first of these numerical examples also demonstrates that the existing derivation of the surface-wave Q from the correspondence principle is inaccurate by 10–20%, whereas the second example illustrates the very high accuracy of predicting the residual geometrical spreading from completely independent data on the structure of the lithosphere.

Most of the results of this chapter represent re-interpretations of well-known datasets, several of which were instrumental in establishing the current frequency-dependent Q paradigm. We selected these classic studies on purpose, in order to convince the reader that even well-established results, broadly considered as canonical, may still deserve a critical re-evaluation based on the more rigorous, attenuation-coefficient approach. Many newer studies use the Q-based approaches criticized here and consequently their findings and interpretations could be questioned in similar ways. As we will see below, in many cases, the changes caused by abandoning the assumptions of the frequency-dependent Q (see Chapter 3) are significant. The attenuation-coefficient viewpoint not only modifies the Q values by factors of up to 20–30 but also often leads to serious changes in the interpretations. One of the most significant general changes of such kind is the disappearance of the frequency-dependent *in situ* Q within the Earth. This frequency dependence of Q is replaced with the recognition of the pervasive presence and variability of the residual geometrical spreading.

The body of observations using the Q-based model and favoring the frequency-dependent Q of the Earth is large and covers several areas and frequency bands. Several of these areas are addressed in this chapter:

1) Long-period and short-period surface waves and L_g;

2) Normal modes;

3) Long-period body waves;

4) Borehole body waves;

5) P_n and "teleseismic P_n";

6) Local and regional-earthquake coda, and total-energy studies; and

7) Lapse-time and temporal variations of seismic coda.

Each of these topics is illustrated on a representative example using previously published data. In considering the examples, we stay within strictly seismological, quantitative, and empirical arguments. Other types of evidence is often advanced in favor of the frequency-dependent Q, such as coming from laboratory studies (*e.g.*, Faul *et al.*, 2004; Romanowicz and Mitchell, 2007). Such evidence was discussed in Chapter 3, where it was shown that the task of correlating the Q values arising from such different types of arguments is always thwarted with difficulties of reconciling the underlying models, and even with the differences of the quantities measured. At the same time, seismological evidence is ample, precise, and significant in itself. The seismological evidence also contains a common physical basis that needs to be carefully analyzed.

6.1 Surface waves

In this section, we consider the observations of Rayleigh waves within ~10–500 s periods and 0.5–1.5-Hz L_g by using the attenuation-coefficient and Q data compiled from Raoof and Nuttli (1984), Mitchell (1995), Durek and Ekström (1996), Weeraratne *et al.* (2007), and also from the Institute of Geophysics and Planetary Physics (IGPP) Reference Earth Model web pages[16]. L_g waves can be interpreted as higher-order surface waves (Knopoff *et al.*, 1973; Panza and Calcagnile, 1975), which allows using them for extending the frequency band of the fundamental-mode Rayleigh waves. Our approach is strictly empirical and quantitative, and reduces to summarizing the observed $\chi(f)$ dependences without using any underlying models beyond the linear expression $\chi(f) = \gamma + \kappa f$. Analysis of this wide frequency band demonstrates an almost amazing commonality in the patterns of $\chi(f)$ and even in the observed values of γ and Q_e. In addition, recognition of the frequency-independent shift γ in surface-wave data leads to a potential explanation for the discrepancy between the surface-wave and normal-mode attenuation measurements at 200–500-s periods, which will be discussed in Section 6.2.3.

Figure 6.1 shows the measured Rayleigh-wave attenuation coefficients at 10–100-s periods in several tectonically active and stable areas around the world. The attenuation coefficients are given in the traditional form, as functions of wave periods. Generally, the values of α are within $0–1 \cdot 10^{-3}$ s^{-1}, comparatively high in tectonic and low in oceanic areas and quickly increasing below ~20-s periods (Figure 6.1).

Note that the variations of α with periods $T=1/f$ in Figure 6.1 appear hyperbolic, which suggests that the dependences on f might actually look simpler. This becomes clear when the same data are plotted against frequency (Figure 6.2). In these plots, $\alpha(f)$ was transformed into the temporal attenuation coefficient, $\chi(f)$, by using eq. (4.13), with dispersion parameters taken from the simplified, linear fundamental-mode group-velocity trend in Figure 6.3, and $\alpha_d = 0.4 \cdot 10^{-4}$ km^{-1}. This α_d value was selected empirically, close to the low bound on the observed α, so that the resulting intercepts of $\chi(f)$ in Figure 6.2b became near-zero. However,

---

[16] URL <http://igppweb.ucsd.edu/~gabi/rem.html>, accessed Sept. 12, 2009.

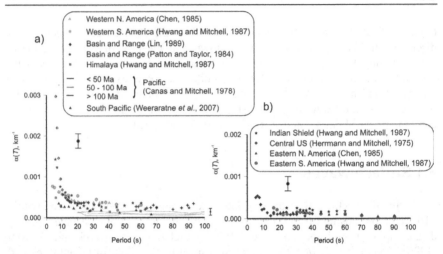

FIGURE 6.1
Fundamental-mode Rayleigh wave attenuation-coefficient data
from Mitchell (1995) and Weeraratne *et al.* (2007): a) tectonically
active (symbols) and oceanic areas (lines) and b) stable areas.
Error bars show typical errors as estimated by Mitchell (1995).
From Morozov (2010b); with permission from Springer.

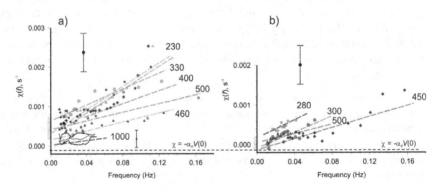

FIGURE 6.2
Fundamental-mode Rayleigh wave attenuation-coefficient data
from Mitchell (1995) and Weeraratne *et al.* (2007): a) tectonically
active (symbols) and oceanic areas (lines) and b) stable areas.
From Morozov (2010b); with permission from Springer.

with some uncertainty it its value, the choice of α_d does not affect the conclusions.
Note that the dispersion curves (Figure 6.3) show significant variations at
frequencies below ~0.03–0.05 Hz, which may contribute to the "spectral
scalloping" of $\chi(f)$ amplitudes observed in the data.

In $\chi(f)$ form, the separation between the tectonic, stable, and oceanic areas becomes clearer in the data , as well as the differences between the several study areas. Several separate linear trends of $\chi(f)$ can be recognized, as indicated by dashed lines in Figures 6.1c and 6.1d. Notably, when extended to the frequency axis, most of these lines cross it at *positive* intercept values, particularly if the correction for α_d is not performed. Similar positive γ values are found in body-wave and coda observations discussed further in this chapter.

From the interpreted linear trends in $\chi(f)$ (Figure 6.2), several important observations can be made:

1) Although lower values of $Q_e \approx 230$ are present in the data from regions of active-tectonics compared to the stable areas, these ranges of Q_e overlap almost completely. Therefore, although Q_e may generally increase with tectonic age (Morozov, 2008), it does not significantly discriminate between the stable and active tectonic types.

2) Nevertheless, the intercept value of $\gamma_D \approx 2 \cdot 10^{-4}$ s^{-1} (gray bars on ordinate axes in Figures 6.2) separates most of the tectonic and active areas. If the dispersion correction is not applied, this threshold should be measured relative to the dashed black line in Figures 6.1c and 6.1d, and equals $\gamma_D \approx 3.2 \cdot 10^{-4}$ s^{-1}. A similar relationship was found for crustal body and coda waves, for which the stable and active areas were separated by the level of $\gamma_D \approx 0.8 \cdot 10^{-2}$ s^{-1} (Morozov, 2008).

3) In relation to the above discriminator γ_D, the oceanic-area data from Canas and Mitchell (1978) (solid lines in Figure 6.2a) generally align with the continental stable-tectonic group (Figure 6.2b), although recent data by Weeraratne *et al.* (2007) (triangles in Figure 6.2a) are close to the edge of the active-tectonic group. Values of γ for oceanic recordings also appear to increase with age (Figure 6.2a), which is an opposite trend compared to the continental lithosphere. In addition, the oceanic data show consistently higher Q_e.

Note that the above observations are empirical and independent of the traditional geometrical-spreading and power-law $Q(f)$ assumptions. However, they reveal several important relationships in the data that have not been noticed in the original interpretations using the Q or correlations of the attenuation coefficient with wave periods (Mitchell, 1995). This shows that raw data representation and classification of observed trends is very important in the analysis of attenuation.

In Figure 6.2, we do not attempt rigorous estimation of the statistical parameter errors and confidence intervals. Unfortunately, reworking published data as attempted here does not allow a complete error analysis in the spirit of the proposed approach. The individual measurement errors are significant (error bars in Figure 6.2); however, amplitude deviations from the interpreted linear trends

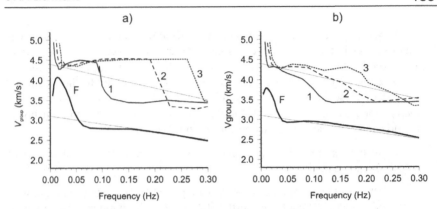

FIGURE 6.3
Group velocities of Rayleigh waves, after Panza and Calcagnile (1975), as functions of frequency: a) for continental structure without a low-velocity channel in the upper mantle and b) for a structure with the low-velocity channel. Labels F, 1, 2, and 3 indicate the fundamental and three higher modes. From Morozov (2010b); with permission from Springer.

are non-random and should be mostly related to wave-mode interferences within the specific structures. In reflection seismology, such frequency-dependent amplitude variations are known as "tuning". A proper inversion for $\chi(f)$ bypassing Q would require revisiting the full raw-amplitude datasets, which are not available to us at present. At the same time, at present we focus only on *the fact* of distinct linear $\chi(f)$ dependences and their characteristic parameters, and consequently can rely on interpretive "visual" analysis and line fitting. Two conclusions nevertheless are quite clear from Figure 6.2: 1) multiple $\chi(f)$ trends exist in the data and 2) these trends may only be viewed as linear at best.

From our procedure of "turning off" the attenuation in order to estimate the geometrical spreading, γ can be viewed as a measure of the residual geometrical spreading or dispersion remaining in the surface-wave amplitudes after their correction. For example, for $\gamma \approx 2 \cdot 10^{-4}$ s^{-1}, this residual geometrical-spreading correction amounts in only $\mu \approx 8\%$ for a 400-s Rayleigh wave propagation time. As we will see below, this relative level of the residual geometrical spreading is similar to that estimated for body waves.

Notably, all values of γ exceed the minimal level of $(-\alpha_d V|_{f=0})$ (horizontal dashed line in Figure 6.2), showing that Rayleigh waves are systematically "under-corrected" by the theoretical geometrical-spreading correction. This is again similar to the observations of lithospheric body and coda waves (Sections 6.3–6.9). The systematic character of this geometrical-spreading term shows that it is caused not only by focusing and defocusing on lateral variations of velocity (Dalton and Ekström, 2006), but also generally deviates from the theoretical

$\Delta^{-\nu}(\sin\Delta)^{1/2}$ dependence (Nuttli, 1973). The variability is also significant between different regions, and particularly within the tectonically active lithosphere (Figure 6.2a).

The relative significance of the residual geometrical spreading in attenuation measurements can be characterized by the cross-over frequency f_c (Section 5.3). For characteristic values of $\gamma = 4 \cdot 10^{-4}$ s^{-1} and $Q_e = 500$, we have $f_c \approx 0.05$ Hz, with some variations for different regions. This frequency corresponds to the ~20-s period below which the apparent attenuation factor $\chi(T)$ starts quickly increasing (Figure 6.1). Physically, below this period, the effects of residual geometrical spreading exceed those of attenuation.

Let us now consider the long-period, ~100- to ~300–400-s Rayleigh waves. Interestingly, the global-average Q^{-1}curve in this range is also hyperbolic (Figure 6.4a), and the corresponding $\chi(f)$ again shows a well-defined linear dependence (Figure 6.4b). By contrast to the shorter-wave case, its $Q_e \approx 84$ is significantly lower, and the intercept $\gamma \approx -8 \cdot 10^{-5}$ s^{-1} is negative, showing that these waves are "over-corrected" by the background geometrical-spreading correction. Taking the same characteristic travel time (400 s), the relative amount of this over-correction is only $|\gamma|t \approx 3\%$. This small value is not surprising, because at such wavelengths, the spherical-Earth model used in accounting for the geometrical-spreading effects is quite accurate. The cross-over frequency for the long-period band equals $f_c \approx 2.1$ mHz. Although the interpretation of this quantity is not as straightforward as in the case of under-corrected geometrical spreading, note that this frequency is close to the transition from the surface-wave to the normal-mode regime (Figure 6.4b).

Note that Q_e still represents an apparent quantity characterizing the attenuation measurements on the surface. For relatively short waves localized within comparatively uniform layers, Q_e should be somewhat greater than the lowest intrinsic Q_i sampled by the corresponding wave. However, it appears that for long surface waves, the above $Q_e \approx 84$ may actually be below the lowest Q_i within the upper mantle and represent the redistribution of wave energy density with changing frequency.

Below ~2.5 mHz, the attenuation-coefficient trend changes with the transition into the low-order fundamental spheroidal modes. As shown in Section 6.2, such a change could be related to the characteristic wavelengths reaching the thickness of the entire upper mantle, and consequently to the transition from predominantly traveling to standing waves.

6.1.1 Crustal model

As shown in Chapter 5, in practically all cases in which a frequency-dependent *in situ* $Q_{\text{in}}(f)$ is interpreted, an useful alternate interpretation can be offered, based on the intrinsic attenuation coefficient, $\chi_i = \pi f / Q_{\text{in}}(f) = \gamma_i + \pi f Q_i^{-1}$. In

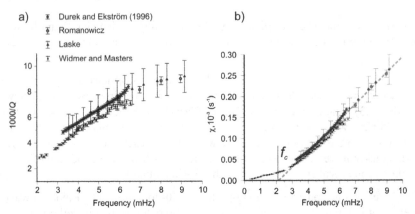

FIGURE 6.4

Spheroidal-mode (black dots) and Rayleigh-wave (other symbols) attenuation data from IGPP reference model website: a) in the original $1000Q^{-1}$ form and b) transformed to $\chi(f)$. Dashed line shows the interpreted linear trend $\chi(f) = -8\cdot10^{-5} + \pi f/Q_e$ [s^{-1}], with $Q_e = 84$. Vertical line indicates the "cross-over" frequency f_c. References to the data by Romanowicz, Laske, and Widmer and Masters are as in the web site. Modified from Morozov (2010b); with permission from Springer.

this expression, Q_i shall be first tried as frequency independent, and some frequency dependence further considered if required by the data.

Surface-wave attenuation models are usually rendered in terms of the frequency-dependent *in situ* Q. However, as we argue that the *in situ* Q actually does not exist but χ_i may be a good approximation to the reality, it would be useful to recast such models in the form of γ_i and Q_i. For example, let us perform such transformation for the models of the tectonically extended crust in the Basin and Range province in the western United States (BR; Mitchell and Xie, 1994) and of the stable eastern United States (EUS; Cong and Mitchell, 1988) shown in Figure 6.5. Both Q models are strongly frequency dependent; however, the power-law $Q = Q_0 f^\zeta$ dependences in them were constructed differently. In the BR model, the upper 15 km of the crust was taken as frequency independent, and ζ was set equal 0.5 everywhere else in the two models.

By approximating the local $Q(f)$ values by depth distributions of γ_i and Q_i, the models become somewhat easier to compare (Figure 6.6). The values of Q_i in the EUS model are high (over 1000) even in the upper crust, and their variations with depth are weak. By contrast, the BR model shows an over ten-times stronger attenuation within the upper crust ($Q_i \approx 60$–200), which quickly drops to ~1000 within the lower crust. Values of γ_i were forced to equal zero in the upper crust of the BR model, which was probably not a very good approximation, particularly in

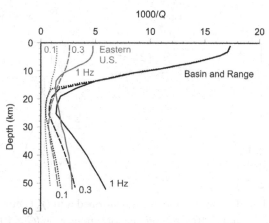

FIGURE 6.5

Frequency-dependent crustal S-wave Q model for the Basin and Range province (black lines) (Mitchell and Xie, 1994) and for the eastern United States (gray) (Cong and Mitchell, 1988), sampled at 0.1, 0.3, and 1.0 Hz (indicated by line dashes). From Morozov (2010b); with permission from Springer.

its extensional tectonic setting. Apart from this contradiction, the γ_i curves are in agreement with the differentiation proposed above (dashed line in Figure 6.6b): γ_i is significantly higher than $\gamma_D \approx 3.2 \cdot 10^{-4}\ \mathrm{s}^{-1}$ within the tectonically active zone (BR); for stable crust (EUS), values of γ are at near or below this level. Note that we use the value of γ_D not corrected for dispersion, because the original model by Mitchell and Xie (1994) also contained no such corrections.

The above comparison was based on an *ad hoc* transformation of the crustal models built within the $Q = Q_0 f^{\zeta}$ paradigm. This transformation results in approximately the same values of Q within the crust, and consequently these models should reproduce the attenuation-data fit numerically predicted by that model. Considering this modeling, an interesting question arises: how can we interpret the γ values in Figure 6.6b, although the forward-modeling approach used by Cong and Mitchell (1988) did not include variable geometrical spreading? The answer to this question is that deviations of the actual structure from their layered crustal models (*i.e.*, γ) becomes interpreted as the "scattering attenuation" parameter Q_s^{-1}. Combined with the true anelastic attenuation Q_i^{-1}, it produces the resulting apparent, frequency-dependent Q_{in}^{-1}:

$$Q_{\mathrm{in}}^{-1}(f) = \frac{\gamma}{\pi f} + Q_i^{-1} \equiv Q_s^{-1} + Q_i^{-1}. \qquad (6.1)$$

Here, $Q_s = \pi f / \gamma \propto f$, which is typical for scattering attenuation (Dainty, 1981).

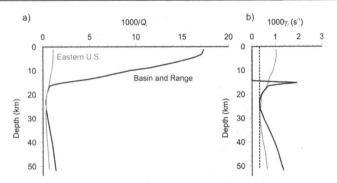

FIGURE 6.6
Crustal models from Figure 6.5 transformed into (γ_i, Q_i) form:
a) $1000/Q_i$ and b) $1000\gamma_i$. Note the low attenuation $(Q > 1000)$
in the eastern United States and strong difference between the
two upper-crustal models. Dashed line indicates the proposed γ_D
threshold separating the active and tectonic structures. From
Morozov (2010b); with permission from Springer.

Thus, when modeling is limited in accuracy, γ can be inverted from the
apparent model $Q_{in}^{-1}(f)$ and interpreted as caused by elastic scattering. Note that
even with such an interpretation, the upper-crust of the BR model with $Q_i^{-1} = 0$
(*i.e.*, non-scattering) but very high Q_i^{-1} (Figure 6.6) appears somewhat
contradictory. In a full and accurate modeling, one would need to start from
$Q_i^{-1} = 0$ and adjust the structure until a correct γ is achieved. After this, Q_s^{-1} should
disappear, and Q_i^{-1} could be inverted from the observed value of Q_e. Such
modeling and inversion needs to be addressed from the raw attenuation-coefficient
data and is beyond the scope of this work.

6.1.2 Love-wave attenuation in layered Earth

Attenuation of long-period surface waves makes a good case study of the
effects of attenuation within the Earth. Surface waves are observed globally, and
their measurements lead to the most uniform sampling of the Earth's mantle.
Because of their low frequencies, surface waves are also sensitive to the averaged
properties of the Earth's medium and are less affected by scattering. At the same
time, the nature of surface waves is well understood, and their propagation can be
precisely described by well-known wave equations.

Nevertheless, it still appears that such good understanding extends only to
the attenuation-free case. As shown below, the problem of surface-wave
attenuation is incorrectly treated by the visco-elastic method, which is the only
method in use today. For example, the well-known visco-elastic solution for the

long-period Love-wave Q_L (Anderson *et al.*, 1965; Aki and Richards, 2002) violates the conservation of energy and overestimates the attenuation. Generally, this violation is not surprising, because the concept of elastic energy is not rigorously supported by the visco-elastic theory (see Chapter 2). In this section, the problem of attenuating surface waves is re-considered from first principles and a new derivation is given, based on an explicit interpretation of the energy balance. To derive the corrected surface-wave solution, we use the equipartitioning approach explained in Section 2.6. The new Q_L in the Gutenberg's continental structure is 10–20% higher than the one derived by Anderson *et al.* (1965) but still shows a similar frequency dependence. The modeled Love-wave $\alpha(f)$ predicts two distinct and near-linear trends within the bands corresponding to the upper-mantle and near-crustal surface-wave modes.

An additional objective of this section is to investigate the origin of the linear dependence of the surface-wave attenuation coefficient. We show that because of the shapes of the Fréchet sensitivity kernels, linear $\alpha(f)$ should be expected from mantle layering. The principal tools of investigation are again the conservation of energy, the Hamilton variational principle, and the Rayleigh-Ritz method. For brevity, we only consider the Love-wave (*SH*) case, and Rayleigh waves can be treated in a similar manner. The conclusion is that again, in a layered Earth, the attenuation coefficient should exhibit piecewise-linear variations with frequency.

Existing model

The first derivation of the surface-wave $Q(f)$ measured on the surface of a layered-Earth model was given by Anderson *et al.* (1965) and is accepted until now (see Aki and Richards, 2002, p. 289–291; Dahlen and Tromp, p. 330–335). The approach is based on two of the theoretical conjectures discussed in Section 3.3. According to the first of them (H1 on p. 80), the attenuation parameter Q^{-1} is interpreted as the complex argument of phase velocity:

$$\arg V = -\frac{1}{2Q_{\text{spatial}}}. \tag{6.2}$$

According to the second conjecture (H2), an analytical dependence of the surface-wave phase velocities V on the wave speeds in the individual mantle layers ($V_{P,i}$ and $V_{S,i}$) is assumed:

$$V = f\left(V_{P,i}, V_{S,i}, \rho_i\right). \tag{6.3}$$

In these expressions, all properties also depend on frequency, ω. It is further assumed that eq. (6.2) applies to both the apparent Q^{-1}, which is observed at the surface, and to the *in situ* Q's, which are included in each of the complex medium velocities $V_{P,i}$ and $V_{S,i}$. This allows extrapolating the partial derivatives of phase velocities, such as $\partial V / \partial V_{P,i}$, into the corresponding complex planes of $V_{P,i}$ and $V_{S,i}$.

The above series of mathematical generalizations allows calculating the surface-wave attenuation simply by using the phase-velocity derivatives. For Love waves, this gives (expression (7.88) in Aki and Richards, 2002):

$$\text{spatial } Q_L^{-1} = \int_0^\infty F_V Q_S^{-1} dz, \qquad (6.4)$$

where z is the depth, and F_V is the S-wave velocity sensitivity kernel,

$$F_V = \frac{\left[k^2 l_1^2 + \left(\frac{dl_1}{dz} \right)^2 \right] \mu}{k^2 \int_0^\infty \mu l_1^2 dz}. \qquad (6.5)$$

In this expression, k is the wave number, μ is the Lamé rigidity modulus, Q_S is the shear-wave attenuation quality factor of the mantle, and l_1 is amplitude of the mode of interest. All values in eqs. (6.4) and (6.5) are depth dependent, and the horizontal SH-wave displacement is given by:

$$u_y(x,z,t) = l_1(z) \psi(x,t), \qquad (6.6)$$

where $\psi(x,t) = \exp(-i\omega t + ikx)$.

However, despite all its elegance and ease of attaining the result, approach (6.2) and (6.3) also leads to significant difficulties with the physical significance of its results. Its key problems are: 1) treating Q^{-1} as a fundamental property of the medium, similar to μ; 2) mixing the notions of wave velocities as parameters of the propagating medium and phase velocities of various wave modes in it; and 3) mixing the "on mass shell" and "off mass shell" regimes in the wavefield. Note that by definition, spatial Q corresponds to the argument of the complex wavenumber (e.g., Aki and Richards, 2002, p.167–169),

$$k \to k + i\alpha = k \left(1 + \frac{i}{2Q} \right), \qquad (6.7)$$

where α is the spatial attenuation coefficient. Positive signs of α for all modes ensure their amplitudes decaying in the directions of propagation. Through its relation to the phase velocity $V = \omega/k$, a positive $\text{Im} k$ corresponds to a negative $\text{Im} V$. However, in a surface wave, a single value of k is common to all depths, and depth-dependent velocities V_P and V_S in formula (6.5) do not serve as phase velocities for any waves. Surface waves are practically always "off mass shell," and therefore there is no reason to take $\text{Im}\,V \neq 0$. The assumptions about negative imaginary parts in V_P and V_S being similar to those of the phase velocities (6.2) represent heuristic extrapolations of the properties of the plane-wave solutions away from the points at which they are valid. Similarly, Q^{-1} describes the argument of the complex wave vector (6.7) and is not an unequivocal medium property.

Finally, considering the properties of the prediction of Q_L resulting from the visco-elastic approach, expression (6.5) is problematic, because it exaggerates the amount of energy dissipation and violates the total energy balance. To see this, note that eq. (6.5) gives the values of Q_L^{-1} as weighted averages of Q_S^{-1} within the layers; however, the weights in the numerator of this ratio are systematically larger than those in the denominator. This flaw can be easily seen on an example of a two-layer model with different velocities but constant *in situ* Q_S. In such a model, assume that the energy in each layer $l = 1,2$ can be subdivided into non-dissipating ($E_{n,l}$) and dissipating ($E_{d,l}$) parts. Let us denote the proportion of $E_{d,l}$ dissipated in time t from each layer by λ; consequently, both $E_{d,l}$ should decrease by factors $(1 - \lambda) = \exp(-\omega Q_S^{-1} t)$. The total relative energy dissipation should therefore occur not faster than λ:

$$\frac{\delta \widetilde{E}}{\widetilde{E}} = \frac{\lambda (E_{d,1} + E_{d,2})}{E_{n,1} + E_{d,1} + E_{n,2} + E_{d,2}} \leq \lambda. \tag{6.8}$$

Therefore, a spatial attenuation factor of $Q_L^{-1} = Q_S^{-1}$ should be expected if the total mechanical energy dissipates in this process, and $Q_L^{-1} < Q_S^{-1}$ if only a part of it dissipates (such as the kinetic energy; see Chapter 2). However, formula (6.5) predicts $Q_L^{-1} > Q_S^{-1}$ on the surface, showing a dissipation of more energy than present in the field, which is unrealizable.

Numerical modeling

Although our model of mantle attenuation is now completely separate from the velocity and cannot be solved by velocity tomography, the general approaches developed for velocity modeling and tomographic inversion remain effective. In particular, the Rayleigh-Ritz method provides efficient numerical solutions of the eigenvalue eqs. (2.112) and (6.18) (Wiggins, 1976). By approximating the functional form of $l_1(z)$ in terms of some basis functions $\phi_i(z)$,

$$l_1(z) = \sum_{i=1}^{N} m_i \phi_i(z), \tag{6.9}$$

where coefficients m_i comprise a discrete model vector **m**, integral eqs. (2.112) become reduced to a matrix eigenvalue problem:

$$kk^* \mathbf{m} = \mathbf{A}_2^{-1} \left(\omega^2 \mathbf{A}_1 - \mathbf{A}_3 \right) \mathbf{m},$$ (6.10)

where the energy matrices consist of depth-averaged pairs of basis functions:

$$A_{1,ij} = \frac{uu^*}{2} \int_0^\infty \rho \phi_i \phi_j dz, \; A_{2,ij} = \frac{uu^*}{2} \int_0^\infty \mu \phi_i \phi_j dz, \text{ and}$$

$$A_{3,ij} = \frac{uu^*}{2} \int_0^\infty \mu \frac{d\phi_i}{dz} \frac{d\phi_j}{dz} dz.$$ (6.11)

Earth-flattening corrections can be incorporated in these integrals in order to account for the Earth's sphericity. By solving this eigenvalue problem, all possible values of $|k|$ and the corresponding eigenfunctions (6.9) are obtained, from which the attenuation spectra (6.14) can be calculated.

For a specific example, consider the Gutenberg continental Earth model in Table 6.1. For this model, 45 cubic polynomial basis functions give a convenient decomposition for $l_1(z)$ in terms of continuous functions, which satisfy:

$$\sum_{i=1}^N \phi_i (z) = 1,$$ (6.12)

and are normalized so that either $\phi_i = 1$ or $d\phi_i/dz = 1$ on the i-th boundary (Figure 6.7). With the shear-wave Q values from the attenuation model MM8 (Anderson *et al.*, 1965; Table 6.1), formula (6.14) yields the apparent frequency-dependent Love-wave Q (Figure 6.8). Note that these Q values are consistently higher than those predicted by the presently used formula (6.5) (gray dashed line in Figure 6.8). This difference is significant and dependent on the underlying velocity and Q distributions within the upper mantle. Therefore, the discrepancy in expression (6.5) should also affect the 1-D, and similarly also 3-D inversions for mantle Q values. These issues will be discussed in Chapter 8.

Model from energy-balance constraints

An alternate expression for the effective spatial Q_L can be derived directly from energy-balance considerations. As argued in Chapter 2, let us view the kinetic energy, E_k, as the dissipating part of the total energy. For weak attenuation, a loss in E_k is continuously replenished from the potential energy according to eq. (2.112). Therefore, for surface-wave propagation, because of the presence of the common factor $uu^* \propto \exp(-2\alpha x)$ in eqs. (2.111), both E_k and E_{el} at any depth

TABLE 6.1 Gutenberg's layered continental
structure model with Q_s values from model MM8
(Anderson *et al.*, 1965).

Layer number	Depth to bottom (km)	ρ(g/cm³)	V_P (km/s)	V_S (km/s)	Q_S
1	19	2.74	6.14	3.55	450
2	38	3.00	6.58	3.80	450
3	50	3.32	8.20	4.65	60
4	60	3.34	8.17	4.62	60
5	70	3.35	8.14	4.57	80
6	80	3.36	8.10	4.51	100
7	90	3.37	8.07	4.46	100
8	100	3.38	8.02	4.41	100
9	125	3.39	7.93	4.37	150
10	150	3.41	7.85	4.35	150
11	175	3.43	7.89	4.36	150
12	200	3.46	7.98	4.38	150
13	225	3.48	8.10	4.42	150
14	250	3.50	8.21	4.46	150
15	300	3.53	8.38	4.54	150
16	350	3.58	8.62	4.68	150
17	400	3.62	8.87	4.85	180
18	450	3.69	9.15	5.04	180
19	500	3.82	9.45	5.21	250
20	600	4.01	9.88	5.45	450
21	700	4.21	10.30	5.76	500
22	800	4.40	10.71	6.03	600
23	900	4.56	11.10	6.23	800
24	1000	4.63	11.35	6.32	800

should decrease with the same logarithmic decrement with travel distance x; we denote this decrement α. The total energy dissipation is a sum of energy losses at each depth:

$$-\frac{d\tilde{E}}{dx} = -\int_0^\infty \frac{d\langle E_k \rangle}{dx} dz = \int_0^\infty 2a_i(z)\langle E_k \rangle dz, \qquad (6.13)$$

where $\alpha_i(z)$ is the local intrinsic spatial S-wave attenuation coefficient at depth z. As shown in Section 5.2.2, with reasonable accuracy, we can regard $\alpha_i(z)$ as a true and local property of the mantle.

Further, we need to decide how α_i at depth z could depend on the wavenumber, k and frequency, ω. Let us assume that this value is the same as for a plane S wave with parameters (k, ω) traveling in a medium with S-wave speed $V_S(z)$. This again represents an *ad hoc* assumption allowing us to relate the Love-wave attenuation problem to a plane S-wave problem, which is tractable

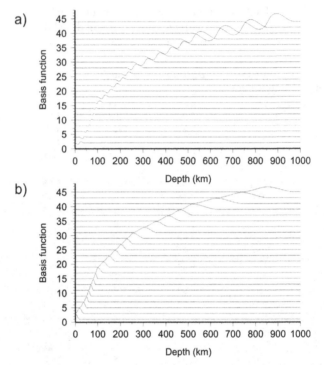

FIGURE 6.7
Basis functions used for modeling Love waves in Gutenberg
Earth model (Table 6.1): a) functions normalized by $d\phi/dz = 1$
at layer boundaries and b) functions having $\phi = 1$ at the
boundaries.

analytically. If it is true, α_i attains a physical meaning that can be compared to
other wave types. Note that this interpretation of α_i is not trivial, because this
plane wave does not satisfy the "mass-shell" equation $\omega/k = V_S(z)$. The surface
wave at depth z does not propagate freely, but is being constantly affected by the
adjoining layers. As also discussed in Section 5.2.2, this "off mass shell" behavior
is an important property of attenuation, and it should be present in this case.

Note that the above assumptions were simply needed to convince ourselves
of the existence of a single parameter αi which only depends on the depth and
frequency and describes both the Love- and S-wave attenuation within the low-
frequency band. In the visco-elastic model, this parameter is introduced as the S-
wave Q (denoted Q_S) and without considering its viability. As we see, the
existence of such a parameter is in fact not automatic and subject to several
approximations, assumptions, and caveats.

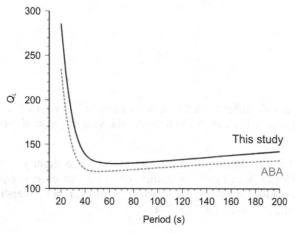

FIGURE 6.8

Love-wave Q_L calculated in the combined Gutenberg/MM8 model by using expressions (6.14) (labeled "This study") and eq. (6.5) (labeled "ABA," after Anderson *et al.*, 1965).

For comparison with the existing solution (6.4) and conventional attenuation measurements, it is convenient to express $\alpha(z)$ through an "S-wave quality factor": $Q_S^{-1} = 2\alpha(z)V_S/\omega$. The observed attenuation coefficient can similarly be expressed as $Q_L^{-1} = 2\alpha/k$, yielding forward model in the form of α-kernels:

$$\alpha = -\frac{d\tilde{E}}{\tilde{E}dx} = \int_0^\infty F_\alpha \alpha_i dz = 2k \int_0^\infty F_Q Q_S^{-1} dz , \qquad (6.14)$$

and Q^{-1}-kernels, if desired:

$$Q_L^{-1} = \frac{2\alpha}{k} = \int_0^\infty F_Q Q_S^{-1} dz , \qquad (6.15)$$

where

$$F_\alpha = \frac{2k}{\tilde{E}} \langle E_{kin} \rangle , \qquad (6.16)$$

and

$$F_Q = \frac{\omega}{2\tilde{E}} \frac{\langle E_{kin} \rangle}{V_S}. \tag{6.17}$$

Note that F_α and F_Q differ from the velocity kernel F_V in eq. (6.5) and, unlike eq. (6.5), expressions (6.14) and (6.15) preserve the sum of the total propagating and dissipated energies.

Note that introduction of attenuation ($\alpha > 0$) also slightly shifts the phase- and group-velocity spectra. To see this, consider the energy equipartitioning variational principle for finding the dependence of $l_1(z)$ on the depth (Section 2.6 and also Aki and Richards, 2002, p. 284):

$$\delta \int_0^\infty \langle L(\mathbf{u}, \dot{\mathbf{u}}) \rangle dz = \frac{1}{2} \left(\omega^2 \delta I_1 - kk^* \delta I_2 - \delta I_3 \right) = 0, \tag{6.18}$$

where $L(\mathbf{u}, \ddot{\mathbf{u}})$ is the Lagrangian density of the elastic field, and I_1, I_2, and I_3 are the energy integrals defined in eq. (2.111). For a fixed ω, the absolute value of the corresponding wavenumber $|k|$ is obtained by solving the eigenvalue problem (6.18). However, integrals (2.111) only depend on α via a common factor uu^*, and therefore $|k|$ is independent of attenuation. Consequently, with non-zero attenuation, the real part of the wavenumber decreases as:

$$\mathrm{Re}\, k = \sqrt{|k|^2 - \alpha^2} \approx |k| \left(1 - \frac{1}{8Q^2} \right), \tag{6.19}$$

which gives a small phase-velocity ($V = \omega/k$) shift due to attenuation. From its stationary-phase definition and the variational principle (eqs. 2.112 and 6.18), the group velocity remains real and changes accordingly:

$$U = \frac{\delta\omega}{\delta\,\mathrm{Re}\, k} = \frac{\mathrm{Re}\, k}{\omega} \frac{I_2}{I_1}. \tag{6.20}$$

6.1.3 Attenuation sensitivity kernels

As shown above, attenuation coefficients nearly linearly depend on frequency in many surface-wave observations. The theoretical examples in Chapter 5 showed four different mechanisms of such linear dependences of $\chi(f)$ arising from wavefront curvatures, incoherent and small-scale reflectivity, and coda. The Love-wave Q_L model considered here illustrates yet another physical effect leading to such linearity for surface waves.

For a fixed velocity/density structure of the mantle, the sensitivity kernels for $\chi(f)$ show generally smooth and near-linear variations with frequency. For example, from eq. (6.14), perturbations in Q_S^{-1} within several depth ranges create frequency-dependent $\chi(f)$ patterns shown in Figure 6.9. Within about two octaves in frequency, all these patterns are close to our linear form $\chi = \gamma + \kappa f$, and because of their similarities, they should strongly trade off during the inversion. Thus, it is likely that the "invertible" amount of information in Love-wave $\chi(f)$ data is roughly equivalent to the four or five parameters required for describing the cumulative $\chi(f)$ curves in Figure 6.10a. Detailed attenuation models, even such as shown in Table 6.1, are over-parameterized and contain significant uncontrolled uncertainties (Anderson *et al.*, 1965). As discussed in Chapter 7, surface-wave attenuation could be principally explained by 1–2 low-Q layers within the mantle in combination with the crust.

Despite what may be commonly thought, for *all* Love-wave modes above ~0.015 Hz, the sensitivity of Q_L^{-1} to the uppermost mantle (<100-km depths) is *much stronger* than to the rest of the mantle. Below this frequency, the sensitivities to the shallow and deeper mantle layers are similar (Figures 6.9 and 6.12). This shows that the variations of crustal and lithospheric thickness and structure should have major effects on the apparent Q_L. For example, note the variation in the sensitivity kernel with crustal thickness changing from 38 to 50 km (dashed line in Figure 6.9).

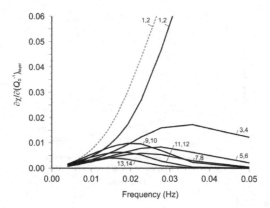

FIGURE 6.9

Sensitivity kernels of Love-wave χ with respect to S-wave Q_S^{-1} of selected pairs of layers in expression (6.14), as functions of frequency. Numbers indicate the pairs of layers in which the S-wave Q values are perturbed (Table 6.1). Dashed line shows the crustal contribution in a model with a 50-km thick crust.

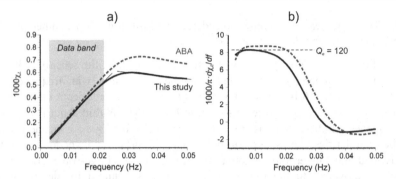

FIGURE 6.10

Attenuation coefficient in Gutenberg/MM8 model: a) 1000χ, and c) the derivative $1000\partial\chi/\partial f$ emphasizing the two Q_e levels. Thin lines in plot a) indicate the interpreted linear trends. Gray box shows the frequency band used for fitting the long-period data. Labels as in Figure 6.8.

Interestingly, the principal effect of the removal of mantle attenuation below the 100-km depth can also be described as approximately "geometrical," *i.e.*, reduction of the intercept γ value at 0.01–0.03 Hz (Figure 6.12). Similar effects of thin attenuative layers within the upper crust were modeled by Morozov *et al.* (2008).

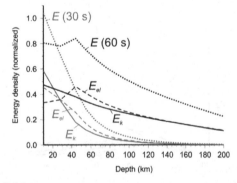

FIGURE 6.11

Normalized distributions of the kinetic (E_k, solid lines), elastic (E_{el}, dashed lines), and total (E, dotted lines) energy density for the fundamental Love-wave modes at 60-s (black) and 30-s (gray) periods. Note that curve E_{el} also represents the velocity sensitivity kernel F_V, and E_k –Q^{-1} sensitivity F_Q in eq. (3.5).

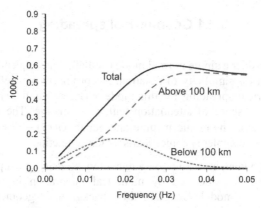

FIGURE 6.12
Total attenuation coefficient (black line, same as in Figure 6.10)
and contributions into it from mantle layers below and above
100-km depth. Note that the solid and dashed lines have similar
shapes but differ by a "geometrical" bulk shift within 0.01–
0.03 Hz range.

6.2 Free oscillations

For free oscillations of the Earth, quality factors have rigorous meanings
analogous to that of the mechanical oscillator, and therefore the whole Q-factor–
based description works correctly. Because each oscillation mode with orders n,l
oscillates at a single eigenfrequency $\omega_{n,l}$, the corresponding attenuation coefficient
$\chi_{n,l} = \omega_{n,l}Q_{n,l}^{-1}/2$ also has a well-defined physical significance of the width of the
spectral peak for this mode. This attenuation coefficient can also be viewed as the
imaginary part of complex $\omega_{n,l}$. Therefore, even for normal modes, a description
using $\chi_{n,l}$ instead of $Q_{n,l}$ could be somewhat more straightforward and insightful.

The current model for the Earth's free oscillations is based on classical
mechanics and is very well developed (*e.g.*, Dahlen and Tromp, 1998). However,
in explaining the physical causes and making quantitative predictions of the
normal-mode Q, this model deviates from mechanics and follows visco-elasticity,
which we heavily criticize in this book. The need for a complete alternate theory
therefore seems indicated; however, we will not undertake this huge effort here.
Instead, we will only illustrate some fundamental-mode observations in terms of
$\chi_{n,l}$, and discuss their implications in terms of the structure of the Earth and the
observed discrepancy between the normal-mode and surface-wave attenuations.

6.2.1 Geometrical spreading

Let us consider a single normal mode oscillating at frequency $\omega_{n,l}$, and drop the mode indices for simplicity of notation. According to our definition (Section 4.1), the geometrical spreading for this mode, $G(\mathbf{r})$, equals the spatial wavefield amplitude in the absence of attenuation within the mantle. The total oscillation amplitude is therefore harmonic in time and equals $G(\mathbf{r},t) = G(\mathbf{r})\exp(-i\omega t)$. This gives the idealized, complete geometrical-spreading function.

However, when inverting for attenuation models, an approximation for $G(\mathbf{r})$ is used, which is derived by numerical integration in some reference, heterogeneous Earth model. Denoting this "background" geometrical-spreading $G_0(\mathbf{r})$, we would now like to see how any difference between $G_0(\mathbf{r})$ and $G(\mathbf{r})$ would affect the measured attenuation level. For simplicity, we illustrate this difference by an analogy with the linear oscillator, to which the normal-mode oscillation is conceptually close.

For a linear oscillator (Section 2.4.1), the time-averaged relative energy dissipation rate equals (see eq. (2.18)):

$$\chi = \frac{\langle D \rangle}{\langle E_k \rangle} = \xi \omega_0 = \frac{\zeta}{m}, \qquad (6.21)$$

where $\xi = Q^{-1}$ is the dimensionless dissipation factor, ζ is the viscous friction constant, m is the mass, and ω_0 is the oscillator's natural frequency. Value χ is the attenuation coefficient, which equals the time-domain logarithmic decrement of oscillations at ω_0 and also the width of the spectral peak at ω_0. Note that according to our "off mass shell" principle, this ratio is the same even for oscillations at $\omega \neq \omega_0$.

Now, note that in normal-mode observations, the natural frequency ω_0 is well-constrained and matched from the observations, whereas m and/or the spring constant $k = \omega_0^2 / m$ are both taken from the reference model. Therefore, for an uncertainty in m or k taken while keeping ω_0 fixed, the corresponding uncertainties in χ are:

$$\delta\chi = -\frac{\zeta}{m^2}\delta m \quad \text{or} \quad \delta\chi = \zeta\omega_0\delta k. \qquad (6.22)$$

This means that if we are observing an oscillator while calibrating its parameters (m,k) by the natural frequency ω_0, errors in the background model should manifest themselves by biases in the dissipation rate, χ. Note that "off mass shell," these shifts are frequency independent.

Another interesting geometrical-spreading–type effect specific to free oscillations is the effect of noise on the measurement, which causes a bias toward higher Q and lower χ (Romanowicz and Mitchell, 2007). This effect can be easier seen in the time domain. According to the definition of χ, it equals:

$$\chi = \frac{\delta E}{E \delta t}, \tag{6.23}$$

where δE is the energy dissipation from level E in time δt. Added white noise would increase E without changing δE, leading to a reduction in the apparent χ. As in (6.22), this effect does not explicitly depend on frequency.

From the above examples, the selected reference model (such as the Preliminary Reference Earth Model, PREM) and measurement procedure act as an equivalent of background "geometrical spreading" for the free oscillations. Errors in the model or measurement noise shift the inferred attenuation coefficients, χ, which further lead to frequency-dependent shifts in Q^{-1}. These effects are quite similar to other types of seismic studies.

As shown in the following section, the above effects for low-order oscillations appear to be comparatively small. For higher modes, the geometrical spreading, as defined in this book, is usually called "focusing" or "defocusing". Focusing is viewed as one of the main issues limiting the resolution of global mantle attenuation tomography (Romanowicz and Mitchell, 2007). Again, because this book is not intended as a thorough study of free oscillations, we refer to the above authors for further detail and only focus on basic observations of the normal-mode χ here.

6.2.2 Frequency dependence of χ

Frequency dependence of normal-mode attenuation is also quite instructive when presented in the form of χ (Figure 6.13). For the fundamental spheroidal modes ($n = 0$), the Q factors stay approximately constant to angular orders l of ~16–17 (frequencies $f_{n,l}$ below ~2.5 mHz) (Figure 6.13a), and therefore the corresponding χ are linear in frequency, with the same $Q_e \approx 325$ (Figure 6.13b). For $l \approx 20$–60, the apparent Q decreases from ~325 to ~140, but χ shows another linear increase, with $\gamma \approx -7 \cdot 10^{-5}$ s^{-1} and $Q_e \approx 93$. This Q_e value is close to the one for long-period surface (see Figure 6.4). However, between 3–7 mHz, the values of χ are slightly shifted from those for surface waves, which is discussed in the next section.

The fact that the variation of the fundamental-mode χ can be described by only three parameters is interesting and should be significant. Note that we should talk not about the "frequency dependence of χ," but only about the implicit correlation between the values of $f_{0,l}$ and $\chi_{0,l}$ for the fundamental modes. Nevertheless, it appears that a significant change in mantle sampling occurs between angular orders of $l \leq 16$ and $l \geq 20$. According to the model proposed below (Section 7.5), linear variations of χ with frequency could indicate certain layers within the mantle dominating the observed attenuation. The level of $Q_e \approx 325$ for low-order overtones suggests that they are dominated by the lower part of the upper mantle (with Q_S of ~150–250), combined with the lower mantle, whose Q_S ranges from ~300 to 350 in the existing models. By contrast, all higher-order modes with l from ~20 to ~60 appear to be dominated by the asthenosperic attenuative layer, which has $Q \approx 70$–100 in the existing models.

The linearity of the $\chi(f)$ pattern for the higher angular-order modes (Figure 6.13b), irrespective of the actual values of l, suggests that the coverage of the attenuative layer is similar for all of these modes, and we have a case of localization as discussed in Section 5.2.2. As argued there, for attenuation occurring within a compact zone, the corresponding χ should increase with frequency and be insensitive to wavefield gradients, and consequently to the normal-mode orders as well. In addition, the attenuative layer within the upper mantle should be highly heterogeneous (see Section 7.5), which may effectively average out its effects on the high-order normal modes, somewhat similarly to coda-wave averaging.

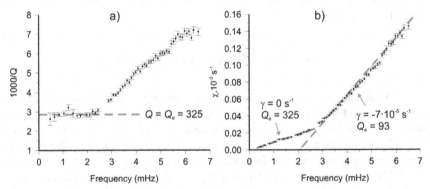

FIGURE 6.13
Spheroidal fundamental-mode attenuation data: a) in $1000/Q$ form and b) transformed to the attenuation coefficient, χ. Dashed gray lines show the apparent linear trends $\chi(f) = \gamma + \pi f / Q_e$. Data from IGPP Reference Earth Model web pages referred to above: http://igppweb.ucsd.edu/~gabi/rem.html, accessed 12 Sept. 2009.

6.2.3 Discrepancy between normal-mode and surface-wave Q^{-1}

The observations of residual geometrical spreading for surface waves and normal modes suggests a potential explanation of the discrepancy between the measurements of the traveling- and standing-wave attenuation noted by Durek and Ekström (1997). This discrepancy consists in systematic, ~15% differences in the attenuation levels measured by the surface-wave compared to normal-mode techniques (Figure 6.4a). In the $\chi(f)$ form, this difference consists in a near-constant, ~10^{-5} s^{-1} upward shift of the surface-wave χ (Figure 6.4b). Note that the amount of this shift is close to the surface-wave $\gamma \approx$ -8·10^{-6} s^{-1} and represents only ~3% of the measured crustal geometrical-spreading effect for surface waves at 100–10-s periods (which is $\gamma \approx$ 3.2·10^{-4} s^{-1} before the correction for dispersion; Figure 6.4). Thus, such a shift could generally be expected from a slightly inaccurate surface-wave geometrical-spreading correction.

As summarized by Romanowicz and Mitchell (2007), long-period surface-wave measurements may be affected by noise and by difficulties in defining the time windows for separating the fundamental modes from the various overlapping wave trains. Durek and Ekström (1997) and Masters and Laske (1997) argued that normal-mode estimates can generally be carried out more accurately and may be more reliable than the surface-wave ones. At the same time, the reduction of the normal-mode χ caused by measurement noise may account for about a half of their gap with the surface-wave estimates (Romanowicz and Mitchell, 2007).

Although the origin of this discrepancy has still not been established, our empirical observations above show that both measurements, and particularly the long-period surface-wave ones allow some room for adjustments by recognizing their geometrical-spreading component. Note that according to the equation:

$$t^* = \frac{\chi^*}{\pi f} = \int_0^t \frac{d\tau}{Q}, \qquad \text{(1.11 repeated)}$$

the geometrical-spreading factor is accumulated along the paths from the source to receiver. Therefore, for example, the predominance of continental surface-wave recordings (which are mostly conducted in tectonically active areas with higher γ_i) from deep-focus earthquakes (also likely with higher γ_i with respect to the overlaying layered mantle and crust) could cause increased γ values when globally averaged for the corresponding wave modes in Figure 6.4b. By contrast, normal-mode measurements are dominated by oceanic areas with presumably lower γ_i.

6.3 Mantle body waves

Long-period body-wave t^* measurements are among the key evidences for the frequency-dependent attenuation factor of the mantle. As explained in Section 5.1, t^* is an "apparent" (*i.e.*, measured) quantity which is usually interpreted as integrals of the *in situ* Q^{-1} over ray paths. However, as argued in this book, the *in situ* Q^{-1} actually does not exist, and t^* has a built-in frequency dependence caused by the uncertainty of geometrical spreading. This uncertainty needs to be well understood and separated from the physics of attenuation.

In this section, we revisit the t^* datasets by Der *et al.* (1982, 1986a, 1986b) and Lees *et al.* (1986) from the attenuation-coefficient point of view. In several extensive studies, these authors, and also Niazi (1971), Der and McElfresh (1976, 1980), Der *et al.* (1985), Der and Lees (1985), and Sharrock *et al.* (1995) examined the dependencies of the *P*- and *S*-wave t^* values, denoted t_P^* and t_S^* respectively, of various teleseismic waves on the frequency and tectonic types of the lithosphere in several areas of the world. Broadly, their key observations showed that:

(i) for body waves beyond distance ranges of ~25°, t^* values are nearly independent of the travel times;

(ii) t^* values increase in tectonically active areas, where the zone of increased attenuation is present; and

(iii) t_P^* decreases with frequency from ~1 s within the long-period band to ~0.2 s at short periods.

The first of these observations indicates that high attenuation is concentrated in the upper part of the mantle, so that body-wave ray paths at ~25° source-receiver distances penetrate below this zone. The second observation suggests that the attenuation is increased in the areas of active tectonics, or otherwise the attenuative zone may be thicker in active tectonic areas.

The third observation above is the most important for our discussion. Until now, variations of t^* with frequency were invariably associated with the *in situ* quality factor of the Earth's mantle decreasing with frequency. The five-time decrease of the long-period body-wave t^* appears to be among the strongest arguments in favor of the frequency-dependent mantle Q hypothesis. Nevertheless, below, we will show that this observation may also be explained by slight variability in the geometrical spreading. The *in situ* attenuation properties should definitely be layered in depth, but are most likely frequency independent.

6.3.1 Frequency dependence of t^*

Der and McElfresh (1980) were among the first to notice that body-wave values of $t_P^* \sim 1$ s and $t_S^* \sim 4$ s observed at long periods could not be the same for all frequencies, and also pointed out the regional variability of the short-period t^*. They concluded that t_P^* and t_S^* should be much lower within the short-period band, because otherwise the energy at 4–5 Hz would not be observed in P waves. Such frequency dependence of t^* is therefore a clearly established fact. However, t^* is still an apparent quantity, and according to the relation:

$$t^* = \int_{\text{Ray path}} \frac{dt}{Q}, \tag{6.24}$$

it also corresponds to the apparent Q. As shown above, the $t^*(f)$ dependence contains contributions from a positive residual geometrical spreading, which increases t^* values at long and intermediate periods. With this geometrical-spreading contribution removed, t_P^* at all frequencies becomes constant and equal ~0.18 s.

To illustrate the point above, Figure 6.14 shows the $\bar{t}^*(f)$ and $t^*(f)$ data summaries from Der et al. (1986b), which were derived from a series of studies analyzing multiple P- and S- body-wave phases at 25°–90° ranges, with travel paths lying within the shield areas of Eurasia. In addition, the $\bar{t}^*$ and t^* dependencies predicted by ray tracing in a layered, frequency-dependent Q model for the eastern United States (EURS) by the same authors are shown by dashed lines (Figure 6.14). An earlier shield-path model (Der et al., 1982; gray dotted line in Figure 6.14) is also overlain on this plot. This model is also close to the absorption-band mantle model by Minster (1978a, 1978b).

Despite the scatter and some discrepancies between different types of observations, it appears that t^* clearly decreases with increasing frequency (Figure 6.14). However, note that the attenuation data were reduced to t^* by using the assumption of perfectly corrected geometrical spreading, which is required to make t^* meaningful (see Section 5.1). By contrast, if not assuming the geometrical-spreading correction to be perfect, the same $\bar{t}^*$ and t^* data can also explained by linear dependence $\chi^* = \gamma^* + \pi f / Q_e^*$, with $\gamma^* \approx 0.06$ and $Q_e^* \approx 5.5\ \text{s}^{-1}$. This dependence corresponds to the thick solid hyperbola in Figure 6.14. Note that this line fits the geometrical-spreading–independent data (P-wave spectral measurement within 1–10 Hz range) even better than the existing model, and it may also be better following the trend of t^* rising at the lower frequencies. A marginal fit of the multi-phase S and rise-time data (Figure 6.14) could be due to poorer reliability of these measurements, and it was similarly problematic in the original interpretation (Der et al., 1986a). Difficulties in fitting multiple data types and the impracticality of formal inversion were emphasized in many t^* studies (e.g., Der et al., 1986a, 1986b; Sharrock et al., 1995).

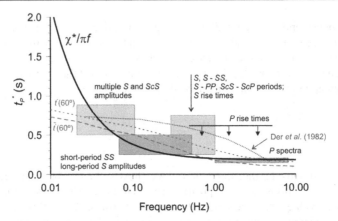

FIGURE 6.14

Data summary from Der *et al.* (1986) with $\bar{t}^*$ and t^* curves modeled by ray tracing at 60° ranges in their Eastern U.S. (EURS) Q model (gray dashed lines). Different data sources, measurement methods, and frequency-t^* value ranges are indicated. The $t^*(f)$ curve for shield areas from Der *et al.* (1982) is shown by grey dotted line. Thick black line corresponds to the attenuation coefficient linear in frequency, with $\gamma^* \approx 0.06$ and $Q_e^* \approx 5.5$.

Figure 6.15 illustrates a simple interpretation technique that can be used to analyze t^* data. Instead of deriving t^* from $\bar{t}^*$ by ambiguous integration of eq. (5.7):

$$t^* + f\frac{dt^*}{df} = \bar{t}^* \cdot \qquad (5.7) \text{ again}$$

we can determine $\bar{t}^*$ from t^* data by using the same equation. The empirical dt^*/df trend for use in eq. (5.7) can be estimated from the same t^* data plot (Figure 6.15a). The resulting $\bar{t}^*$ values simulate spectral measurements within the long-period band, and they are independent of the uncertainty of geometrical spreading. Therefore, $\bar{t}^*$ values can be reliably interpreted and modeled. Note that the present data can be satisfied with a constant value $\bar{t}^* \approx 0.18$ s mentioned above (Figure 6.15b).

The estimated value of the general trend Q_e^* allows presenting the cumulative attenuation coefficient χ^* in "reduced" form $\left(\chi^* - \pi f / Q_e^*\right)$ (Figure 6.16). As with time reduction in refraction seismology, using a shift linearly increasing with f allows highlighting the detail in $\chi^*(f)$ dependences. As we see, all of the data align horizontally, which shows that they are consistent with $\gamma^* \approx 0.06$

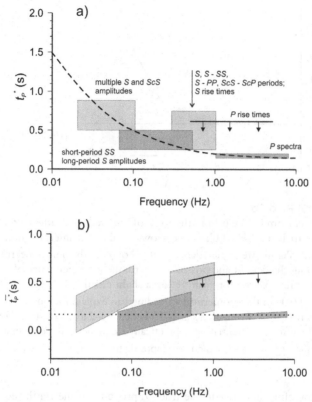

FIGURE 6.15
Geometrical-spreading independent t^* interpretation technique with: a) the same data as in Figure 6.14 with the $t_P^*(f)$ trend estimated from the data by eye (dashed line) and b) the trend removed by using eq. (5.7) and producing a geometrical-spreading–independent $\bar{t}^*$. Note that after this correction, the attenuation can be considered as frequency independent, with $\bar{t}^* \approx 0.18$ s (dotted line).

and $Q_e^* \approx 5.5$ s^{-1}. Note that a curve with a slight "absorption band" (increased χ^* from ~0.1 to ~5 Hz, similar to the dashed black curve in Figure 6.16) might fit the data a little better; however, this conclusion still does not seem to be warranted by the data scatter and quality (Der *et al.*, 1986b).

The above analysis shows that, similar to the frequency-dependent Q, the traditional t^* represents an "apparent" quantity because of its sensitivity to the geometrical-spreading correction, which is a subjective part of the measurement procedure. Values of t^* change whenever the real geometrical spreading differs from the postulated theoretical level. By contrast, $\bar{t}^*$ is measured from the spectral

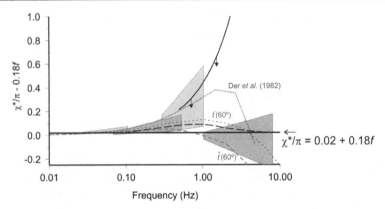

FIGURE 6.16

Data from Figure 6.14 in the form of the "reduced" attenuation coefficient. Dashed black line shows an alternate interpretation with slightly frequency-dependent Q_e^* or γ^*. With some concerns about the data fit and possible mild frequency dependence of Q_e, the principal result remains clear: a slight positive shift of $\gamma^* \approx 0.06$ in the geometrical spreading can explain the observed increase of t^* from ~0.2 s at 1–10 Hz to ~1–2 s at 0.01–0.02 Hz. The effective attenuation is practically frequency independent and equals $Q_e^* \approx 5.5$ s^{-1}, which corresponds to $t_e^* = 1/Q_e^* \approx 0.18$ s.

ratios and therefore is closer to the true Q_e^* property of the Earth (eq. (5.5)). For this reason, spectral-ratio measurements should be more reliable in constraining the attenuation.

Our derivation of χ^* from published t^* values still inherits all the limitations and approximations of these data, such as: 1) the P- and S-wave data are tied together by simple scaling $t_S^* = 4t_P^*$; 2) t^* is considered to be independent of t; 3) the same geometrical-spreading rate γ^* is assumed for P- and S-waves, their reflections and multiples; and 4) the free-surface refection coefficient for ScS_n multiples is assumed to equal one (Der et al., 1986a, 1986b; Lees et al., 1986). None of these limitations are necessary in the χ^* approach, and the corresponding parameters can be included in the model. Inverting the raw data spectra, amplitudes, and pulse shapes directly within the χ^* model would certainly reduce the errors and improve the data fit (Figure 6.14) and quality of our interpretation. However, this would require revisiting a large, multi-vintage, and complex dataset, which is not available to the present analysis.

As shown above, body-wave Q_e is practically frequency independent from the present data (Figure 6.14). After an ~6% geometrical-spreading correction, long-period t^* values approach those observed at short periods, and equal ~0.18 s for P waves. From eq. (5.20), this means that all the data in Figure 6.14 can be explained by a frequency-independent κ_i within the Earth. Inversion for the depth and regional variations in κ_i is a complex problem certainly impregnated with further uncertainties, which will be addressed elsewhere. However, from the attenuation data reduction to the (γ^*, Q_e^*) form and from eqs. (5.19) and (5.20), we see that the *in situ* κ_i model *should be frequency independent* unless additional data provide evidence to the contrary in the future.

6.4 Body-wave measurements in boreholes

In crustal body-wave observations, $Q = Q_0 f^\eta$ typically increases with frequency ($\eta > 0$), and large values of $\eta \sim 0.5$–1 are not uncommon in surface wave, $L_g Q$, L_g coda Q, and borehole studies (for an overview, see Abercrombie, 1998). Very large values of $\eta \sim 1.3$–1.8 were reported from borehole measurements at frequencies below 10 Hz (Adams and Abercrombie, 1998). Above this frequency, η decreases to a moderate $\eta \sim 0.2$–0.3, and Q_0 increases. Adams and Abercrombie (1998) called these changes the "10-Hz transition" and cautioned against extrapolating $Q(f)$ dependencies across this frequency. Based on the above discussions, we are now prepared to explain this transition by the changes in the geometric properties of the wavefield.

High-quality downhole seismometers offer good possibilities for measuring the upper-crustal Q. However, Sams and Goldberg (1990) and White (1992) noted that in these environments, even frequency-independent Q measurements are not without problems. For example, ultrasonic *in situ* borehole measurements using the spectral ratio technique resulted in P-wave Q_P values that were more than five times larger than those measured under the corresponding pressure conditions in the lab (Goldberg and Zinszner, 1989). Such discrepancies were attributed to mode conversions and scattering. Note that such factors play the role of the "geometrical spreading" around the borehole, in agreement with our general argument. A specific example of such near-borehole scattering is also found in the dataset considered below.

For a specific case example, let us consider the measurements of body-waves Q_P and Q_S in the Kanto region, Japan, by Kinoshita (2008). In this experiment, P and S waves from local earthquakes were recorded by seismometers in two deep boreholes. The observation frequency band was 1–10 Hz. To derive the Q values, each direct arrival was compared to the corresponding reflection from the free surface (Figure 6.17). Because of the simple geometry, known velocity structure and depth of the geophone, the time lags between the direct and reflected arrivals were accurately known, and by taking spectral ratios of their amplitudes, frequency-dependent Q values were estimated, as shown in Figure 6.18a. The reported values of Q_0 were ~11 and 33 for P and S waves, respectively. Such low values seem more appropriate for water-saturated sediments and invite some scrutiny in this particular case. Parameter η varied from 0.76 to 1.34; as in

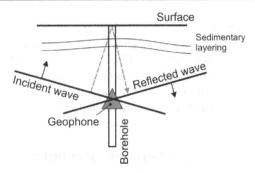

FIGURE 6.17
Borehole body-wave attenuation measurements using direct
waves from local earthquakes and their reflections from the free
surface (Figure 3.14b, p.100).

other examples in this book, such high positive η suggests strong residual
geometrical spreading and make these data an interesting example to revisit.

The key to the reported values of Q_0 and η above is in using the following
model for the spectral powers of the direct and surface-reflected arrivals, denoted
$F_i(f)$ and $F_o(f)$, respectively:

$$\left|\frac{F_o(f)}{F_i(f)}\right| = |G(f)|e^{-2\pi f t / Q(f)}, \tag{6.25}$$

where t is the two-way time lag between these two waves. By considering a half-
space reflection problem, the theoretical free-surface transfer function $|G(f)|$ was
estimated as equal to one, and after fixing this value for $|G(f)|$, the above
parameters of the frequency-dependent Q were obtained.

However, as we saw in Chapter 5, plane waves are subject to non-trivial
geometrical spreading (which can also be called scattering in this case) caused by
short-scale reflectivity. Therefore, $|G(f)|$ should likely be below one, and it would
be safer to determine it from the data rather than making theoretical assumptions
about its level. Re-plotting the data in the $\chi(f)$ form reveals quite different
conclusions from those by Kinoshita (2008) (Figure 6.19). Well within the
measurement errors, the frequency dependence of χ can be considered linear, and
the "geometrical" limit ($f \to 0$) is about $|F_o(f)/F_i(f)| = e^{-0.85} \approx 0.4$ to the P-wave
energy ratio and $e^{-0.5} \approx 0.6$ for S waves (Figure 6.19). Note that the frequency
ranges of this linear fit extend to 10 Hz for P-wave data and to 6 Hz for S-wave
data, which is much broader than the corresponding ~3 and ~2 Hz high-cut limits
to which the power-law $Q(f)$ model could be fit (compare Figures 6.18 and 6.19).
The values of effective Q_e estimated directly from the plots give $Q_P \approx 700$ or
higher for P waves (when viewing the lowest-frequency point as an outlier; Figure

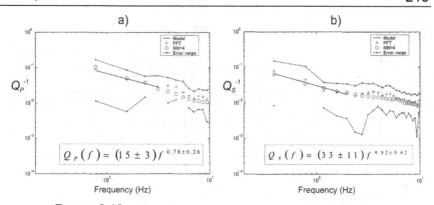

FIGURE 6.18
Log-amplitude spectral ratio data from station SHM in Kanto area (Japan) from Kinoshita (2008) for a) P- and b) S-waves. Values of $Q(f)$ by Kinoshita (2008) are given in the insets, and the data fits for these models are shown by solid lines. From Morozov (2008); with permission from John Wiley and Sons.

6.19a) and $Q_S \approx 100$ for S waves (Figure 6.19b). These values are much higher than Q_0 based on the assumption of $|G(f)| = 1$ (Figure 6.18, insets).

Similar to the cases of surface and L_g waves above, within the error bounds on the spectral ratio values, there is no need to invoke a frequency-dependent Q_e. Interestingly, from the basic theory of Q as "imperfect elasticity," Q_S is expected to be always lower than Q_P. This relation was violated by the values $Q_P \approx 15$ and $Q_S \approx 33$ given by Kinoshita (2008), which therefore appeared to be an important anomaly. However, the present analysis shows that the effective Q_e for these waves clearly satisfy the $Q_P \gg Q_S$ relation, but the apparent anomaly corresponds to a ~50% higher geometrical transmission loss for P waves. This suggests P waves reflectivity is stronger than for S-waves, which could be due to various reasons, such as variations in the porosities of sedimentary rocks.

Cases of $\eta > 1$ are always particularly disturbing, because they suggest a relative increase in high-frequency energy with propagation time and do not fit even in the visco-elastic model. Kinoshita (2008) also gave such an example (Figure 6.20). However, as the corresponding $\chi(f)$ plot shows, there actually is little evidence for attenuation in this case, i.e., the effective Q_P^{-1} can be set equal to zero. At the same time, the "geometric" amplitude ratio is near $e^{-0.75} \approx 0.5$ (Figure 6.20b), causing the large value of η.

The example in Figure 6.20 also illustrates another important point which can be easily missed in the log-log plots of $Q^{-1}(f)$. Note that the lower-frequency band of ~0.8–3 Hz in which the $Q(f)$ fitting was performed (solid line in Figure 6.20a) actually corresponds to the range in which the spectral ratios increase with frequency (Figure 6.20b). Such an amplitude increase is likely related to

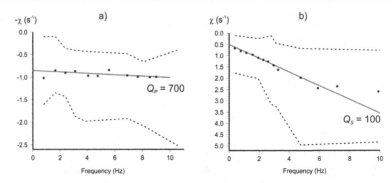

FIGURE 6.19
Temporal attenuation coefficients for: a) P and b) S waves at
station SHM, respectively, derived from Q values in Figure 6.18.
Effective attenuation values are labeled Q_P and Q_S. From
Morozov (2008); with permission from John Wiley and Sons.

transmission amplitude variations ("tuning" or scattering) and *should not* be
attributed to attenuation.

Thus, Kanto borehole data show that the attenuation in this area is
moderate to low, and is at least not detectably frequency dependent. This is
contrary to the reported very high $\eta \approx 0.8–1.2$ and very low values of $Q_0 \approx 11–33$
(Kinoshita, 2008). Such values were caused by an assumption of a complete free-
surface reflection, which was far from reality, and attributing the first-order effect
of reflectivity to Q.

Instead of being equal to one, the effective upcoming-to-downgoing
transfer amplitudes determined from the data were about 0.6–0.7 for P-P and 0.8
for S-S reflections. Such deviations from the theoretical level of (-1) could be
caused by multiple effects, including transmission losses, non-vertical incidence,
and mode conversions. All such factors should be strong in the near surface, with
the strongest one being the reflection at the basement-sediment contact. In the
available velocity structure (Kinoshita, 2008), the reflection coefficient for P-
waves at normal incidence is about 0.4. Two transmissions through this contact
should reduce the free-surface reflected wave energy to ~0.70 of the incident one,
which is close to the level estimated from the data above.

A number of other factors may additionally increase the reflectivity of the
near-surface. For example, seismic waves from local source-receiver distances of
~10–50 km arrive at ray parameters of ~0.2–0.3 s/km, which corresponds to the
incidence angles on the sedimentary boundaries of up to 40° (Figure 6.17). At
such angles, amplitude variations and P/S mode conversions should be strong, and
they should further increase the reflection amplitudes. Further, at oblique
incidence, the seismometer samples different parts of the wavefield on its upward
and reflected downward propagation, and therefore they also should have different

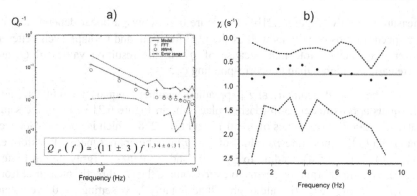

FIGURE 6.20
Interpretation of borehole data from station FCH in Kanto area (Japan): a) P-wave results with error bounds by Kinoshita (2008). Inset shows $Q(f)$ inferred from these data for ~0.8–3 Hz frequency range (solid straight line). b) Attenuation coefficients derived from the same data points, and their estimated error bounds. Note the lack of detectable attenuation but significant deviation of the amplitude ratio from 1.0 ($\gamma \approx 0.5$; horizontal line). From Morozov (2008); with permission from John Wiley and Sons.

amplitudes. Finally, velocity heterogeneity and reflectors surrounding the boreholes may cause additional variations in the amplitude ratios. All of these factors include certain "geometrical" (frequency-independent) components, although their frequency-dependent effects should also be significant. All of the above show that accurate theoretical predictions of $|G(f)|$ may be very difficult even in well-constrained, borehole recording. Nevertheless, the cumulative effect of these factors can be measured from the data, as described previously.

6.5 P_n

To illustrate how the attenuation-coefficient model works for regional body waves, consider the P_n Q results from a study by Xie (2007) from PASSCAL[17] seismic experiments in the Tibetan Plateau (Figure 6.21). The quantity plotted along the vertical axis in Figure 6.21a is the stacked spectral ratio, which is proportional to χ (see Section 5.1.2). However, when, χ is plotted against a linear frequency scale, the frequency dependence of the attenuation coefficient can be

---

[17] Program for Array Seismic Studies of the Continental Lithosphere, one of the core programs of the Incorporated research Institutions for Seismology (IRIS) consortium.

measured directly (Figure 6.21b). As Figure 6.21 shows, a linear dependence fits the spectral ratios equally as well as the $Q_0 f^\eta$ function, and perhaps even better at lower frequencies. Note that because of $\eta > 0$, the resulting values of Q_e are systematically higher than the corresponding Q_0.

The best-fit geometrical P_n spreading parameter equals $\gamma \approx 0.002$ s^{-1}, and Q_e equals approximately 340 (dashed black line in Figure 6.21b). From the same data, Xie (2007) gave values of $\eta = 0.14$ and $Q_0 = 278$, which is about 25% lower than our Q_e. For moderate η, values of (Q_0, η) can be approximately transformed to (γ, Q_e) (Section 5.4), leading to $Q_e = 400$ and $\gamma = 0.005$ s^{-1} (gray line in Figure 6.21b). This transformation seems to over-estimate the Q_e, directly measured from the $\chi(f)$ plot (Figure 6.21b), although still acceptably[18]. By setting $\gamma = 0$, we obtain $Q_e = 310$ (dotted line in Figure 6.21b). As expected, this value of Q_e is the closest to $Q_0 = 278$ by Xie (2007).

Figure 6.21a also shows two versions of the background geometrical-spreading model in the form t^{ν}, one of which is frequency dependent. Note that when using the (γ, Q_e)-type attenuation parameters (Figure 6.21b), there is no need for re-plotting the data for the different background models. With the use of frequency-dependent geometrical spreading (gray line and symbols in Figure

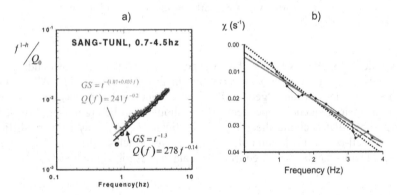

FIGURE 6.21

P_n attenuation parameters in Tibet: a) Stacked spectral ratios from Xie (2007) for two (Q_0, η) models inverted by using different geometrical spreading (black and gray lines and symbols). Two selected geometrical-spreading laws and the corresponding $Q(f)$ are given in the labels. b) The same data in $\chi(f)$ form in linear frequency scale. Lines represent different options for linear fitting (see text). From Morozov (2010a); with permission from Springer.

---

[18] Note that this is the only real data example in this book in which a solution with $\gamma = 0$ is acceptable.

6.21a), a part of the slope of $\chi(f)$ would simply be attributed to the geometrical-spreading dependence, leading to the corresponding changes in γ and Q_e. This is only an interpretational change, which affects neither measurements nor modeling.

6.6 Teleseismic P_n

Along with contrasting t^* and Q values arising from long- and short-period body-wave measurements, a completely different argument is sometimes advanced in favor of the mantle Q increasing with frequency, namely the observations of an anomalous, high-frequency "teleseismic" P_n (also often called long-range P_n) from Peaceful Nuclear Explosions (PNE) in Russia (Ryberg and Wenzel, 1999). Seismic waves from the PNE's at frequencies above 5 Hz were found to travel at P_n velocities to over 3000-km distances within the uppermost mantle of the East European Platform. Because these waves were also followed by extensive codas, scattering processes were required for their explanation. A multiply scattering waveguide mechanism was proposed by Ryberg *et al.* (1995), which favored preferential propagation of high-frequency waves (SWG in Figure 6.22). It is still unclear whether such strong scattering would actually increase or decrease the uppermost-mantle attenuation at these frequencies; nevertheless, the fact of efficient propagation of waves at 5 Hz which appear much weaker at 1–2 Hz suggested that high-frequency Q should be high (Ryberg and Wenzel, 1999).

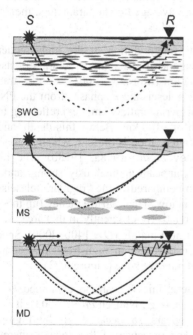

FIGURE 6.22
Three models of the uppermost mantle explaining the high-frequency teleseismic P_n phase and its coda. S is the source, R – receiver. SWG – strong multiple scattering within uppermost ~80 km of the mantle; MS – moderate scattering near the Lehmann discontinuity (~120–150-km depth); and MD – reflections and refractions on from mantle discontinuities with scattering occurring within the crust.

The subject of the "scattering waveguide" within the uppermost mantle is among the most spectacular, and also strongly popularized findings from regional PNE seismology. Three contrasting models were proposed to explain teleseismic P_n observations: 1) strongly scattering, fine-scale, stochastic mantle immediately below the Moho (Ryberg *et al.*, 1995; labeled SWG in Figure 6.22); 2) scattering below ~100 km, near the interpreted Lehmann discontinuity (Thybo and Perchuc, 1997, Nielsen *et al.*, 2003; MS in Figure 6.22); and 3) "conservative" model resolving only reflections and refractions within the upper mantle and attributing scattering to the crust (Morozov *et al.*, 1998a; MD in Figure 6.22). Note that the rationale for models MS and MD was based on the "structural" constraints derived from PNE data (*i.e.*, detailed variations of the travel times and amplitudes), whereas the argument for model SWG was based entirely on the general features of its spectral content and coda. In its focus on explaining solely the spectral-amplitude and averaged coda parameters, the SWG model is also similar to the observational Q studies.

However, the scattering-waveguide interpretation of the high-frequency teleseismic P_n also represents a notable example of mistaking structural effects for a frequency dependence of mantle attenuation (Morozov, 2001). Careful travel-time and amplitude analysis shows that the teleseismic P_n phases from the PNE's consist of series of P-wave multiples ("whispering gallery" modes) reflecting from the free surface traveling above the depth of ~100 km. Below this depth, and to ~150–220-km depths, a strong increase in attenuation is present, which causes the longer-range phases to attenuate (Morozov *et al.*, 1998a, 1998b). The high-frequency amplitude of the teleseismic P_n appears anomalously strong, and the coda appears anomalously long only when compared to the "regular" teleseismic P waves penetrating the attenuative layers below. The high-frequency (up to ~10–15 Hz) P waves and their multiples travel relatively efficiently in this area merely because of the low attenuation in the uppermost mantle ($Q \approx 1400$–2000). Spectra of the various phases show a progressive decrease in high-frequency content, proportional to the time spent within the attenuative zone (Figure 6.23).

The characteristic coda of the teleseismic P_n can be explained crustal scattering, similarly to codas of other phases, (*e.g.*, Nielsen *et al.*, 2003). It is not anomalously high-frequency, but only appears so because of the high-pass filtering used by Ryberg *et al.* (1995). The presence of the coda is therefore unrelated to the efficiency of the teleseismic P_n propagation. This propagation itself is not anomalously efficient, merely because this phase is not a P_n, but represents two or three multiple P-wave reflections from the free surface. Thus, the SWG model, along with its inferences of preferential high-frequency properties of the uppermost mantle, finds no support in the data, but rather illustrates the freedom of model-driven arguments in attenuation studies. A combination of models MD and MS with normal, frequency-independent Q and weak scattering (if any) within the upper mantle (Figure 6.22) remain in good agreement with PNE observations.

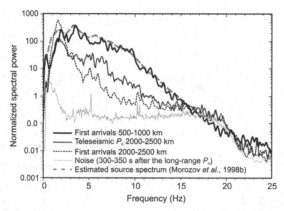

FIGURE 6.23

Spectrum of teleseismic P_n phase compared to those of other phases from PNE Quartz-4. Note the absence of anomalously high teleseismic P_n amplitudes. Progressive changes in the amplitude spectra of the arrivals are related to the presence of an attenuating zone below ~150-km depth.

6.7 L_g

Measurements of regional phase attenuation, and particularly of the L_g phase, are most important in seismic nuclear test monitoring research. Despite the generally applied and empirical character of most of such studies, in many of them, Q_0 and η values are used in interpretative sense, *i.e.*, by correlating them with geological structures, performing seismic regionalization, or comparing them to other seismic phases. L_g attenuation parameters also show strong variability, which is associated with major geological structures. For example, several researchers, such as McNamara *et al.* (1996), Phillips *et al.* (2000), Xie (2002), Fan and Lay (2003), Xie *et al.* (2004), revealed high and variable L_g attenuation within the Tibetan Plateau ($Q_0 \approx 120$–150), with localized areas having $Q_0 \approx 60$–90. In many studies, large values of $\eta \approx 0.37$–0.5 were reported. Such cases are of particular interest to the present discussion, because they suggest a significant residual geometrical attenuation and a much higher Q_e.

Figure 6.24a shows the stacked spectral ratios from Xie (2002), from which this author interpreted the frequency-dependent L_g attenuation parameters of $Q_0 \approx 126$ and $\eta \approx 0.37$. Transformation of the same points into the attenuation-coefficient form again reveals a linear trend of χ, whose slope gives $Q_e \approx 230 \pm 15$ (Figure 6.24b). As above, this value is significantly higher than Q_0. The character of $\chi(f)$ data fit suggests no deviations from a linear trend, apart from some

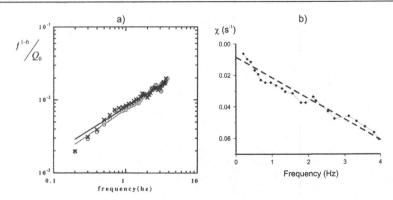

FIGURE 6.24
L_g attenuation in the Tibetan plateau: a) Stacked spectral ratios from Xie (2002), with $Q_0 = 126$ and $\eta = 0.37$ (straight lines). b) The same data [crosses in a)] re-plotted as $\chi(f)$ in linear frequency scale.

undulations related to data errors and spectral effects of the lithospheric structure (probably, "tuning") discussed in Section 2.8.

Note the clear non-zero intercept value of $\gamma = 0.008$ s^{-1} (Figure 6.24b). Interestingly, this value of γ falls on the boundary, γ_D, separating the stable and active tectonic regimes for body and coda waves, as summarized in Section 7.3.

6.8 Coda

Seismic coda is among the best objects to study attenuation both empirically and theoretically. Because of its averaging of multiple wave modes in the vicinity of the source and/or receiver, coda shape is usually simple, and its time-frequency dependence can be described by only a few parameters. Coda Q, often denoted Q_c, is typically used as such a parameter, to which the dependences of Q_c on frequency and sometimes on coda lapse time may be added. When viewed as an empirical parameter only describing the coda shape, such Q_c is quite adequate, although maybe having a redundant frequency dependence. However, Q_c often appears to be too easily attributed to the scattering properties of the subsurface. As illustrated on other wave types above, such an association requires very accurate modeling of the effects of the structure within which the propagation takes place, and such modeling is particularly difficult for codas.

In Chapter 3, we already considered an example of long-range seismic coda decaying with time and frequency (Figure 3.9 on p. 90). The notable observation was the geometrical-spreading, corrected-coda amplitudes decayed with frequency slower than with time, thereby violating the expected

$A(t,f) \propto \exp(-\pi f t/Q_c)$ character and leading to a Q_c increasing with f. Here, we continue revisiting the coda problem from two points of view. First, we review one of the classic local-earthquake coda examples by Aki (1980), and show that its apparent frequency-dependent Q_c transforms into linear frequency dependences of the attenuation coefficient. Second, we attempt a numeric modeling of long-range codas in a realistic lithospheric structure which also illustrates how frequency-dependent Q_c arises from a "normal," frequency-independent Q rheology combined with refractions and reflectivity within the subsurface.

6.8.1 Local-earthquake coda

Figure 6.25a shows the results of local-earthquake coda measurements in central California, Hawaii, and in central and western Japan (Aki, 1980). These observations were key to establishing the coda-Q concept, and also to illustrate its frequency dependence. To produce these Q values, Aki (1980) used a theoretical geometrical spreading of t^{-1} in eq. (2.119) to compensate the time-domain log-amplitude coda slopes and converted $\chi(f)$ into $Q^{-1}(f)$ by using the inverse of eq. (1.10). Two main observations were made from the resulting curves (Aki, 1980; Figure 6.25a): 1) three of the four $Q^{-1}(f)$ dependencies appear to converge at the higher frequencies whereas the fourth (TSK) does not and 2) therefore, station TSK in central Japan suggests a different attenuation mechanism from the other areas. Although the observed $Q(f)$ dependences were identified as apparent and

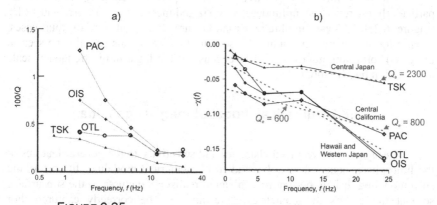

FIGURE 6.25
Local-earthquake coda attenuation data from Aki (1980): a) in a) $1000/Q(\ln f)$ form and b) $[-\chi(f)]$ form. Labels indicate seismic stations: PAC – central California; OIS – western Japan; TSK – central Japan; and OTL – Hawaii. Note the interpreted linear $\chi(f)$ trends (dashed lines and labels showing values of Q_e). From Morozov (2010a); with permission from Springer.

did not allow discrimination between the anelastic and elastic-scattering contributions, small-scale scattering was nevertheless viewed as viable attenuation mechanism (Aki, 1981). Even with allowing for the Q_i/Q_s uncertainty, the frequency-dependent Q was still attributed either to the scattering or rheological properties of the medium.

However, returning the data to the attenuation-coefficient form reveals linear patterns of $\chi(f)$ and suggests different conclusions (Figure 6.25b). The convergence of the inferred $Q^{-1}(f)$ curves near ~25 Hz (Figure 6.25a) could be principally due to dividing the values of χ by the frequency. The difference of station TSK at high frequency is principally due to low attenuation (Q_e^{-1}) in this area, which still does not mean a difference in its mechanism. Division by f also somewhat exaggerated the difference between the results from stations OIS and OTL below 5 Hz (Figure 6.25a).

By contrast to the $Q(f)$ interpretation, the intercepts and slopes of the $\chi(f)$ trends (dashed lines in Figure 6.25b) reveal significantly more information from these data. Comparing these parameters, we see that: 1) all geometrical-spreading terms γ are positive; 2) in Hawaii and for both areas in Japan, γ values are lower, from (0.02–0.03 s^{-1}) than in central California (0.06 s^{-1}); 3) attenuation values in Hawaii and western Japan are quite similar, with $Q_e \approx 600$; 4) in central Japan, attenuation is low, with $Q_e \approx 2300$; and 5) in terms of both parameters, central California is distinctly different from the other three areas, with its $\gamma \approx 0.06$ s^{-1} and $Q_e \approx 1250$.

On top of the linear $\chi(f)$ trends, some "spectral scalloping" is also clear, particularly the reduced amplitudes near 6 Hz and increased at ~1–2 Hz and 12 Hz (Figure 6.25b). These amplitude variations are consistent for all four cases, suggesting their common origin. These amplitude variations can also be seen in the $Q^{-1}(f)$ form, although less clearly because of the logarithmic frequency scale and division by f (Figure 6.25a).

6.8.2 Numerical model of long-range coda

Because of its averaged character and sensitivity to the general features of the lithosphere, coda can also be studied by numerical modeling. In performing such modeling, it is important to implement realistic structures of the subsurface, so that the effects of geometrical spreading can be correctly captured. For simplicity and ease of interpretation, we use long-range P-wave coda in this example, in which the effects of the incident and scattered waves can be reasonably separated.

In Morozov et al. (2008), we generated four finite-difference models of short-period coda wavefields resulting from crustal and upper-mantle velocity models. Four 1-D velocity models were constructed (Figure 6.26), including the global IASP91 model (Kennett and Engdahl, 1991) and three models derived from

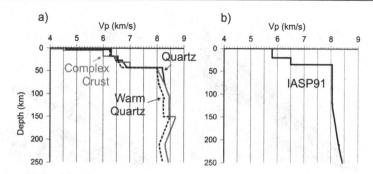

FIGURE 6.26
Upper 250 km of 1-D V_p velocity models used in coda
modeling. a) Models based on PNE studies and b) the global-
average IASP91 model.

PNE studies in northern Eurasia. These models were derived by careful travel-
time and amplitude analysis of refracted and reflected waves from reversed PNE's
and numerous chemical explosions (Morozova *et al.*, 1999). These models still
represent the most detailed and accurate pictures of the crustal and upper-mantle
layering within the 1–5 Hz seismic frequency band.

The first of the PNE-based models is called "Quartz" after the namesake
PNE profile from which it was derived. The model contains a three-layer crust
overlain by a 3-km–thick sediment layer and complex mantle with low velocity
zones at 110- and 210-km depths (Figure 6.26a). Another model, "Complex
Crust" has a high-contrast, five-layer crust with a somewhat exaggerated
attenuative layer with $Q_S = 10$ on its top, and the mantle structure identical to that
of Quartz (Figure 6.26a). A model called "Warm Quartz" (Figure 6.26a) simulates
a high-heat flow regime in the Quartz mantle, which is done by applying a
temperature-related negative velocity gradient within it (Christensen and Mooney,
1995). Note the presence of strong velocity gradients, contrasts, and low-velocity
zones within these models, as evidenced by observations of numerous reflections
and refractions in PNE studies. In order to collect data on the sensitivity of the
results to crustal velocities, we repeated all Quartz simulations after applying ±5%
perturbations to the crustal velocities (Figure 6.26a) while keeping the mantle
fixed.

Because of their complexity, all three Quartz-based models show
significantly more complex wavefields compared to the IASP91 model (Figure
6.27). In this figure, three-component wavefields were simulated by using the
reflectivity method (Fuchs and Müller, 1971). This method accurately accounts for
all reflections, multiples, and mode conversions within a vertically layered
structure. Realistic models with reflective crust and uppermost mantle produce
strong crustal L_g phases in addition to the P and S phases (Figure 6.27).

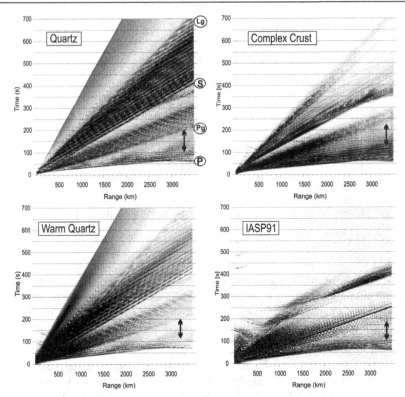

FIGURE 6.27
Synthetic seismograms created using the four different velocity
models of Figure 6.26. These records serve as the Green's
functions during modeling the coda in eq. (6.29). Note the
similarities between the realistic (Quartz) models and a contrast
with the wavefield in IASP91 model. Arrows indicate the
positions of the receiver and coda time ranges used in coda
modeling. Key regional phases are labeled in the first plot.

To examine the frequency dependence of coda attenuation, we selected a
coda time window following the teleseismic P wave arrivals, marked with arrows
in Figure 6.27. These windows are relatively arrival-free in the direct-wave
seismograms (Figure 6.27); however, the near-source or near-receiver scattering
would introduce exponentially decaying coda amplitudes in them. The choice of
the P-wave coda can also be justified by several additional reasons: 1) S- and L_g
phases are less pronounced in the PNE records; 2) their codas lie on top of the P-
wave coda, complicating the observation and requiring their decomposition; and
3) the L_g phase is significantly more difficult to model accurately. Similarly, the
Rayleigh wave (R_g) is not examined here because it decays quickly in the near-
surface sediments and is not viewed as a significant factor in coda formation

(Dainty, 1985). R_g is also not seen beyond ~200 km in PNE records. Finally, the use of P-wave coda at offsets of ~2900 km provides a sufficient coda time window before the onset of the S wave.

To fully address the crustal scattering problem, 3-D modeling of crustal features including topography, shorelines, mountain ranges, basins, and faults, is required. However, full 3-D modeling is still not practical due to limited knowledge of the structural detail and scattering properties of the crust. Instead, we use a heuristic simulation of teleseismic P-wave coda-amplitude decay modeled as scattering from heterogeneities near the Earth's surface, by combining 1-D reflectivity synthetics with the single-scattering approximation of coda intensity. The use of near-surface crustal scatterers simplifies the computations, and it is also justified by the increased heterogeneity of the shallow crust and supported by several array observations (*e.g.*, Greenfield, 1971; Dainty, 1985; Bannister *et al.*, 1990; Gupta *et al.*, 1991) and also by previous coda modeling (*e.g.*, Dainty and Schultz (1995)). As this is not a true 3-D crustal model, it cannot provide a completely realistic picture of the resultant coda decay; however, it captures the key mechanism of coda generation, provides improved estimates of crustal attenuation, and elucidates the problem of the frequency-dependent apparent-coda Q_c.

By approximating crustal heterogeneities as point scatterers distributed within the Earth's crust, the coda intensity (energy density) U recorded at a receiver at time t is an integral over volume V containing all scatterers:

$$U(\mathbf{r},t) = \int dt_s \iiint_V d^3\mathbf{r}_s \lambda(\mathbf{r}_s) U_{source}(\mathbf{r}_s,t_s) G(\mathbf{r}_s,t_s \mid \mathbf{r},t), \qquad (6.26)$$

where t_s is the time of the direct arrival at the scatterer, $\mathbf{r}_s$ represents the scatterer positions, λ is the scattering amplitude describing the amount of energy reflected at each scatterer, U_{source} is the seismic source function representing the energy density arriving at the scatterer from the source, and $G(\ldots)$ is the Green's function describing the propagation of the scattered energy from the scatterer to the receiver. This expression is similar to (5.62) but does not assign any specific form to the Green's function. Because the upper crust is considered as the primary contributor to the seismic coda, this volume integral is replaced with a surface integral:

$$U(\mathbf{r},t) = \int dt_s \iint_S d^2\mathbf{r}_s \lambda(\mathbf{r}_s) U_{source}(\mathbf{r}_s,t_s) G(\mathbf{r}_s,t_s \mid \mathbf{r},t). \qquad (6.27)$$

This approximation also allows us to reduce the volume of simulations by computing only the surface-to-surface Green's functions and to use a near-surface source modeling algorithm. In the 1-D case, the Green's function G is also translationally invariant in time and space:

$$G(\mathbf{r}_s,t_s;\mathbf{r},t) = G(\mathbf{r}-\mathbf{r}_s,t-t_s),\qquad\qquad (6.28)$$

and therefore expression (6.27) further simplifies to:

$$U(\mathbf{r},t) = \int dt_s \iint_S d^2\mathbf{r}_s \lambda(\mathbf{r}_s) U_{source}(\mathbf{r}_s,t_s) G(\mathbf{r}-\mathbf{r}_s,t-t_s).\qquad (6.29)$$

For a source located at a large distance, the primary P wave travels at a significantly higher apparent velocity (8–9 km/s) compared to the scattered waves (such as $V = 2.9$–3.5 km/s for a scattered L_g). Therefore, for a given lag time t, the coda consists of scattered waves originating from an elliptical ring surrounding the receiver (Figure 6.28). The scattering area within the ring increases with t due to its increasing size and partly compensates the energy decay due to the geometrical attenuation. For example, for a plane of constant scattering potential and no intrinsic attenuation, the coda energy would stay constant, provided that only surface-wave modes contribute to the Green's function (6.28).

In the 1-D model considered here, the source and receiver contributions to the coda are equal and inclusion of the near-source scattering would double the resulting singly scattered coda intensity. However, this intensity is also controlled by factor λ in eq. (6.29), which is set equal to a frequency-independent constant for simplicity. The value of this constant can be estimated by matching the modeled coda amplitudes to those observed from real PNE's (Morozov and Smithson, 2000), and consequently it already incorporates the averaged contributions from both the source and receiver areas.

To evaluate integrals (6.29) numerically, we pre-computed the Green's

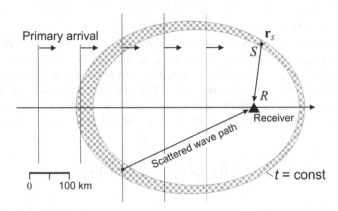

FIGURE 6.28
Near-receiver scattering area (shown by pattern) contributing to the long-range P-wave coda recorded within a narrow lag-time interval near $t = $ const.

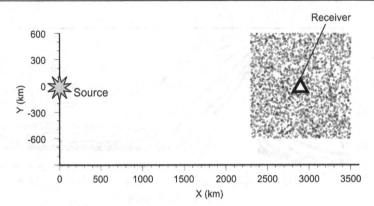

FIGURE 6.29
Monte-Carlo sampling in long-range coda modeling. Scatterers
are located near the surface, centered on the receiver, and
uniformly and randomly distributed.

function (Figure 6.27) and used Monte-Carlo sampling to integrate over the near-receiver area (Figure 6.29). Random sampling helped avoid spurious coherency that could be caused by spatial aliasing when using a regular integration grid. The resulting P-wave coda was measured at near-teleseismic, 2900-km distances to allow a significant time interval for measurements before the onsets of P_g and S waves. The size of the scattering region size was chosen to be a 1200-km square on a side, allowing 600 km from the central receiver to the nearest edges (Figure 6.28). For scattered waves of L_g-type velocity, this gives a ~200-s window in which to reliably measure the coda amplitude decays. As the Q_S value in the reflectivity synthetics is constant throughout the crustal section, and crustal scattering does not depend on the source-receiver offsets, the receiver's offset value should not influence coda Q_c and Q_e values measured from the output traces.

Note that by using the primary field $U_{source}(\mathbf{r},t)$ in eq. (6.27), we only consider the near-receiver scattering (Figure 6.28). Due to reciprocity, the same scattered amplitudes should be produced by near-source scattering in the 1-D model. Dainty (1990) showed that this was also true in real NORESS records from Semipalatinsk nuclear explosions in Kazakhstan. In the single-scattering approximation of this study, the source- and receiver-end codas will add up, and the resulting dependence of coda slopes on the frequency should not change.

Sixty-four realizations of the random grid with 2000 points in each were computed using a cluster computer, and the resulting amplitude envelopes were stacked to produce the final synthetic coda records (Figure 6.30). As expected, the modeled log-amplitude coda envelopes are linear within the time intervals during which the scattered waves stay within the modeled scattering region. The resulting logarithmic decrements (χ) can therefore be accurately measured for seven frequency bands frequencies centered at 0.4, 0.79, 1.58, 3.16, 4.48, 6.32, and 8.94 Hz (Figure 6.30).

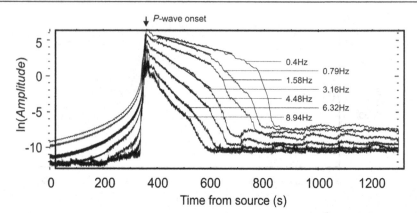

FIGURE 6.30

Synthetic coda envelope for model Quartz (see Figure 6.26) with crustal Q_s = 300, filtered within seven narrow frequency bands (labeled). Straight line segments indicate measurements of χ.

Multiple simulations were performed as described above for each model, with *P*- and *S*-wave quality factors within the layers in the crustal part of the model set proportional as $Q_P = 2Q_S$, in which values of Q_S ranged from 100 to 1000. This range of Q_S corresponds to $L_g\,Q$ values of ~300–1000 measured from Russian PNE's by Li *et al.* (unpublished) and extends to low-Q values in order to investigate the crustal effects. Mantle attenuation in each velocity model was fixed by using the mantle Q_p values derived from *P*-wave Quartz PNE data. These Q_S values ranged from Q_S = 750 below the Moho to 600 above the 410-km discontinuity, with an attenuative zone of Q_S = 200 in the region of the Lehmann discontinuity at 110–150-km depths (Morozov *et. al.*, 1998).

Having set up the forward problem, our objective is now to investigate the relation of the apparent frequency-dependent χ and Q_c measured from the log-amplitude slopes in Figure 6.30 to the actual crustal Q input in the models. Once again, this can be much better accomplished by checking the $\chi(f)$ dependences than by transforming them into the $Q_c(f)$ form. Figure 6.31 shows clear linear dependencies of $\chi(f)$ in two of the models to the frequencies of about 7–10 Hz. Above these frequencies, the fidelity of modeling appears to drop, and the attenuation is replaced by frequency-independent numerical noise. Straight lines fit to the points within the 0–7 Hz frequency give the "geometrical" non-zero intercepts at f = 0 Hz: $\gamma \approx 0.8 \cdot 10^{-2}$ s^{-1} and $\gamma \approx 1.6 \cdot 10^{-2}$ s^{-1} for models Quartz and Complex Crust, respectively. From the frequency dependences of χ, we can also obtain the effective-coda Q_e equal to ~800 and 600, for the two respective models (Figure 6.31). Note that these γ's are clearly positive, showing that the geometrical spreading in the numerical models occurs faster than prescribed by the simplified

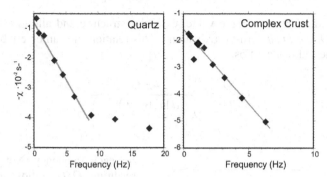

FIGURE 6.31
Logarithmic coda amplitude decrements measured from
synthetics in two realistic models with crustal Q_S = 500. Note the
linearly increasing χ with frequencies increasing to ~7–10 Hz.
Solid gray lines indicate linear regressions using: γ = 0.8·10⁻² s⁻¹
and Q_c = 800 for Quartz, and γ = 1.6·10⁻² s⁻¹ and Q_c = 600 for
Complex Crust models.

theoretical model. Some outliers visible at lower frequencies should likely be
related to tuning within the crustal layers (Figure 6.31).

In the traditional interpretation (*e.g.*, Figure 6.25a), when converted to
frequency-dependent $Q_c = \pi f/\chi$, the values of coda amplitude slopes could also be
fit by using a dependence of type $Q_0 f^\eta$, with $Q_0 \approx 250$ at 1 Hz and $\eta \approx 0.5$ (Figure
6.32). Although this interpretation could be acceptable as explaining the coda
shape, our model contains no rheological or frequency-selective scattering
mechanisms to support an *in situ* frequency-dependent attenuation. Therefore the
observed coda $Q_c(f)$ (Figure 6.32) is a purely "apparent" quantity, or an
observational artifact which may not be very helpful for understanding the true
properties of the crust and mantle.

Relation of Q_c to crustal S-wave Q_S

Coda modeling also allows correlating the coda Q_c values measured on the
surface to the S-wave Q_S within the subsurface. For a crude general estimate, we
can utilize the fact that zones with the lowest Q_S (*i.e.*, highest χ_S) contribute most
to the observed coda attenuation. The highest values of both geometrical and
dissipative χ should be present at shallow depths. Therefore, it appears reasonable
to try looking for a relationship between the near-surface intrinsic attenuation for
S waves and coda Q_c.

Interestingly, modeling discussed above shows that when using
velocity/density models close to the correct platform structure (*i.e.*, the Quartz-
type models), values of γ stay nearly constant. This relative stability shows that

that γ is indeed related to the velocity-density structure and allows relating the observed Q_c the *in situ* crustal Q_S's alone. This relationship can be established by forming the following ratios:

$$\beta = \frac{Q_c}{Q_S} = \frac{-\pi f}{Q_S \left(\dfrac{d \ln A(t,f)}{dt} + \gamma \right)}, \qquad (6.30)$$

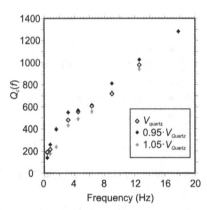

FIGURE 6.32

Apparent coda $Q_c(f)$ for model Quartz with crustal velocities scaled by factors 0.95, 1.0, and 1.05 (legend). Note the strong frequency dependence ($\eta \approx 0.5$) of $Q_c(f)$ despite the constant-Q_S crustal rheology.

for each frequency. Note that the resulting Q_c/Q_S ratios are nearly frequency independent, always exceed one. Therefore, coda Q_c always under-estimates the attenuation, because attenuative areas usually occupy smaller volume than the one traversed by seismic rays (see Figure 5.12b). For higher Q_S, β values approach one for realistic models (Figure 6.33); however, the accuracy of our assumptions of the low Q_S zone dominating coda attenuation decreases in this limit. Note that for IASP91 model, Q_c exceeds Q_S by more than a factor of two even at the highest crustal Q_S.

Because of the nearly constant γ, curves $\beta(Q_S)$ become nearly frequency independent as shown by lines in Figure 6.33. This shows that the frequency dependence of Q_c is apparent and the geometrical spreading is approximately frequency independent in these models. Therefore, these curves can be estimated by numerical modeling in each model, and then utilized for deriving the average crustal Q_S from the observed Q_c. Interestingly, functions $\beta(Q_S)$ are very similar (within only ~20% differences) for all three realistic Quartz-based models, despite the strong difference in γ for the Complex Crust model (Figure 6.33). However, the $\beta(Q_S)$ curve for the IASP91 model, which is definitely inadequate for this area, strongly differs from those derived for the Quartz model (Figure 6.33).

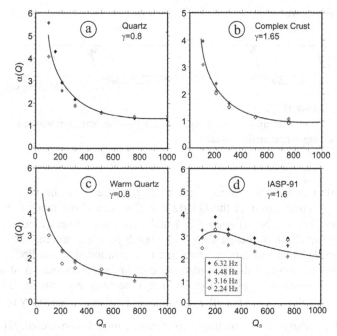

FIGURE 6.33

Ratios $\beta = Q_c/Q_S$ of synthetic coda Q_c factors to the "true" S-wave Q_S within the crust modeled for four lithospheric models in Figure 6.27.

Q_c and composition of coda wavefield

In a heterogeneous lithosphere, coda wavefield is complex and far from any simple symmetry. Generally, crustal S waves should be the primary contributors to the coda, and consequently we relate the observed Q_c to the crustal S-wave Q_S. However, the values of Q_S and Q_c differ. In models, because a part of the seismic energy travels through the mantle with typically higher Q_S, the observed values of Q_c are greater than crustal Q_S (Figure 6.33). In regard to the geometrical attenuation parameter γ_i, it can be either negative or positive, depending on the position of the source within the lithospheric structure, as discussed in Chapter 4. Values of $\gamma > 0$ correspond to the scattered energy decaying with distance faster than surface waves, which should be a common case due to: 1) high attenuation of the surface waves in the weathered zone (note the Complex Crust model, Figure 6.33) and 2) multiple reflections and scattered seismic waves leaking into the mantle as suggested, for example, by Gupta *et al.* (1991).

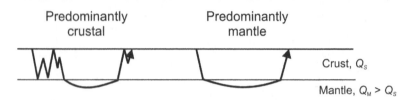

Predominantly   Predominantly
crustal   mantle

FIGURE 6.34
Two end-member wave propagation types for scattered waves in
a long-range arrival coda.

Both γ and the functional form of $\beta(Q_S)$ depend on the velocity/density structures. A comparison of the Quartz and Complex Crust models shows that coda decays are steeper (Figure 6.31) and $\beta(Q_S)$ are lower (Figure 6.33) for Complex Crust. Both of these effects likely result from the presence of the low-Q sedimentary layer in this model. Note that in a detailed study of PNE's L_g Q in Russia, Li *et al.* (unpublished) pointed out a quantitative correlation of low L_g Q with the presence of young (low-velocity) sedimentary basins. From this correlation, a value of $Q_S \approx 60-70$ was estimated for these sedimentary rocks.

Interestingly, in the "realistic" crust/mantle models considered, $\beta(Q_S)$ is not constant and decreases from ~3–4 at $Q_S = 100$ to about ~1–1.4 at $Q_S = 1000$ (Figure 6.33). As expected, the measured values of $\beta(Q_S)$ are also fairly frequency independent. By contrast, in the IASP91 model, $\alpha(Q_S)$ is roughly constant within ~2.5–3.5 range and somewhat less stable with respect to the frequency variations (Figure 6.33). This means that only in the IASP91 model, the observed Q_c is proportional to the Q_S of the crust, and it equals the Q_c of the realistic models only when $Q_S \approx 100$.

The above observations can be explained by considering two end-member wave types contributing to the modeled codas (Figure 6.34). For the first type, scattered waves propagate predominantly through the crust, possibly with some contributions from mantle paths (Figure 6.34). For the second type, only a fixed portion of the path lies within the crust, and the principal contribution to attenuation comes from mantle paths. The mantle in our models has high Q_S values of ~700–1000, and therefore the increased mantle paths correspond to lower apparent Q_c. With these definitions, Morozov *et al.* (2008) observed that:

1) The amount of decrease of Q_c relative to Q_S is controlled by the relative time seismic waves travel within the mantle, and therefore waves of "predominantly crustal" type have higher β than those of the "predominantly mantle" type (Figure 6.34).

2) Because of its simple crustal structure, the IASP91 model is relatively depleted in crustal waves (also see Figure 6.27d), and therefore the contribution from mantle paths in it is significant for all values of crustal Q_S. For such paths, β is determined by the time

seismic waves travel within the crust, which stays approximately constant.

3) Waves of the "crustal" type are abundant in Quartz-based models (see Figure 6.27a to 6.27c). With an increase in crustal Q_S, their role in the coda should increase, for example, more multiple reflections occurring within the crust. This should lead to reduced values of β. Figure 6.33 shows that β decreases with increased crustal gradients and reflectivity, and it reaches ≈ 1 for the Complex Crust model.

4) For average crustal $Q_S \leq 100$, seismic waves dissipate quickly, and therefore long-range coda wave propagation occurs predominantly through the mantle. Consequently, coda Q_c values are similar in all models and close to Q_S. Nevertheless, such low whole-crust Q_S values appear unlikely.

6.8.3 Observations of long-range coda

The use of the attenuation-coefficient model (γ, κ) allows us to resolve a hitherto puzzling observation from the PNE profiles (Morozov et al., 2008). Within the East European Platform, earlier measurements of the P-wave coda from the Quartz profile resulted in Q_c values of ~380 at about 2 Hz and ~430 near ~5 Hz (Morozov and Smithson, 2000). This corresponded to the power-law (1.13) of $Q_c(f) \approx 270 \cdot f^{0.3}$ for the apparent Q_c. However, within the Siberian Craton (PNE Kimberlite-3), practically frequency-independent, time-domain coda amplitude decays were observed. In the Q_c-type paradigm, these decays corresponded to $Q_c(f) \approx 1050 \cdot f^{1.0}$. Although the intrinsic attenuation was expected to be low within the Siberian Craton, such strong frequency dependence appears to be vastly different from the East European Platform and was viewed as surprisingly high. On the other hand, for surface and L_g waves, Mitchell (1995) and Mitchell et al. (1997) suggested that high η should in fact be expected in cratonic areas. Such contrasting relationships appear to be difficult to classify and understand in the form of a frequency-dependent Q.

However, in the attenuation-coefficient form, the explanation of the above observations is quite simple. From Quartz-4 data, the $\chi_c(f)$ dependences are linear to about ~2 Hz and show intercepts $\gamma \approx 0.75 \cdot 10^{-2}$ s^{-1} and frequency-independent slopes corresponding to $Q_e \approx 850$, with the estimated range of uncertainty of 780–960 (Figure 6.35). Above ~2 Hz, the $\chi_c(f)$ trend from this PNE changes to $\gamma \approx 1.4 \cdot 10^{-2}$ s^{-1} and $Q_e \approx \infty$ (long-dashed gray line in Figure 6.35). The nature of such change is unclear; it is likely that the S waves dominating the coda below 2 Hz gradually die out and are faster-spreading, but slower-attenuating waves (such as body P-waves) overtake them in the coda.

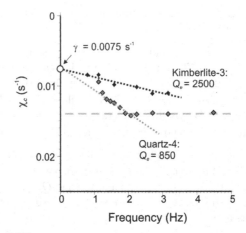

FIGURE 6.35
P-wave coda-attenuation coefficients measured in PNE Quartz-4 (recorded at 2600 km from the source) and Kimberlite-3 (1100 km from the source). Lines indicate the interpreted trends.

Most interestingly, from Kimberlite-3 records, γ is nearly the same, but Q_e is much higher than in the lower-frequency portion of Quartz-4 coda: $Q_e \approx 2500$ ±300 (Figure 6.35). Such high values of Q_e within the Siberian Craton agree with observations of P_g waves propagating to over 1600 km from these PNE's (Figure 6.36). These are likely the longest-propagating short-period P_g waves observed on Earth.

Note that the values of $\gamma \approx 0.75 \cdot 10^{-2}$ s^{-1} from both PNE's agree remarkably well with numerical simulations based on the velocity model derived from inverting the Quartz profile (preceding Section 6.8.2). This suggests that γ values could be quite stable, even in cases of strong apparent frequency dependences ($\eta \approx 1$). Although Mitchell (1995) and Mitchell et al., (1997) noted that high η values were typical for stable cratons, we consider such strong frequency dependence as abnormal. The abnormality consists in treating the high apparent η as a material property of the cratonic lithosphere. The observations only show that the apparent coda Q_c is nearly proportional to the frequency within the Siberian Craton, and consequently, $\chi \approx \gamma =$ const. Therefore, coda waves spread in a frequency-independent manner and slightly faster (by factor exp($-\gamma t$)) than the surface waves in the geometrical-spreading model adopted for coda background. This means that body waves make a significant contribution to scattering, but the attenuation is in fact quite low and frequency independent, with $Q_e \approx 2500$. Therefore, at least from this example, we argue that compared to the (Q_0, η) combination, γ provides a more stable, model-independent, and transportable criterion for correlating the observations of attenuation with crustal tectonic types.

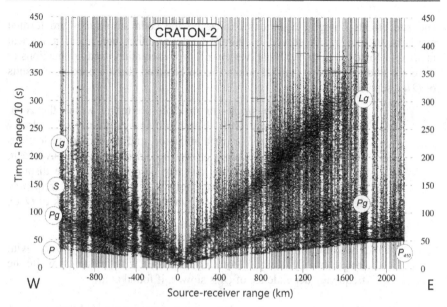

FIGURE 6.36
Seismic section from PNE Craton-2 showing major regional phases. Note the P_g and L_g phases propagating to 1600–1700 km to the east within the Siberian Craton.

6.8.4 Separation of anelastic attenuation and scattering

The elastic and anelastic attenuation processes lead to distinctly different distance and frequency dependences of seismic-wave amplitudes. Back-scattering increases seismic amplitudes near the source, whereas the anelastic attenuation causes the characteristic exponential decay $\exp(-\pi f t/Q)$ dominating at far distances. Similar observations also apply to the singly or multiply scattered coda, in which case t becomes the coda lag time. Based on these observations, inversion methods allowing differentiation between the anelastic- and scattering-attenuation parameters of the medium were proposed, based on the radiative transfer theory (Wu, 1985; Hoshiba, 1991; Sato and Fehler, 1998).

Although we disagree with the use of Q variables, and particularly Q_s for describing the medium (Section 3.6), and argue that scattering cannot be unambiguously isolated in principle (Section 5.7), let us still adhere to conventional Q terminology. We therefore address the problem of whether and how Q_i and Q_s can be separated by analyzing the time-frequency variation of coda amplitude. Owing to the difficulty of this problem, the techniques for separating the Q_i and Q_s are relatively complex. After correcting for geometrical spreading and site effects, integral equations expressing the coda amplitudes are formed similarly to (5.62) and solved by Monte-Carlo or other non-linear numerical methods. However, with the broad variety of implementations and applications to different study areas, two general points should be noted about the existing results of Q_i/Q_s separation:

1) While using either single- or multiple scattering within the coda, forward models are still based on assumptions of velocity uniformity, isotropy, S-wave dominance, and absence of reflections and mode conversions from crustal boundaries. As we have seen, inaccuracies of such assumptions may strongly affect the distance dependences of amplitudes and the values and frequency dependences of the inverted Q's, which may thwart the task of Q_i/Q_s separation.

2) The resulting Q_i and Q_s values usually differently vary with frequency. Typically, Q_s quickly increases with frequency while the frequency dependence of Q_i is slower, if found at all.

A good example illustrating these points was given by Mayeda *et al.* (1992). These authors used the radiative transfer model in the form of multiple lapse-time window analysis (MLTWA) to separate the effects of Q_s^{-1} and Q_i^{-1} by inverting coda Q_c^{-1} measured in central California, Long Valley (eastern California), and Hawaii (Figure 6.37). Despite the complexity of the inversion by Mayeda *et al.* (1992), their results can be interpreted quite easily in the spirit of our attenuation-coefficient view (Figure 6.37). The frequency-dependent Q_s in

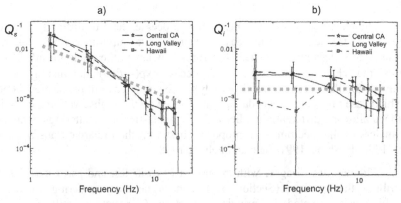

FIGURE 6.37

Coda Q^{-1} in Central California, Long Valley, and Hawaii (Mayeda *et al.*, 1992): a) Scattering, Q_s^{-1}, and b) anelastic, Q_i^{-1}. Thick dashed gray lines indicate simplified interpretations for a) $Q_s \propto f$ and b) $Q_i = $ const.

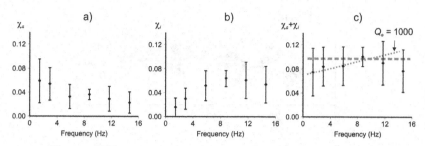

FIGURE 6.38
Central California coda data from Figure 6.37 in attenuation-coefficient form: a) scattering attenuation, b) anelastic attenuation, and c) their sum. Dashed gray lines in plot c) indicate two possible interpretations.

these three areas of active tectonics shows high values of $\eta > 1$, and Q_i shows $\eta \approx 0$ at frequencies below ~4–8 Hz (Figure 6.37). Note that practically everywhere within the estimated 90% confidence levels shown by error bars, we can reduce the interpretation to a more "conservative" Q_s dependence with $\eta = 1$, and setting $Q_i = $ const (gray dashed lines in Figure 6.37). With these modifications, the "scattering" factor $\exp(-\pi f t/Q_s)$ represents the geometrical attenuation, $\gamma = \pi f/Q_s$ (see Section 3.6).

As in the previous examples, restoring the attenuation coefficients from Q^{-1} data presentation allows estimating the true attenuation intensities. The attenuation coefficients derived from the scattering and anelastic Q's by Mayeda *et al.* (1992) show complementary variations with frequency (Figures 6.38a and 6.38b). The total attenuation coefficient shows no visible frequency dependence within the whole frequency band, or at most a dependence with $Q_e \approx 1000$ between 1–10 Hz (Figure 6.38c). Therefore, our interpretation is that the dissipation within the study area is low, and the attenuation is mostly geometrical (*i.e.*, coda amplitude decay is different from $1/r$ assumed in the background model). Partitioning of the attenuation coefficient into the "scattering" and "anelastic" ones may depend on the type of convergence achieved by the iterative inversion algorithm. Given their scatter, it appears unlikely that coda-χ data in Figure 6.38c should allow resolving the scattering and anelastic components. In particular, the total attenuation coefficient shows little evidence for anelastic attenuation, and therefore the entire attenuation could be attributed to "scattering."

Reconstructing the χ data from the reported Q_i^{-1} and Q_s^{-1} of course does not give a complete picture of all the data involved in the inversion. Figure 6.39 shows an example of how the complete dataset of source-, site-, and geometrically corrected logarithms of coda amplitudes are fit in MLTWA. This figure also illustrates the effects of the two attenuation factors on the model. In the absence of scattering ($Q_s^{-1} = 0$), the gray lines would merge at large distances, approaching a

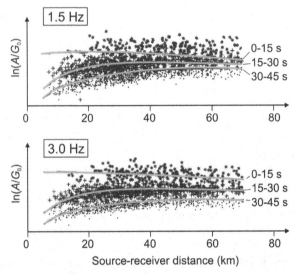

Figure 6.39
Coda-amplitude data fits (gray lines) at 1.5 and 3.0 Hz for
Central California, from Mayeda *et al.* (1992). The amplitudes
were measured within three lag-time windows of 0–15, 15–30,
and 30–45 s (labeled) and shown by asterisks, crosses, and dots,
respectively.

straight line sloping to the right proportionally to Q_i^{-1}. By contrast to Q_i^{-1}, back-scattering causes the amplitudes to increase to the distance of about 30–60 km, after which they decay faster. However, note that the scatters in the corrected amplitude values are comparable with the separations between the modeled trends. A comparison of the data points (not the model curves) in Figure 6.39 at 1.5 and 3.0-Hz observation frequencies suggests that these data could indeed be modeled with frequency-independent attenuation, for example as shown by thick dashed line in Figure 6.38c.

6.8.5 Lapse-time dependence

Similarly to MLTWA, lapse-time coda Q analysis utilizes the combined dependence of coda attenuation on frequency and coda lag-time to derive additional detail about the attenuation structure of the subsurface. In the more traditional local-earthquake setting with closely located sources and receivers, the logarithmic decrement of coda amplitudes often decreases with increasing coda lapse (lag) times. This dependence on the lag time is often interpreted as caused by the variations of the intensity and scale-lengths of lithospheric heterogeneity with depth. The bottom of the scattering ellipsoid z_{max} (Figure 6.40) is interpreted as the

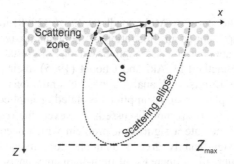

FIGURE 6.40
Single-scattering ellipsoid for local-earthquake recording (for
mathematical detail, see Figure 5.13 on p. 170). S is the source
(local earthquake) and R is the receiver. Note that the ellipsoid is
only a kinematic constraint. Although the ellipsoid reaches depth
z_{max}, scattering should likely be concentrated near the surface
(gray pattern), and the coda χ_c relates to this area.

"depth of investigation," and Q_c measured at increasing lags is interpreted as
scattering Q at the corresponding depths.

For a specific case example, let us illustrate the above interpretations using
the results from a recent paper by Mukhopadhyay *et al.* (2008) (Figure 6.41).
These authors studied the frequency-dependent coda attenuation in ~60-s codas
recorded from the Chamoli earthquake in India. They provided a very clear,
detailed, and well-documented presentation using the traditional *ansatz*
$Q_c(f) = Q_0 f^\eta$ and found Q_c to be strongly and positively dependent on both the
frequencies and coda lapse times.

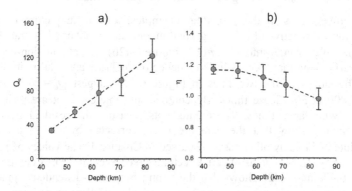

FIGURE 6.41
Coda lapse-time dependence of: a) Q_0 and b) η from
Mukhopadhyay *et al.* (2008).

As above, the study by Mukhopadhyay *et al.* (2008) interests us not as a specific case example but as an illustration of the general caveats of the Q-based approach to coda attenuation. Although based on long-established approaches, such as the classic method by Aki and Chouet (1975), their paper lacks critical evaluations of the results and analysis of the limitations of the underlying assumptions. A complex set of assumptions is stated or implied, but their validity and impact on the results are not discussed. However, the resulting values of η themselves already indicate a significant problem which needs to be addressed. These values are significantly greater than one for all depths (Figure 6.41b), suggesting a very unrealistic increase of high-frequency energy with propagation time, which presumably occurs over large volumes of the lithosphere.

The "depth of investigation" labeled in the abscissas of both plots in Figure 6.41 is defined as the lowest point of the scattering ellipsoid (z_{max} in Figure 6.40). However, if scattering decreases with depth (and this should generally be true), the lowermost point of the ellipsoid should have the smallest effect on Q_c. As discussed in Chapter 5, most scattered waves should originate closer to the surface for all lapse times (Figure 6.40). Therefore, the dependence of Q_c on lapse times is unlikely caused by the increasing sampling depth. Q_c simply increases with lag times, and as shown below, it is hardly related to the scattering properties of the crust or mantle. Perhaps the term "depth" could be replaced with "pseudo-depth," as it is customary in electrical resistivity imaging.

Conclusions drawn by using the various models should be commensurate with their degrees of accuracy. In the present case, the dependence of Q_c on coda lag time shows that the uniform half-space model of Aki (1969) is insufficiently accurate. This model does not predict this lag-time dependence. If variations in the probability density of heterogeneities are introduced in order to correct this inaccuracy, it is only reasonable to try a heterogeneous velocity/reflectivity structure first. The difficulty of considering heterogeneous structures does not justify interpretations reaching beyond their ranges of validity.

However, as in the preceding examples, a realistic picture of seismic attenuation irrespective of the background model can be obtained by transforming the values of Q_c values into $\chi_c = \pi f Q_c^{-1}$ (Figure 6.42b). The attenuation-coefficient data clearly show that there is little evidence of anelastic attenuation in the study area. The slopes of the two lines in Figure 6.42b suggest $Q_e = \infty$ at 10-s and $Q_e \approx 10,000$ at 30-s lapse times. By contrast, the intercepts, γ, are positive and decrease with lapse times. These intercepts measure the residual geometrical spreading and show that the data are under-corrected by the t^{-1} geometrical correction, as in many other cases discussed in Chapter 4. The values of $\gamma \approx 0.02$–0.06 s^{-1} are typical for areas of active tectonics (see Section 7.3). A comparison of Figures 6.42a and 6.42b shows that the strong frequency-dependent–apparent Q_c represents the nearly constant residual geometrical spreading: $Q_c \approx f / \pi \gamma$.

Decreasing γ values with increasing lapse times are not surprising and are commonly observed. Coda envelopes often show flattening slopes (*i.e.*, increasing

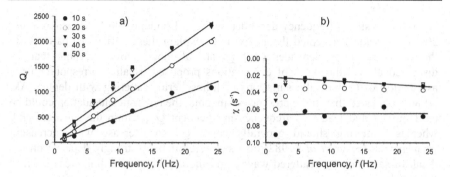

FIGURE 6.42

Coda parameters as functions of frequency and lapse times.
a) Coda Q_c *versus* frequency for several lapse times (symbols given in the legend) and b) the same data in the form of χ. Note that χ is positive, practically independent of the frequency, and decreases with lapse times. The range of possible $\chi(f)$ trends is indicated by lines; however, both of them correspond to $Q_e > 10,000$.

apparent Q_c) with increasing lapse time, caused by the extinction of quicker decaying scattered-wave modes. In the case of local earthquake data, such modes could be the near-source or receiver reflections and surface waves. However, this varying slope is still a coda property, and attributing it to the subsurface requires great caution.

The $\chi(f)$ plot (Figure 6.42b) also shows that below ~5 Hz, the amplitudes are systematically reduced for all lapse times. The reason for this reduction is unclear and deserves some investigation; this could be related to over-estimated spectral signatures of the earthquakes $A_0(\omega)$ at these frequencies.

6.8.6 Temporal variations

During the past 30 years, numerous studies addressed spatial and temporal variations of coda-wave attenuation, particularly in relation to major earthquakes and volcanic eruptions (*e.g.*, Chouet, 1976; Fehler *et al.*, 1988; Londono, 1996; Moncayo *et al.*, 2004, Novelo-Casanova *et al.*, 2006; Del Pezzo *et al.*, 2004). In these cases, the key observed coda characteristics usually are the temporal variations of the values of Q_c at selected frequencies, and also the dependences of Q_c on frequency.

There appears to exist a significant difference between temporal coda Q_c variations measured in volcanic and non-volcanic regions. In non-volcanic regions, strong frequency dependences are common ($\eta \approx 1$; *e.g.*, Roecker *et al.*, 1982), whereas volcanic Q_c values vary from weak (*e.g.*, Chouet, 1976; Fehler *et al.*, 1988) to strong frequency dependences (*e.g.*, Londono, 1996; Moncayo *et al.*, 2004). No widely accepted theory exists to explain these variations, particularly the weak frequency dependence of Q_c at volcanoes. However, note that, as discussed above, the types of explanations proposed usually correspond to the association of Q_c with the variations of lithospheric scattering with depths. Aki (1980) suggested that in the presence of magma, the anelastic attenuation could be predominant and lead to a frequency-independent Q^{-1}, within the seismic band, whereas scattering should generally cause Q^{-1} to decrease with increasing frequency. Roecker *et al.* (1982) also suggested that Q_c strongly depends on the depth of sampling by scattered waves, and the depths of sampling were inferred from coda lag times.

As one can see, the above arguments are based on contrasting the scattering to anelastic Q and the frequency-independent to frequency-dependent Q_c. However, as we have seen, both the values of Q_c and their lag-time and frequency dependences are controlled by the selected theoretical model for the background structure. From this observation, an alternate model for temporal variations of Q_c immediately arises. The crustal and uppermost-mantle structures should be significantly different in the volcanic and non-volcanic areas, and this difference can in principle explain the observed differences in Q_c values and their temporal variations. Thus, the first reason for changing the apparent frequency dependence of Q_c should be sought in varying γ, which describes the effects of the upper-crustal structure. Values of Q_e are also likely to respond to the variations of the fluid content, opening pores, and related properties within the upper crust. Scattering properties should also change as a result of such variations; however, elastic scattering still does not appear to be a proven factor which would be comparable to γ and the anelastic Q.

In addition, another difference between volcanic and non-volcanic areas exists in the characters of the events causing temporal changes in the coda. Non-volcanic studies are usually conducted over time spans of several years, over which the variations of γ due to seasonal effects are likely to be the strongest. In volcanic studies, the temporal changes are often more rapid and associated with eruptions. In this case, deeper sources of temporal variations can be expected. Both γ and the average intrinsic Q_i can be affected by the changes in the volcano plumbing system caused by movements of magma, gases, and hydrothermal fluids.

To illustrate the above model, we briefly revisit two well-known datasets from the literature. First, we consider the non-volcanic study by Chouet (1979) which pioneered the temporal coda-variation research. In the second example, we review the data from Mount St. Helens eruption by Fehler *et al.* (1988). Similar to the study by Chouet (1979), this dataset also contains clear (but very different) temporal and frequency-dependent effects, allows comparing the geometrical-spreading and Q^{-1} contributions, and leads to interesting observations of their variations during the process of volcanic activity.

The two examples above show that the frequency-dependent temporal attenuation coefficient $\chi(f)$ contains more information than usually recovered by using the conventional Q-factor interpretation. Attenuation-coefficient plots allow grouping the observed dependences into consistent patterns, which turn out to be linear within significant segments of the frequency band. These $\chi(f)$ segments provide wider and clearer separations between the responses from different structures. The resulting measurement reveals more detail in the attenuation patterns, which allows detecting temporal trends in geometrical-spreading and Q_e variations that have not been noticed before.

Seasonal variation of local-earthquake coda Q

To illustrate the relation of the frequency-dependent Q-type models to $\chi(f)$, let us consider the data from an aftershock study of an earthquake swarm along the San Andreas Fault (SAF) in central California. From July 17, 1973 to June 24, 1974, 185 earthquakes of magnitudes 0.9 to 3.3 were recorded at the temporary station STC located in Stone Canyon, located approximately 1 km west of the SAF (Chouet, 1979). Note that the presence of such a major, linear crustal structure as the SAF zone in addition to the sedimentary and crustal layering makes this area particularly interesting for studying the effects of the geometrical spreading on attenuation measurements.

Figure 6.43a reproduces the frequency-dependent $Q^{-1}(f)$ at Stone Canyon stations given by Aki (1980). The observations were grouped in four time

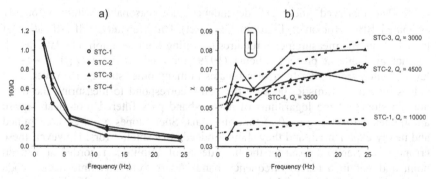

FIGURE 6.43
Frequency dependence of local-earthquake coda attenuation at Stone Canyon (California) from Aki (1980). Labels indicate the consecutive time intervals. a) Data in $100/Q$ form. b) The same data in $\chi(f)$ form. Note that the separation of periods STC-2 through STC-4 is much clearer, and $\chi(f)$ shows linear dependencies on f (dashed lines; Q_e values given in labels). A reference 10% error bar on χ is shown in inset.

intervals: July 18, 1973 to September , 1973 (STC-1); September 12, 1973 to November 10, 1973 (STC-2); November 11, 1973 to February 10, 1974 (STC-3); and February 11, 1974 to June 24, 1974 (STC-4). From these data, $Q^{-1}(f)$ appeared to increase at all frequencies sometime from mid-September to mid-October 1973, as noted by Aki (1980). However, beyond this general observation, more detailed variations of attenuation are difficult to analyze from this plot. Chouet (1979) and Aki (1980) described this change as intriguing but unclear, and broadly attributed it to some rapid but strong variations of lithospheric heterogeneity within a large volume (tens of kilometers in radius) of the lithosphere.

Similar to the examples of the previous sections, re-plotting the $Q(f)$ data in terms of $\chi = \pi f Q^{-1}$ reveals linear attenuation patterns, which show detailed relationships between the datasets (Figure 6.43b). The convergence of all $Q^{-1}(f)$ near ~25 Hz in Figure 6.43a appears principally related to dividing the values of attenuation coefficients, χ, by the frequency. By contrast, the intercepts and slopes of the attenuation-coefficient trends (dashed lines in Figure 6.43b), show that: 1) the geometrical-spreading factors are slightly below $\gamma \approx 0.04$ s^{-1} for STC-1 and equal about 0.05–0.06 s^{-1} for STC-2 through STC-4 and 2) the attenuation-factor Q_e values are high, with $Q_e \approx 10{,}000$ for STC-1 dropping to $Q_e \approx 3000$–4500 for STC-2 through STC-4. Such high γ values appear characteristic for central California (Morozov, 2010a) and may be related to the proximity of the San Andreas Fault. At the same time, the values of Q_e appear unusually high compared, for example, to $Q_e \approx 800$ in central California, estimated in Section 6.8.1, which may indicate a well-consolidated and dry crust in the area of Stone Canyon.

The observed linear $\chi(f)$ dependences are reasonable within a roughly estimated 10% error on χ (Figure 6.43b, inset). Unfortunately, full estimation of the errors in $\chi(f)$ measurements requires revisiting the raw amplitude data, which are not available at present. Chouet (1979) reported errors in $Q_c(f)$ at ~1–3% levels, which are similar to error estimates in many other studies. However, such values may be optimistic, because they only correspond to measuring the time-domain slopes of the logarithms of narrow-band pass filtered root-mean square amplitudes across long (~30-s) time intervals. Such slopes are well constrained and nearly exponential, and they should indeed have small errors. However, these errors only refer to measuring the time decay of amplitude recorded at a given point and within a narrow frequency band. Apart from this time decay, coda amplitudes are also affected by the spatial and frequency amplitude variations (e.g., resonances, tuning), which are visible from the systematic amplitude variations in Figure 6.43b. Note the "spectral scalloping" resulting in reduced amplitudes at 1.5, 6 (with one exception for STC-1), and 24 Hz and increased at 3 and 12 Hz in all recordings (Figure 6.43b). Because of such amplitude variations, the scatter of the resulting Q is much stronger than would be expected from a single-scattering model with such small errors in Q (see Figure 6.44).

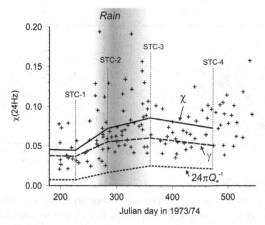

FIGURE 6.44

Temporal variations of local-coda χ in Stone Canyon at 24 Hz. Crosses are coda data from Chouet (1979), and black lines show the interpreted γ values (dashed), contribution from Q_e^{-1} (short dash), and total χ (solid) from Figure 6.43. Gray dashed lines indicate the middles of measurement intervals. The rainy season, which started on October 7, 1973, is indicated schematically.

The uncertainties in the resulting (γ, Q_e) values can be examined by using the standard linear regression analysis. Figure 6.45 shows cross plots of (γ, Q_e) for the four time intervals, with their optimal parameter values shown by diamonds and areas of likelihood levels exceeding e^{-2} contoured. The stronger amplitude oscillations at lower frequencies (Figure 6.43b) cause most of the errors for STC-1 and STC-2 periods (Figure 6.45a), and by excluding these frequencies, significantly tighter estimates of γ and Q_e can be obtained (Figure 6.45b). However, dropping the single 1.5-Hz reading from STC-1 shows that its Q_e^{-1} can in fact be set equal zero (Figure 6.45b), which means that the attenuation in this interval may be too low to be detectable.

With using the frequencies below 6 Hz or not, the variation of attenuation parameters with observation time are clear and well constrained (Figure 6.45). The attenuation increases from STC-1 to STC-3 and later reduces at STC-4. However, note that the contribution of γ dominates that of Q_e even at the largest observation frequency (i.e., $\gamma > \pi f/Q_e$ with $f = 24$ Hz; Figures 6.43b and 6.44). For example, for STC-2, the geometrical-spreading and Q^{-1} contributions would only become equal at the "cross-over" frequency $f_c = \gamma Q_e/\pi \approx 86$ Hz, which is far above the observation band. Thus, it appears that the "geometrical" effect should be the main cause of both the apparent coda Q_c and of its temporal variation.

By comparing the temporal trends of γ and Q_e to the $\chi(f)$ data derived from $Q(f)$ for $f = 24$ Hz by Chouet (1979), we can further evaluate the significance of

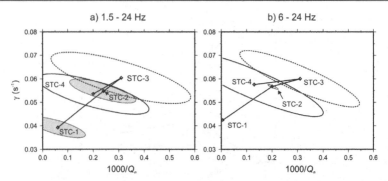

FIGURE 6.45

Cross-plot of $(\gamma, 1000Q_e^{-1})$ corresponding to linear $\chi(f)$ trends identified in Figure 6.43b. Ellipses indicate areas of squared data misfit less than twice higher than for the optimal (γ, Q_e) values (diamonds). Fitting using a the entire 1.5–24 Hz frequency band and b) using frequencies 6–24 Hz. Note the temporal variations in both γ and Q_e (lines).

these trends. Despite their wide scatter, the individual χ_c data are consistent with the inferred temporal trends. Notably, there actually appears to be no increase in χ_c until the start of the rainy season (October 7, 1973; Figure 6.44). Therefore, contrary to the conclusion by Aki (1980), it still appears likely that the increase of both γ and Q_e^{-1} was associated with a change in the near-surface conditions during the rain season.

Three significant differences of our empirical $\chi(f)$ model compared to the traditional coda-scattering model allow us to look for the source of coda in the near subsurface. First, as determined above, the geometrical factor, γ, not scattering, dominates coda attenuation at Stone Canyon. Geometrical spreading is most sensitive to the near-surface structure, where the strongest velocity contrasts and reflectivity are present. Numerical waveform coda modeling (Section 5.6) showed that introduction of a low-velocity, high-attenuation layer in the upper crust can significantly increase γ, and therefore the effect of a rising water table could be qualitatively consistent with the observed increase of γ (Figure 6.44). Second, in our model, values of Q_e are significantly higher (3000–10,000 compared to 100–1000 by Chouet, 1979), and this smaller amount of attenuation could be attributed to the upper crust. Third, the attenuation factor (Q_e^{-1}) is frequency independent, and therefore it does not have to be explained by scattering, but could be caused by anelastic losses within the wet and weathered near-surface layers.

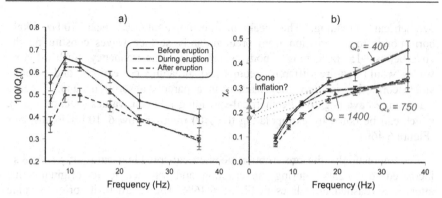

FIGURE 6.46

Coda attenuation data at Mount St. Helens. a) Q_c^{-1} data from Fehler *et al.* (1988). Note the peak at 10 Hz and decreasing Q_c^{-1} after the eruption. b) The same data in $\chi_c(f)$ form. Dashed grey lines show the interpreted linear trends in $\chi_c(f)$. Note the linear $\chi_c(f)$ dependences at $f > \sim18$ Hz, with slope decreasing after the eruption.

Variation during 1981 Mount St. Helens eruption

Another interesting example of temporal variations of coda attenuation was given by Fehler *et al.* (1988). Three groups of measurements were conducted at Mt. St. Helens: 1) prior to the September 3 to 6, 1981 eruption; 2) during the eruption; and 3) after it. In the coda quality-factor form, the attenuation (Q_c^{-1}) relatively weakly varied with frequency within each of these intervals and decreased by 20–30% after the eruption (Figure 6.46a). Fehler *et al.* (1988) noted the attenuation peak near 10 Hz, which might correspond to the peak at 0.5 Hz hypothesized by Aki (1980) for non-volcanic regions. Aki (1980) interpreted this peak as caused by scattering and estimated the corresponding scale length of lithospheric heterogeneity as ~4 km. Following the same argument, Fehler *et al.* (1988) suggested that a heterogeneity scale length of ~500 m could be responsible for the 10-Hz Q_c^{-1} peak at Mount St. Helens.

However, if we do not assume scattering from the very beginning, but re-plot the same data as $\chi_c = \pi f Q_c^{-1}$, the interpretation changes (Figure 6.46b). Above ~18 Hz, $\chi_c(f)$ shows linear behavior similar to that in Figure 6.43. The intercept values $\gamma \approx 0.18$ s^{-1} are three to four times higher than those in Figure 6.43, likely because of the stronger positive geometrical spreading (defocusing) caused by the cone of the volcano. Similar to the case of Figure 6.43, but somewhat less dramatic, the effect of "non-geometrical" attenuation (Q_e^{-1}) represents only about a half of the observed coda attenuation, and the rest is represented by the residual geometrical spreading. The peak in the apparent Q_c^{-1} near 10 Hz likely corresponds to a transition from predominantly surface waves constituting the coda below ~15 Hz to mostly body waves above this frequency. Such different waves would require different geometrical-spreading corrections, and when a single correction is used, they lead to an apparent "absorption band." Surface waves may have negative geometrical-spreading intercept values (Section 7.6.5), which can be noted by extrapolating the $\chi_c(f)$ trends below 6–10 Hz frequencies (Figure 6.46b).

Interestingly, the pre-eruption geometrical-spreading value of $\gamma \approx 0.18$ s^{-1} increased to ~0.25 s^{-1} during the eruption and returned to its original value afterwards (dashed gray lines in Figure 6.46b). The return to its original value could be expected, because γ is a property of the structure (shape of the mountain and its subsurface), which was generally restored after the eruption cycle. The co-eruptive increase of γ was likely related to the inflation of the cone of the volcano (Dzurisin et al., 1981). Attenuation decreased from $Q_e \approx 400$ before the eruption to ≈750 after it (Figure 6.46b). This change could be explained by the removal of gas and magma from the volcano chambers and reduction of anelastic attenuation. Note that this is a significantly stronger change (about 50%) compared to the 20–30% change in the apparent Q_c^{-1}. At the same time, the values of Q_e are about twice higher than those of Q_c (Figure 6.46b), and therefore a smaller volume of attenuative material needs to be removed in order to account for this change.

The character of the co-eruptive $\chi_c(f)$ curve in Figure 6.46b is less clear, suggesting an increase in γ, which could be due to cone inflation. The value of Q_e is ~1400, which may appear surprisingly high, compared to the values of 150–300 in Q-factor based measurements (Figure 6.46a). Two possible reasons could lead to the increased values of both of these parameters: 1) migration of the earthquake sources to the top of the mountain during the eruption and 2) increased high-frequency noise during the eruption, causing over-estimated Q_c values. This second reason appears quite likely, judging by the near-constant $\chi_c(f)$ values at 20–40 Hz during the eruption (Figure 6.46b). These and other explanations could be examined further by revising the raw data and by modeling.

Extents of temporal variations in the subsurface

From geological considerations, the largest values of both γ_i (low values and high gradients of the velocities) and Q_i^{-1} (high attenuation) should be concentrated in the near surface and/or in the area of the magma chambers, chimneys and cones of the stratovolcanos. The same areas are also the least consolidated and most likely to be affected by rapid changes in the environmental conditions (seasonal variations or eruptions). Because of the relatively small changes in the observed γ and Q_e^{-1} and high contrasts in γ_i and Q_i^{-1} in the subsurface, only a small portion of the crust needs to be affected by the changes in order to produce the observed effects. For example, recalling the relation between the intrinsic and apparent (γ, κ) properties for fixed ray paths (5.17):

$$\gamma = \frac{1}{t} \int_{path} \gamma_i d\tau, \text{ and } \quad Q_e^{-1} = \frac{1}{t} \int_{path} Q_i^{-1} d\tau, \tag{6.31}$$

we see that with Q_e changing from infinity to 3000 and typical values of $Q_i \approx 10$–20 in wet near-surface sediments, only 0.3–0.7% of the total ray path needs to be affected by attenuation. Similar estimates should apply to molten lava, gas, and fluids beneath the volcano. Note that low-velocity near-surface layers (and presumably also the volcano chimney and cone) provide efficient waveguides trapping surface and guided waves known as "ground roll" in reflection seismology. All this shows that near-surface scattering should be the main mechanism producing the coda.

As an end-member model, scattering from the free surface but accounting for realistic scattered wavefield could make a good quantitative model of the coda, such as L_g coda model in Section 5.6.4. The free surface, including topography, weathered zone, and sedimentary sequences, represents by far the strongest reflector package, which also contains the strongest attenuation and is traversed by all waves reaching the receivers (Figure 6.47a). In volcanic environments, the magma-source region is characterized by low velocities, high attenuation, and extreme complexity that enhance scattering (Chouet, 2003; Lin *et al.*, 2005). Thus, the principal difference of volcanic settings appears to be in an enlarged zone of heterogeneity, in which both the sources and areas of temporal variations may be

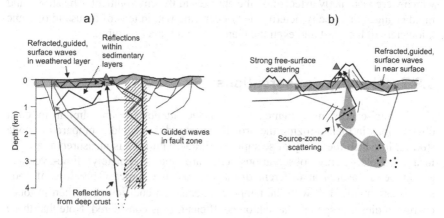

FIGURE 6.47
Schematic models of scattering responsible for coda formation:
a) non-volcanic case (vicinity of San Andreas Fault) and
b) volcanic case. Gray indicates zones subject to rapid temporal variations. Black dots are the earthquake sources, gray triangles are seismic stations. Note that in both cases, scattering is concentrated near the free surface and potentially within other, relatively compact zones of increased heterogeneity.

located (Figure 6.47b).

The above difference in the shapes of high-χ_i structures explains the difference in the observed characters of temporal changes of γ and Q_e (Figures 6.43b and 6.46b). It appears that in the volcanic case (except in the co-eruptive phase), the change occurred within the subsurface and consisted in the removal of the low-Q_i gas and magma, which led to a reduction of Q_e^{-1}. At the same time, γ remained unchanged because ray shapes did not change near either the source or the receiver (Figure 6.46b). By contrast, in the non-volcanic case (Figure 6.43b), the change related to seasonal precipitation (as interpreted here) occurred near the opposite (receiver) end, which also corresponded to the zone of strongest scattering. Therefore, the high- γ_i, high-Q_i^{-1} near-surface layer affected both the resulting γ and Q_e^{-1}, as shown in Figures 6.43b to 6.44. An additional difference was in the observed seasonal cycle in Figure 6.44 being incomplete; with longer recording, both γ and Q_e^{-1} in this case would have likely returned to the original values.

To recapitulate the approach, abandoning the uniform-scattering coda model at the early stages of interpretation allowed us to reveal important data relationships which were expressed in the linear patterns of $\chi(f)$ in Figures 6.43b and 6.46b. This observation allowed us to measure two new parameters (γ and Q_e) and detect their variations with the time of observation. Further, scattering and temporal changes were shown to occur in significantly more localized areas, which were also likely affected by the changes in the environment. The strong and rapid changes in velocity heterogeneity occurring within tens of thousand of cubic kilometers of the crust and even the mantle are no longer required.

6.9 Total-energy observations

Total-energy measurements of seismic attenuation are similar to coda observations in their utilizing the broadly averaged wavefield. Compared to coda studies, in which the decay of seismic amplitude with time is measured at a fixed location, total-energy observations compare the completely time-averaged amplitudes measured at different distances from the source. Therefore, if coda studies essentially deal with the temporal attenuation coefficient, χ, then in total-energy studies, the spatial attenuation coefficient, α, is considered. Note that these coefficients are no longer related by a simple multiplication by wave velocity, $\chi \neq V\alpha$, because they do not correspond to the same traveling wave. The geometrical-spreading factors, $G(t,f)$ for coda and $G(\mathbf{r},f)$ for total-energy, both represent averaging of numerous wave modes and propagation paths and are extremely difficult to model.

Because of the above similarities, total-energy and coda studies also share the problems related to using simplified geometrical-spreading models for detailed inversion scattering, which is prone to "over-fitting" the data. Total-energy

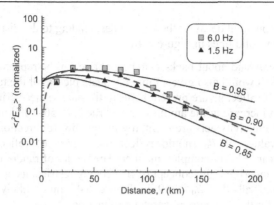

FIGURE 6.48

Averaged observed total energy curves $<E(r,f)>$ *versus* hypocentral distance in southern California corrected for source and receiver effects (Mayeda *et al.*, 1991), at two frequencies (labeled symbols). Theoretical curves by Wu (1985) for seismic albedos of $B_0 = 0.85$, 0.90, and 0.95 are shown (black lines). Gray dashed line shows the empirical geometrical-spreading dependence of eq. (4.3) with $\gamma = 0.02$ km^{-1} and $\delta v = 0.45$.

measurements are often used for differentiating between scattering and anelastic attenuation, and for estimating the seismic albedo (Section 5.8). As in other applications discussed in this book, scattering Q derived from such measurements can often be interpreted in terms of more accurate, specific, and productive effects of the lithospheric structure.

As an example, consider Figure 6.48 summarizing the interpretation of the total seismic energy from local earthquakes in southern California by Mayeda *et al.* (1991). By fitting the data (symbols) with theoretical curves resulting from the scattering theory (Wu, 1985), these authors estimated the seismic albedo as $B_0 \approx 0.8$–0.9. If B_0 were indeed related to scattering, such high values would mean that scattering greatly exceeds the anelastic attenuation in the study area: $Q_s^{-1}/Q_i^{-1} = B_0/(1 - B_0) \approx 4$–9. However, it is important to understand how these values of Q_s^{-1} and Q_i^{-1} were constrained from this dataset, which can be summarized as follows. First, the total-energy values in Figure 6.48 were multiplied by $4\pi r^2$, where r was the hypocentral distance. Assuming a uniform and isotropic background geometrical spreading and no scattering or attenuation, this correction would make all observed energies constant with distance. With respect to this constant value, scattering can principally be recognized by the amplitudes increasing away from the hypocenter, similarly to what we discussed in the coda MLTWA example (Figure 6.39). This effect should be significant at ~30–50-km distances from the source. By contrast, the anelastic attenuation causes a continuous, exponential decay of the amplitudes, leading to the distance-amplitude curves shown by solid lines in Figure 6.48.

Everything said about back-scattering above can be repeated about crustal reflections, which would also give a more adequate picture of the true character of the wavefield. If reflectors are present beneath the source, near-critical reflections would cause an increase in the amplitudes recorded at ~50–100-km distances (see Section 7.4). Reflectivity acts preferentially in the upward direction, and therefore the inverted values of B_0 should reflect the depths of reflectors and their amplitudes, but in a fairly complex manner. The model of random scattering in a uniform space underlying the concept of B_0 strongly restricts its applicability. The selection of the "albedo" $B_0 \approx 0.9$ in Figure 6.48 more likely represents an approximation for the geometrical-spreading function.

Mayeda *et al.* (1991) pointed out that the anelastic attenuation Q_i^{-1} in this dataset is weak. This can indeed be seen from the absence of a significant frequency-related amplitude decrease in Figure 6.48, and particularly at greater distances. This shows that most of the total-energy variation with distance is caused by the effects of the subsurface structure, which we can describe by a phenomenological geometrical-spreading function.

The functional form of such phenomenological geometrical spreading can be very general and constrained only by its expected general behavior and the symmetry of the problem. For example, the amplitudes could be expected to behave asymptotically as $r^{-1+\delta v}$ at $r \to 0$ and $r^{-1}e^{-\gamma r/V}$ at $r \to \infty$. Both of these asymptotics can be combined in the following "*ansatz*" for geometrical spreading:

$$G_{res}(r) = \frac{2r^{\delta v}}{e^{\gamma r/V} + 1}.$$ (6.32)

The gray dashed line in Figure 6.48 shows another empirical function related to eq. (4.3):

$$G_{res}(t) = t^{-\delta v}e^{-\gamma t},$$ (6.33)

with $t = r/V$, $\gamma = 0.08$ s^{-1}, $\delta v = 0.45$, and $V = 3.5$ km/s This function fits the data more closely than scattering models with different values of seismic albedo (Figure 6.48). In the absence of independent, detailed information about the background structure, this empirical geometrical spreading model appears satisfactory.

In conclusion of this example, significant random scattering again seems to be unnecessary in order to explain these observations. On the contrary, reflectivity within the middle and lower crust, the Moho and crustal/mantle velocity gradients should have significant influence on the observed total-energy values.

6.10 Attenuation-coefficient tomography for L_g coda Q

In the preceding examples, we showed that many observed seismic attenuation processes can be consistently interpreted in terms of parameters γ and κ describing the dependence of the temporal attenuation coefficient $\chi(f) = \gamma + f\kappa$ on frequency f. Up to this point, we derived χ from published Q^{-1} data by using relation $\chi = \pi f \, Q^{-1}$. This transformation removes the common ambiguity of the frequency-dependent Q, and parameter γ becomes most useful for interpretation. However, when working with direct seismic amplitude measurements, this transformation is undesirable and unnecessary. During the inversion for γ and κ, this transformation also causes difficulties with error estimation and assigning balanced weights to the data points. A direct procedure bypassing Q^{-1} and resolving these values from the seismic-amplitude data is required. Moreover, the values of γ and κ discussed in all examples above were *apparent*, *i.e.*, related primarily to presentation of the observed data. In this section, we take a further step and develop a tomographic procedure inverting for the *in situ* γ_i and κ_i.

The tomographic method is quite general. Minimal model parameterization and the constrained Simultaneous Iterative Reconstruction Technique (SIRT) are used for regularization-free inversion. In addition to attenuation parameters, scattering amplitudes are also inverted for and found to be highly variable and strongly correlated with geological structures. The approach is illustrated on 2-D mapping the surface-consistent L_g coda attenuation measured from five PNE profiles in Russia. Although somewhat overestimated compared to the two detailed PNE's point studies in Section 6.8.3, coda Q_e^{-1} values also show pronounced correlations with geological structures.

As any tomographic inversion, the procedure includes several key steps: 1) definition of a discrete parameterization of the unknown field; 2) forward model; 3) inversion; and 4) analysis of resolution. We will only be discussing steps (1) to (3). For several reasons, the resolution analysis does not appear very important in this example. First, the choice of the forward model is open to some questions and is made from somewhat heuristic, illustrative considerations discussed below. Second, the regularization-free inversion, which is achieved by a judicious choice of model parameterization, reduces the trade-off during the inversion. Third, due to recording limited to linear profiles while mapping an extended area (Figure 6.49), model parameterization is quite non-uniform, and the uncertainty of the spatial detail in the resulting model is mostly caused by subsequent interpolation.

6.10.1 Surface-consistent parameterization

The overall problem consists in inverting the observed, frequency-dependent, L_g coda attenuation coefficients, χ_c, for the spatial distributions of

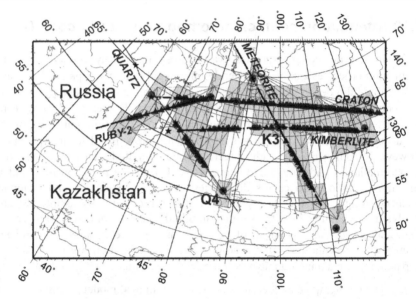

FIGURE 6.49

Sources (circles) and receivers with measured L_g coda χ_c values (triangles) in northern Eurasia. Black stars are the PNE's. The area was subdivided into 3° grid cells, of which those containing sources or receivers were used in the inversion (gray). Gray lines show the Delaunay triangulation used for producing interpolated maps (Figure 6.51). Two PNE's used in Section 6.8.3 are labeled: Q4 – Quartz-4 and K3 – Kimberlite-3 (also see Figure 6.52).

parameters γ_i, κ_i, and λ describing the residual geometrical spreading, anelastic dissipation, and scattering intensity, respectively. For the underlying regional coda model, see Section 5.6.4. The values of χ_c are defined at the positions of receivers (triangles in Figure 6.49), whereas the positions of model parameters γ_i, κ_i, and λ are distributed on a grid covering the study area. Because all three types of parameters are viewed as physical properties of the subsurface, we can make the approximation of "surface-consistency," in which these parameters depend only on their geographic locations. Therefore, collocated sources and receivers have identical γ_i, κ_i, and λ values in this model. Note that because the PNE sources were buried at 500–700 m below the surface, this approximation may not be entirely justified; however, it is likely adequate at regional scales and helps achieve a simpler, more tightly constrained solution.

To define the spatial inversion grid for parameters γ_i, κ_i, and λ, some care needs to be exercised in order to not over-parameterize the model excessively. To achieve a minimal parameterization, we define a spatial grid with increments of 3° in both latitudes and longitudes covering the entire area of profiling. Fifty-three of these cells containing the actual source or receiver points are included in the inversion (Figure 6.49). For each of these selected cells, a bilinear basis function (finite element) $\psi_i(x,y)$ is defined, where x and y are the longitude and latitude of the observation point, and i is the cell number. Each basis function is defined so that $\psi_i(x,y) = 1$ at its center and $\psi_i(x,y) = 0$ at the centers of the adjacent cells. "Blocky" basis functions typical for travel-time tomography (with $\psi_i(x,y) = 1$ within the entire cell) were also tried, with no significant differences in the results.

With thus-defined basis functions, the following identity holds at any point (x, y):

$$\sum_{k=1}^{N_g} \psi_i(x,y) = 1, \qquad (6.34)$$

where the summation includes all cells used in the inversion. With the help of this functional basis, values of γ_i, κ_i, and λ at any point (x, y) can be expressed through the corresponding values at the nodes; for example,

$$\gamma_i(x,y) = \sum_{k=1}^{N_g} \psi_k(x,y)\gamma_{i,k}. \qquad (6.35)$$

This expression restricts all possible fields of γ_i, κ_i, and λ to continuous functions given by bilinear splines in (x,y). In the resulting discrete parameters $\gamma_{i,k}$ and $\kappa_{i,k}$, subscripts 'i' will be further dropped for simplicity of notation. Finally, perturbations of all these $3N_g$ parameters are included in a single vector $\mathbf{p}$, which is further used in the inversion. In this study, we do not consider the dependence of parameters κ_i, and λ on frequency.

6.10.2 Forward model

The purpose of the forward model is to predict the observed values of χ_c by using the model $\mathbf{p} = (\gamma_k, \kappa_k, \lambda_k)$. Determination of χ_c requires specifying a background "geometrical" model $G_0(\mathbf{r},t)$ for coda amplitudes. This model is included in both the forward model for χ_c and in the measurement of χ_c values from recorded PNE amplitudes. Because of its presence in both the predicted and measured quantities, the requirement of accuracy of this geometrical-spreading model is not critical. However, selecting $G_0(\mathbf{r},t)$ reasonably close to the correct geometrical limit of coda amplitudes would increase the accuracy of linear approximation on which the inverse method is based.

A general, single-scattering, 2-D forward model for L_g coda was developed in Section 5.6.4. However, in this PNE example we use a different model similar to the one used for long-range coda in Section 6.8.3. The advantage of this model

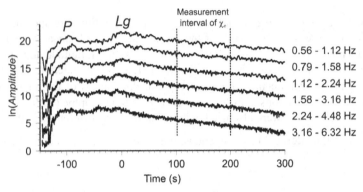

FIGURE 6.50
Stacked amplitude envelopes from six recordings within 660–760-km distances from the PNE Kimberlite-3 in Siberian Craton (Figure 3.9a modified). Records are filtered within several frequency bands (labeled). Regional phases and the coda attenuation measurement intervals are indicated. Time zero corresponds to the L_g arrival, and log-amplitude envelopes are shifted in order to separate the curves.

is in its simplicity in view of the relatively sparse data coverage concentrated at the profiles (Figure 6.49). This model represents the coda by a sum of near-source and near-receiver contributions and avoids tracing rays to scatterers located broadside of the profiles. This model is appropriate for P-wave codas; however, note that P- and S-wave codas still strongly contribute to χ_c measured in the L_g-coda time windows (Figure 6.50). Also, the observation distances of L_g from the PNE's are only about two to three grid cells used in the inversion, and therefore the presented inversion represents a relatively crude estimate, in which we emphasize the stability rather than the detail of the solution. Finally, the most significant advantage of this model is in the background geometrical spreading for the coda reducing simply to $G_0(\mathbf{r}, f) = 1$ (Morozov and Smithson, 2000). With this background model, χ_c equals the negative time derivative of the logarithm of coda amplitude:

$$\chi_c(f) = -\frac{d \ln A(t, f)}{dt}, \qquad (6.36)$$

which is measured by fitting straight lines to the logarithmic coda envelopes in each individual record (Figure 6.50).

To derive a model for the observed γ and Q_e, let us use an approximation for event energy envelopes by Morozov and Smithson (2000), in which the intensity of the primary event was described by a short pulse of a parabolic shape arriving at time t_0, with amplitude P_0 and duration τ. The coda of this arrival was described by an amplitude exponentially decaying after the initial pulse,

$$P_{\text{coda}}(t,\lambda,\chi) = \begin{cases} 0, & t < t_0, \\ \lambda P_0 \tau e^{-2\chi(f)t}, & t \geq t_0. \end{cases} \quad (6.37)$$

In this expression, λ is the scattering amplitude factor, defined as the relative coda intensity at the time of the primary event. In general, this parameter may also depend on the frequency, but such dependence is not considered in the present analysis.

From records from PNE Quartz-4 (Q4 in Figure 6.49), Morozov and Smithson (2000) estimated the duration of the primary onsets $\tau = 1.25$ s for the teleseismic P, and $\lambda = 0.22$ for all events. The relative coda intensity at the time of the primary event therefore was $\lambda\tau \approx 0.27$ (eq. 6.37), which can be considered as significantly smaller than 1. For L_g, such an estimate is more difficult to make; however, the appearance of coda records suggests that L_g coda is similarly weak compared to the onset. Therefore, the total wavefield intensity at the receiver can be approximated as:

$$P_{SR}(t) = \left[P_{primary}(t) + P_{coda}(t,\lambda_S,\chi_S) \right] * \left[P_{primary}(t) + P_{coda}(t,\lambda_R,\chi_R) \right], \quad (6.38)$$

where subscripts S and R correspond to the source and receiver locations, respectively, and the asterisk denotes the convolution in time. To the first order in $\lambda\tau$, this expression reads:

$$P_{SR}(t) = P_{primary}(t) * P_{primary}(t) + P_{primary}(t) * \left[P_{coda}(t,\lambda_S,\chi_S) + P_{coda}(t,\lambda_R,\chi_R) \right], \quad (6.39)$$

and the resulting coda intensity is given by the second term in this expression:

$$P_{SR,coda}(t) \propto \lambda(x_S,y_S,f) e^{-2\chi(x_S,y_S,f)t} + \lambda(x_R,y_R,f) e^{-2\chi(x_R,y_R,f)t}, \quad (6.40)$$

where we also assumed the surface consistency of λ. As almost everywhere in this book, $\chi(x,y,f)$ is approximated by a linear function of frequency:

$$\chi(x,y,f) = \gamma(x,y) + \kappa(x,y)f. \quad (6.41)$$

On the left-hand side of eq. (6.40), taking into account the source- and receiver-site effects, the observed coda intensity can be approximated as:

$$P_{SR,coda}(t) \approx \zeta(f) e^{-2\chi_c t}, \tag{6.42}$$

where $\zeta(f)$ is a time-independent amplitude normalization factor of the specific seismic record. With these approximations, eq. (6.42) leads to a system of non-linear equations,

$$d_{n,t,f} = \lambda_S(f) e^{-2[\chi(x_S,y_S,f)-\chi_c]t} + \lambda_R(f) e^{-2[\chi(x_R,y_R,f)-\chi_c]t} - \zeta_n(f) = 0, \tag{6.43}$$

for all t, f, and record numbers n. This system needs to be solved within the class of bi-linear functions given by eq. (6.35).

6.10.3 Tomographic inversion

Solving eqs. (6.43) is equivalent to minimizing the following objective function:

$$\Phi(\gamma,\kappa,\lambda,\zeta) = \frac{1}{2} \sum_f \sum_n \int_{T_n} \left[\Lambda_S(f) + \Lambda_R(f) - \zeta_n(f) \right]^2 dt, \tag{6.44}$$

where the source and receiver terms are, with P denoting S or R, respectively:

$$\Lambda_P(f) = \lambda_P(f) e^{-2\chi(x_P,y_P,f)t - 2\chi_c t}, \tag{6.45}$$

and T_n is the coda observation time window. The summation in eq. (6.44) takes place over all records and frequencies of interest, and the minimization is performed in terms of all parameters in vector **p**.

Within the class of piecewise-bilinear functions given by the finite-element basis (6.35), the objective function (6.44) becomes a non-linear function of model parameters: $\Phi(\gamma_i, \kappa_i, \lambda_i, \zeta_n)$. However, unconstrained degrees of freedom are still present in this parameter space, such as the arbitrary scaling of parameters λ and ζ in eq. (6.44) and source-receiver trade-off caused by non-uniform spatial sampling of the model. Such degrees of freedom are usually removed by regularization, which is performed by adding constraint terms to the objective function (6.44). For example,

$$\tilde{\Phi}(\gamma_i,\kappa_i,\lambda_i,\zeta_k) = \Phi(\gamma_i,\kappa_i,\lambda_i,\zeta_k) + \sum_k w_k \Psi_1(\gamma_i,\kappa_i,\lambda_i), \tag{6.46}$$

where functions $\Psi_1(\gamma_i, \kappa_i, \lambda_i)$ penalize various undesirable types of solutions. However, this regularization method has two disadvantages: 1) it requires a non-trivial selection of weights w_k and 2) it shifts the solution from minimizing the objective function (6.44) and therefore from satisfying eqs. (6.43). Most importantly, properties of the resulting constraints may be difficult to assess in order to evaluate their effects on the results. Therefore, a simpler regularization scheme with readily interpretable constraints is desirable.

To obtain a stable solution unbiased from the minimum of the original objective function $\Phi(\gamma_i, \kappa_i, \lambda_i, \zeta_n)$, the iterative nature of the SIRT solver can be utilized. We seek the solution iteratively, with increments performed in the directions of gradients of the objective function (6.44) as follows. After each iteration, the solution is corrected to satisfy the following criteria:

1) Clipping the values of λ_i and κ_i to make them non-negative at all points. These constraints correspond to non-negative scattering amplitudes and Q_e, respectively. Note that unlike κ_i, the geometrical-spreading term γ_i can take both negative and positive values.

2) Normalization: $\sum_i \lambda_i = N_{grid}$, where N_{grid} is the total number of grid points involved in the inversion. This constraint is enforced by re-scaling all values of λ_i and removed the scaling invariance of the minimum of expression (6.44).

3) Clipping the deviations from a smoothed model: $\left| \lambda_i - \lambda_i^{smooth} \right| < \beta \lambda_i^{smooth}$, where λ_i^{smooth} is the smoothed value of λ_i derived by averaging the adjacent points but excluding the i-th one, and β is the tolerance parameter selected equal to 0.05 in our inversion. This criterion guarantees that adjacent cells differ by no more than β in terms of λ_i. Similar constraints were implemented for γ_i and κ_i.

4) Similar constraints, for example: $\left| \lambda_i - \lambda_i^{apriori}(x, y) \right| < \beta' \lambda_i^{apriori}(x, y)$, could be added to keep the solution in the vicinity of some *a priori* model. Such constraints could be useful to remove instabilities at isolated points near the circumference of the model, or for extrapolation of the solution outside of the area of coverage. It us useful to parameterize the *a priori* model at user-specified support points between which the values of $\lambda^{apriori}(x, y)$ are linearly interpolated by using a Delaunay triangulation. However, such a constraint is not used in the solution presented below.

To derive the perturbations of parameters γ_i, κ_i, λ_i, and ζ_n during each SIRT iteration, eqs. (6.43) need to be linearized, giving:

$$d_{f,n,t} = d_{f,n,t}^0 + \delta d_{f,n,t} =$$

$$= d_{f,n,t}^0 + \left[\delta\lambda_S - 2t\left(\delta\gamma_S + f\delta\kappa_S\right)\right]e^{-2(\alpha_S - \alpha_c)t} + \qquad (6.47)$$

$$\left[\delta\lambda_R - 2t\left(\delta\gamma_R + f\delta\kappa_R\right)\right]e^{-2(\alpha_R - \alpha_c)t} - \delta\zeta_n$$

$$= 0.$$

Quantities $\delta\gamma$, $\delta\kappa$, and $\delta\lambda$, evaluated at the source and receiver locations, are further expressed as combinations of the corresponding values at the nearest grid nodes by using eqs. (6.35). By combining all parameter perturbations of $\delta\gamma_i$, $\delta\kappa_i$, $\delta\lambda_i$, and $\delta\zeta_n$ into a common model vector $\mathbf{p}$, the resulting equations form a linear system:

$$\mathbf{A}\delta\mathbf{p} - \mathbf{B} = 0 . \qquad (6.48)$$

This system is mixed determined (Menke, 1989), because we have 7986 data points for 885 unknowns, yet some parameter combinations remain unconstrained as described above. A formal solution to this system in the least-squares sense is:

$$\delta\mathbf{p} = \left(\mathbf{A}^T\mathbf{A}\right)^{-1}\mathbf{A}^T\mathbf{B} , \qquad (6.49)$$

where T denotes the matrix transpose. In back-projection methods, matrix $\mathbf{A}^T\mathbf{A}$ in this equation is replaced with its diagonal, whose inverse is non-singular and can be easily calculated. With this replacement,

$$\delta\mathbf{p}_{SIRT} \approx c\left[diag\left(\mathbf{A}^T\mathbf{A}\right)\right]^{-1}\mathbf{A}^T\mathbf{B} . \qquad (6.50)$$

This method requires storage of only two vectors: $\mathbf{A}^T\mathbf{B}$ and the diagonal of $\mathbf{A}^T\mathbf{A}$, and consequently it can be applied to very large problems. In our study, the inverse (6.50) was calculated repeatedly by scanning through the entire $\chi_c(f)$ dataset, which was terminated when model updates became sufficiently small. Between these iterations, "trimming" operations were applied to the model in order to satisfy the constraints above. To suppress parameter oscillations during this process, a scaling factor $c < 1$ ($c = 0.5$ in our inversion) was applied during stepping (6.50) and gradually reduced when data error increases were detected. Our final model (Figure 6.51) resulted in a ~90% data error reduction from the starting model, which used $\gamma_i = 0$, $\kappa_i = 0$, and $\lambda_i = 1$ at all nodes.

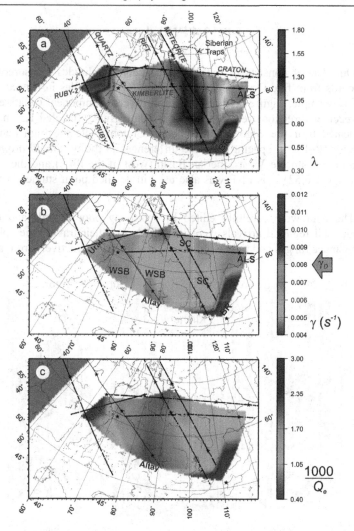

FIGURE 6.51

PNE L_g coda inversion results for: a) coda-scattering amplitude λ; b) residual geometrical spreading γ; and c) effective coda attenuation $Q_e^{-1} = \kappa/\pi$. In plot b), note the level of $\gamma_D = 0.008$ s^{-1} which appears to discriminate between stable and active tectonic areas (Section 7.3). Major tectonic regions are indicated by: ALS – Aldan Shield; BR – Baikal Rift; SC – Siberian Craton; and WSB – West Siberian Basin.

6.10.4 PNE data

Our dataset consists of log-amplitude coda amplitudes measured in the PNE records from five Deep Seismic Sounding (DSS) profiles in Russia (Figure 6.49). For example, Figure 6.52 shows the transverse-component (relative to the source-receiver direction) record from PNE Kimberlite-3 (labeled K3 in Figure 6.49) located near the edge of the Siberian Craton. Note the high density of recordings (10–15-km spacing) and the differences in the wavefield propagating within and beneath the West Siberian Basin (west of the PNE) and the Siberian Craton (Figure 6.52).This PNE was also used in detailed point studies in Section 6.8.3.

The data were carefully edited by removing poor records and clipped amplitudes, all regional arrivals were identified, and their travel times picked[19]. The L_g distance ranges and time windows were selected interactively from all PNE

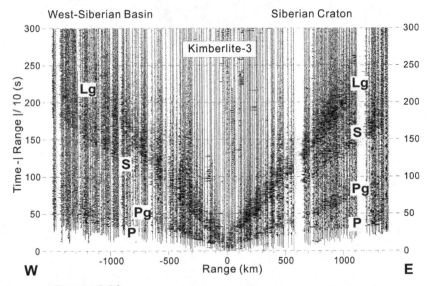

FIGURE 6.52
Transverse-component record from PNE Kimberlite-3 (K3 in Figure 6.49), filtered within 0.5–8.0 Hz frequency band. The regional phases (labeled P, P_g, S, and L_g) are clear and observed to far source-receiver distances. Note the difference between the branches of S and L_g waves in two directions from this PNE.

---

[19] Data preparation, processing, and measurements of χ_c by H. Li, a Ph.D. student at the University of Wyoming. For more amplitude- and attenuation-related studies using PNE, see her dissertation, Li (2006).

record sections. Coda windows started 20 s after the picked L_g onsets and extended for 50–100 s. Pre-L_g noise windows (*i.e.*, the *P*- and *S*-wave coda; *cf.* Morozov and Smithson, 2000) were also picked. By using these windows, L_g to pre-L_g amplitude ratios were calculated, and traces with such ratios below 1.1 were discarded.

Edited records were further band-pass filtered within four overlapping frequency bands of 1–2, 1.5–3, 2–4, and 3–5 Hz. Within each band, three-component trace envelopes were formed, and $d\ln(Amplitude)/dt$ derivatives and their standard errors were measured by using the "robust fit" technique in Matlab. The resulting values of frequency-dependent log-amplitude coda slopes and their estimated uncertainties were saved in a database which was used in subsequent inversion.

6.10.5 *Lg* coda χ in northern Eurasia

The resulting maps of λ, γ_i, and κ_i within the area of L_g-coda Q data coverage from five PNE profiles are shown in Figure 6.51. To produce continuous maps, we used a linear interpolation within a Delaunay triangulation constructed on the centers of the 53 grid cells used in the inversion (gray lines in Figure 6.49). This explains the shapes of some of the features located near the edges of the coverage area (such as the Baikal Rift zone and the Urals; Figure 6.51).

As expected, scattering amplitudes λ show strong variations, which remarkably well correlate with several tectonic areas (Figure 6.51a). High λ values are found in the Uralian and in the Sayan-Baikal (BR in Figure 6.51b) fold belts; such high λ should be related to stronger surface topography and complex crustal structures created by tectonic movements. High scattering amplitudes are also present in the western part of the Siberian Craton. By contrast, the eastern part of the Siberian Craton (the Aldan Shield; Figures 6.51a and 6.51b) and most of the West Siberian Basin show low λ. Such low scattering is again in general agreement with earlier observations of a relatively uniform crust in these areas (Morozova *et al.*, 1998). Interestingly, the zone of increased scattering within the shield areas is close to the Siberian Traps (Figure 6.51a).

The geometrical-spreading exponent, γ_i, also correlates with tectonic features (Figure 6.51b). Note that the level of $\gamma_D = 0.008$ s^{-1} (marked in Figure 6.51b,) separates the stable cratonic areas (with $\gamma_i < \gamma_D$) from the tectonically active Baikal Rift, for which γ_i is distinctly higher than γ_D. As shown in Section 6.8.2, this difference could be caused by stronger velocity/density contrasts within the tectonically active crust. Within the stable parts of the study area, γ varies only moderately, from about 0.004–0.006 s^{-1} (Figure 6.51b). The Urals show increased values of γ_i, and the West Siberian Basin shows lower values, which is likely related to the variations of crustal thickness and mantle velocity gradients (Morozova *et al.*, 1998).

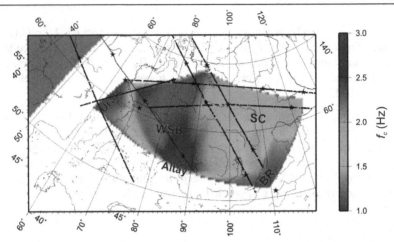

FIGURE 6.53
Cross-over frequency f_c values inverted from PNE L_g coda.
Labels as in Figure 6.51.

The effective surface-consistent coda attenuation factor $Q_i^{-1} = \kappa_i/\pi$ values also show distinct correlations with tectonic structures and reasonable stability across the study area (Figure 6.51c). Coda Q_i is low in the Baikal Rift zone ($Q_i \approx 250$–300) and near the Urals ($Q_i \approx 130$) and higher within the Siberian Craton ($Q_i \approx 400$). Within the eastern part of the West Siberian Basin, the attenuation is the lowest ($Q_i \approx 1400$). However, although indicating reasonable correlation with geological structures, the values of Q_e^{-1} appear somewhat increased (up to ~1.5 times) compared to point measurements performed in the records from two individual PNE's in Section 6.8.3. Potential reasons for such over-estimation will be considered in the following sub-section.

The map of cross-over frequency $f_c = \gamma_i Q_i/\pi$ also shows an interesting correlation with tectonic structures (Figure 6.53). This frequency increases when both γ_i and Q_i are high, *i.e.*, for a high-quality crust with a pronounced sedimentary cover (see numeric modeling in Section 6.8.2). The result in Figure 6.53 corroborates this prediction: f_c is low in the Urals and Baikal Rift areas (~1 Hz), higher within the Siberian Craton (~2 Hz), further increases to ~2.5 Hz in the West Siberian Basin, and reaches ~3 Hz in the Altay foreland. However, the last of these points may be less reliable, because it is only constrained by a single, non-reversed PNE Quartz-4 (Figure 6.49).

As a measure of the relative significance of the attenuation and geometrical-spreading effects, f_c is a useful parameter for describing the attenuation structure and its impact on the measurements. Taking the characteristic value of $f_c \approx 2$ Hz in the study area, and the useable frequency band for most DSS PNE data from approximately 0.5 to 3–4 Hz, we see that f_c is located in the middle of the observation band. Therefore, neither γ_i nor Q_i (*i.e.*, κ) can be neglected in the analysis of attenuation. Within the cratonic areas with high $Q_i \approx 1000$, the effects of geometrical spreading become progressively more important. Note that the same also applies to most attenuation studies using L_g and short-period body-wave recordings. Disregarding the residual geometrical spreading in such cases results in strong positive frequency dependence of the apparent Q, which is often observed in cratonic areas (Mitchell, 1995; Mitchell and Cong, 2005).

Although indicating generally good correlation with tectonics, the effective coda Q_i^{-1} values appear somewhat over-estimated in Figure 6.51c. This could likely be related to the parameter trade-off still present in the model. With the regularized SIRT inversion followed by spatial interpolation, comprehensive analysis of inversion uncertainties and their presentation are difficult tasks which will be considered in future studies. At this stage, we only suggest several factors that could affect the reliability of Q_i^{-1} results (Figure 6.51c).

The general difficulty of measuring the logarithmic decrements of amplitudes is well known. Spectral amplitudes often show "scalloping" which makes measurements of spectral slopes problematic because peak amplitudes and not their root-mean squares need to be followed during slope-fitting. In our case, κ effectively measures the double time-frequency derivative of the logarithm of geometrical-spreading–corrected spectral amplitudes:

$$\kappa = -\frac{\partial^2}{\partial t \partial f} \ln \left[\frac{A(t,f)}{G_0(t,f)} \right]. \qquad (6.51)$$

As the second derivative in t and f, κ should be most affected by data noise and parameter trade-off. Further, because of κ being small and non-negative, "leaking" of the noise and inversion uncertainties could increase its values, particularly in areas where κ is smaller.

Another factor that was not considered in this study, that could cause increased values of Q_i^{-1} is the frequency dependence of the scattering amplitudes, λ. If scattering amplitudes decrease with frequency, they could increase the values of Q_i^{-1}, and vice versa. In addition, the L_g coda in PNE data is located on top of the codas of the preceding P-wave arrivals, which also contribute to an apparent increase of Q_e^{-1} by steepening the coda decay with time. This could be the most likely reason for the increased Q_e^{-1}.

With some doubts about the potentially over-estimated values of Q_e^{-1}, the values of λ and γ appear to be reliable (Figures 6.51a and 6.51b). They represent lower-order parameters, correlate well with the tectonics, and lead to significant (~90%) reductions in data errors. In particular, the map of γ (Figure 6.51b) represents the most important result of this study. This map quantitatively correlates with other world-wide studies summarized in Morozov (2008) and almost perfectly corresponds to the numerical modeling and analysis of two selected PNE's in Section 6.8.2.

To summarize the above example, it demonstrates that mapping of spatially variable γ_i definitely needs to be performed whenever Q measurements are attempted, particularly in large areas with contrasting crustal and lithospheric structures. Once γ_i is incorporated, only frequency-independent Q_i remains to be measurable using the present-day seismic data. However, both γ_i and coda Q_i can be inverted directly from the attenuation coefficient, $\chi_c(f)$, data. These quantities are assumption- and model-ambiguity free, and therefore they represent a solid basis for interpretation. Both of these quantities can also be modeled by realistic numerical waveform simulations, which allow the derivation of useful links of γ and Q_e to the lithospheric structure and to the *in situ* crustal S-wave attenuation.

Chapter 7

Results

Throwing pebbles into the water, look at the ripples
they form on the surface. Otherwise, such occupation
is an idle pastime.

Kozma Prutkov, Fruits of Reflection (1853-54)

Several general results arise from the preceding analysis and data examples. Methodologically, we found no compelling support for Q being a physical attribute of the propagating medium. On the contrary, theoretical evidence and uncertainties in its determinations show that this parameter is insufficient for describing attenuation. This is the most important point, because the concept of Q is critical for most existing attenuation data interpretations. In addition to the examples used in Chapter 6, numerous other datasets may need to be reconsidered in light of the new approach.

The proposed dismissal of Q fundamentally affects the theory of attenuation. Visco-elasticity, in its specific sense of a mathematical theory formulated in terms of local memory variables and equivalent models, was criticized for its lack of physical basis and reducing the attenuation problem to phenomenological descriptions. As explained above, a full theory of attenuative elastic waves should be formulated specifically for each mechanism of seismic-wave attenuation. Such theory would certainly be far more difficult than visco-elasticity, but it could be developed by using the approach outlined in Section 2.4.

For presenting and contrasting our models, we intentionally selected old, already well-discussed and long-accepted results. This data selection emphasizes the generality of the proposed approach and universality of its principal claim. In brief, this claim is that the available attenuation data can be explained more naturally if we do not assume a Q-based model of Earth materials and avoid using the associated models during seismic-amplitude measurements and data presentation. Scattering is also relieved of its dependence on theoretical models and included as a part of geometrical spreading or intrinsic (macroscopic) energy dissipation. This approach also appears to be more productive in terms of quantitative conclusions and correlations with geology.

Clearly, much work remains to be done before the subject of attenuation is well understood. Some directions of this future analysis are outlined in Chapter 8. Below, we summarize the key general observations that appear to be established by the attenuation-coefficient analysis in this book.

7.1 Frequency dependence of attenuation coefficient

In most of the theoretical and experimental observations above, we saw that the temporal attenuation coefficient obeys the quasi-linear law hypothesized in Section 2.8,

$$\chi = \gamma + \kappa f + \chi_{err},$$ (2.128 again)

in which κ is a non-negative constant. The constancy of κ is not surprising and corresponds to expression (2.128) which simply represents a Maclaurin-series approximation for $\chi(f)$. The essence of the debate about preferring the attenuation-coefficient model (2.128) or the frequency-dependent Q,

$$\pi f \chi^{-1} \equiv Q = Q_0 \left(f/f_0 \right)^{\eta},$$ (1.13 modified)

which implies a power-law form for χ,

$$\chi = \pi f_0 Q_0^{-1} \left(f/f_0 \right)^{1-\eta},$$ (7.1)

is about allowing a non-zero limit of χ at frequencies approaching zero, $f \to 0$. Because this limit is physically unrealizable, the contradiction between these two expressions can always be amended by making $Q \to 0$ at lower frequencies, $i.e.$, by using positive η values. For example, for about five-fold variations in f, dependences (2.128) and (7.1) are nearly equivalent for any values of f and f_0 (Figure 7.1). From the range of observations, approximation (2.128) simply

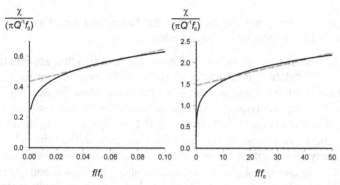

FIGURE 7.1

Comparison of a $\chi(f)$ dependence in eq. (7.1) with $\eta = 0.8$ (black lines) with linear dependence (2.128) (dashed gray lines) for two intervals of f/f_0.

extrapolates the $\chi(f)$ trend to $f = 0$ whereas (7.1) considers only curves passing through point $\chi|_{f=0} = 0$ (Figure 7.1). Note that in real data (see examples in Chapter 6), the range of frequencies for which $\chi(f)$ rolls off to zero in approximation (7.1) is always below the observation range, and the non-linearity shown by black lines in Figure 7.1 is not observed.

There exist several reasons why χ can be non-zero at zero frequency (Chapter 5). The causes of $\gamma = \chi|_{f=0} \neq 0$ include ray bending and multi-pathing, scattering, layer interactions in surface-wave propagation, and general uncertainty of the background model. In fact, $\gamma = 0$ is only found in several very special cases involving perfectly known (uniform-space) geometrical spreading or perfectly coherent scattering. Therefore, despite its using three parameters instead of two in (2.128), parameterization (Q_0, η, f_0) in (1.13) and (7.1) is problematic, because it misses the important case of $\gamma \neq 0$ and imposes sharp variations of χ at $f < f_0$ (Figure 7.1).

Thus, from both theoretical and observational standpoints, we find the linear $\chi(f)$ form (2.128) to be most suitable. Let us now see how this form helps us in the classification of global observations of seismic attenuation.

7.2 Summary of global attenuation-coefficient observations

The attenuation-coefficient data discussed in Chapter 6 can be grouped into three sets of nearly linear branches within 100–400 s, 10–100 s, and ~0.5–100 Hz period/frequency bands (Figure 7.2). These frequency bands correspond to distinct wave types and sampling depths within the Earth, and therefore discontinuous transitions between them were also expected.

The first of these groups of $\chi(f)$ branches, at 100–400 s, appears to be well defined and is "global" in character in our model, owing primarily to its wave length of up to 2000 km and our use of global compilation for deriving it (Figure 7.2). The two higher-frequency groups are sensitive to the tectonic types and geologic structures (Figure 7.2). Both γ and Q_e values vary regionally for both Rayleigh and body waves. As shown in the preceding section, the values of γ are systematically lower within stable regions and tend to reduce with tectonic ages. Within oceanic areas, γ shows lower, continental-type values, and Q_e values are high (~1000; see Figure 6.2a, p. 186). These observations generally agree with our suggestion above that for short-period waves, γ should be principally controlled by the upper-crustal structure. The upper portion of the crust is more heterogeneous in active continental environments (Christensen and Mooney, 1995; Mitchell, 1995) and is virtually absent in the oceanic crust.

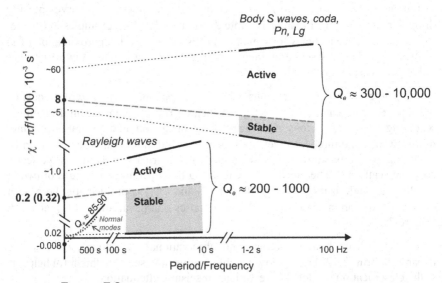

FIGURE 7.2

Summary of observed $\chi(f)$ dependences from studies in Chapter 6. "Reduced" χ values are shown, in which the linear dependences corresponding to $Q_e = 1000$ appear horizontal. Typical values of Q_e and γ indicated. Long-dashed gray lines and bold symbols show levels $\gamma = \gamma_D$ discriminating between stable- and active-tectonic environments. The values of surface-wave γ_D not corrected for dispersion are given in parentheses. From Morozov (2010b); with permission from Springer.

Note that the transition between the two Rayleigh-wave branches occurs nearly continuously at the attenuation-coefficient levels of $\chi \approx 3 \cdot 10^{-4}\,s^{-1}$ at $f = 0.01$ Hz (see Figures 6.2 and 6.4b on p. 186–190). By contrast, with the transition to lithospheric and crustal wave modes, χ jumps upward by over ~(3–6)$\cdot 10^{-3}\,s^{-1}$ (Figure 7.2). This once again suggests that the crust (and likely the upper crust for higher-frequency waves) should be the cause of such increased γ values.

Between the two higher-frequency groups of $\chi(f)$ branches, there is a broad gap from ~1–2 s to ~10 s, in which no attenuation measurements are available (Figure 7.2). This frequency range is characterized by well-known difficulties of measurement and interpretation. At these frequencies, the wavefield changes from predominantly surface- to body-wave types, and the complexity of the lithospheric structure is felt most acutely by both of these types of waves. In addition, the depth and type of sampling changes significantly across this gap.

Several general conclusions can be made from the observed distribution of $\chi(f)$ segments: 1) similar levels of Q_e are observed for most wave types; 2) the key distinction between these wave types appears to lie in γ; and 3) γ increases when switching to progressively higher-frequency modes. These observations suggest that the frequency-dependent dissipation mechanisms, *i.e.*, anelastic dissipation and small-scale elastic scattering, should be similar for most waves. This supports the long-wave approximation we made in this book (Figure 5.2 on p. 148). On the contrary, geometrical-spreading mechanisms are specific to each wave, and consequently the values of γ_i, which lead to γ, must be different. Finally, higher-frequency waves encounter stronger velocity/density contrasts, their background geometrical spreading is usually faster, and consequently their γ's are higher as well. For the same reason, the range of Q_e values is also broader for high-frequency waves (Figure 7.2). A less trivial but important point is that the values of γ for higher-frequency phases are positive, which was discussed in Chapter 4.

7.3 Crustal ages, tectonic types, and γ

As stable and measurable quantities, γ, Q_e, and also f_c derived from them, should be the best candidates for correlating with geological structures and tectonic types. Many authors performed such correlations in terms of Q_0 and η parameters, and therefore, in order to produce a classification in terms of γ and Q_e, we have to use the transformation above.

Attenuation data in (η, Q_0) form suggest correlations with tectonics, although these correlations may sometimes be intricate. From numerous local-earthquake coda studies, active tectonic regions are generally characterized by low Q_0 and high η, while stable cratons are characterized by higher Q_0 and lower η (*e.g.*, Aki, 1980; Benz *et al.*, 1997). However, in Peaceful Nuclear Explosion

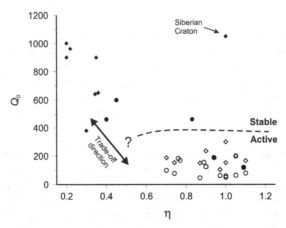

FIGURE 7.3

Correlation of S-wave attenuation parameters from coda and L_g Q compiled by Morozov (2008) with crustal tectonic types. Black symbols correspond to stable tectonic areas; open symbols to active tectonic areas; and the dashed line indicates a separation between them. Modified from Morozov (2008); with permission from John Wiley and Sons.

(PNE) studies in the Siberian Craton, we found high values of $\eta \approx 1$ in combination with a very high Q_0 (Morozov, 2008; Figure 7.3). It appears that tectonically active zones occupy a region of lower Q_0 and increased η, and stable areas may be found within the rest of the (η, Q_0) domain. Although different tectonic types can be separated in the (η, Q_0) plane as shown by the dashed line in Figure 7.3, the position and shape of this boundary is somewhat uncertain. This uncertainty is particularly difficult to describe because of the trade-off between parameters (η, Q_0) and the background geometrical spreading, which could be different in these areas. This trade-off could potentially intermix the points belonging to the stable- and active-tectonic populations in Figure 7.3. Also note that from surface-wave and L_g studies, stable cratons show not low but high η values (Section 6.1), further complicating this classification.

Most importantly, the above picture still does not explain the physical reasons for the difference in the attenuation properties within the active- and stable-tectonic regions. Mitchell (1995) suggested that this difference could be due to the lithosphere becoming drier and higher-Q_0 with age; however, this still does not explain the differences in η, and the whole argument is hampered by the dependence of both Q_0 and η on the selected reference frequency f_0 and assumed geometrical spreading.

By contrast, as shown in Chapter 4, the geometrical attenuation parameter γ should be closely related to the crustal structure and therefore could be the key to

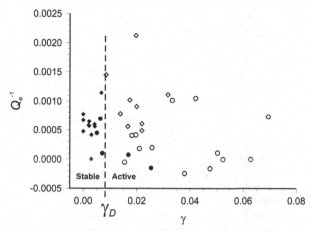

FIGURE 7.4

The same data as in Figure 7.3 in (γ, Q_e^{-1}) form. Note that the tectonically active and stable areas are separated by the level of parameter $\gamma = \gamma_D$ alone. Modified from Morozov (2008); with permission from John Wiley and Sons.

differentiating between crustal types. A transformation into the (γ, Q_e^{-1}) parameterization by using eq. (5.25) indeed shows this to be the case (Figure 7.4). In this form, the tectonically stable and active areas can be separated by the level of geometrical spreading of $\gamma_D \approx 0.008$ s^{-1} for body, coda, and L_g waves (dashed line in Figure 7.4). Apart from two outliers, all tectonically stable areas lie below this threshold. Such reduced γ values within colder, drier, and more uniform lithosphere could be related to lower gradients and smaller velocity contrasts within its structure.

The values of S-wave or L_g Q_e lie within the 1000–3000 range for both stable and tectonically active areas, although the stable-area data tend to cluster more tightly near $Q_e \sim 1500$ (Figure 7.4). Thus, Q_e does not appear to be a significant discriminator. The geometrical-attenuation values, from shallow borehole data using surface reflections (Section 6.4) cannot be directly compared to γ from surface measurements (Figure 7.4) because they use different mechanisms. The resulting values of S-wave $Q_e^{-1} \approx 0.01$ expected for sedimentary rocks should be comparable to those shown in Figure 7.4, and they would fall above the range of Q_e^{-1} values shown in this plot.

Because parameters γ and Q_e are more "physical," it is important to correlate them with geological properties, and particularly with the ages of the most recent tectonic activity. Mitchell and Cong (1998) performed such comparison in terms of the apparent frequency-dependent Q factors taken at frequencies of 1 and 5 Hz. These authors found that these Q's, and primarily $Q_0 = Q(1\text{Hz})$, increase with tectonic ages. By converting the data of Mitchell and

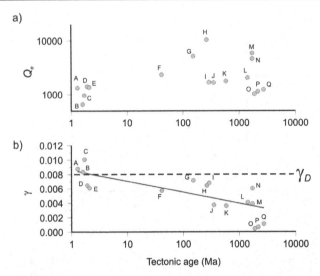

FIGURE 7.5
Attenuation-coeffcient parameters as functions of time since the
most recent tectonic activity: a) Q_e and b) γ. Data from Mitchell
and Cong (1998). Dashed line is the $\gamma_D = 0.008$ s^{-1} threshold
from Figure 7.4, and the solid line represents the inferred
$Q_e(Age)$ trend transformed from Mitchell and Cong (1998).
Labels indicate the tectonic regions: A – The Andes Mountains;
B – Basin and Range; C – Tethys region (the area of
convergence of the Eurasian, African, Arabian, and Indian
plates); D – the Arabian Peninsula; E – the East African Rift;
F – the Rocky Mountains; G – northeastern China; H – the
eastern Altaid belt of Eurasia; I – the Tasman Province of
Australia; J – the Atlantic Shield of South America; K – the
African Fold Belts; L – the North American Craton within the
United States; M – the Australian Craton; N – Eurasian cratons;
O – African shields; P – the Brazilian Shield; and Q – the Indian
Shield. From Morozov (2008); with permission from John Wiley
and Sons.

Cong (1998) into the (γ, Q_e) form, we also see that Q_e values tend to increase, but
also become more scattered with age (Figure 7.5a). However, this difference could
be due to limited sampling of the active-tectonic areas in this study (as well as
throughout the world). At the same time, γ appears to consistently decrease with
age (Figure 7.5b), and additional sampling of active areas might only strengthen
this observation. With the exceptions of the Arabian Peninsula and the East
African Rift, all regions with older than 10-Ma tectonics fall beneath the
$\gamma_D = 0.008$ s^{-1} level of geometrical attenuation (Figure 7.5).

7.4 Causes of positive γ and η from local earthquakes

In different types of attenuation measurements (*e.g.*, body-wave, L_g, coda, or total energy), different seismic phases and time windows are selected for measuring the amplitudes. The peak and averaged whole-record amplitudes shown in Figures 4.2 to 4.4 are likely most relevant to *P*-wave, coda, and total-energy studies. However, our goal here was not to model any specific case with sufficient accuracy — this would be difficult. Instead, we demonstrated that the geometrical spreading generally:

1) does not fit into any simple models such as $t^{-\nu}$;

2) is variable and sensitive to the lithospheric structure; and

3) is measurable and to some extent can be modeled by ray-based and waveform synthetics.

In particular, numerical modeling in Sections 4.2 and 6.8.2 show that because of the upper-crustal reflectivity, geometrical spreading from local earthquakes is *commonly faster* than expected from an isotropic, uniform-space, body-wave t^{-1} model. These conclusions are practically invariant with respect to the choices of time windows and types of amplitude measures.

Several common features of γ and Q_e can be seen in the synthetics. For most crustal models without strong mid-upper crustal reflectivity (such as in Figures 4.2 to 4.4), numerical modeling showed values of $\gamma > 0$. Such structures are likely abundant around the world, and therefore the geometrical spreading should most often be under-corrected by the standard body-wave correction r^{-1}. In the traditional attenuation measurements based on transforming $\chi(f)$ into $Q(f)$ (eq. (1.9)) and taking $Q(f) = Q_0 f^\eta$, positive γ values lead to $\eta > 0$. As shown in Section 5.4, parameter η roughly equals $\eta \approx \pi\gamma/(f_{obs}Q_e)$, where f_{obs} is the central observation frequency. Therefore, the fact of $\gamma > 0$ could explain widespread observations of positive and high η's, especially at lower observation frequencies f_{obs}, and in higher-γ, lower-Q_e settings. Note that both of these last conditions correspond to tectonically active areas (Chapter 6).

The values predicted for γ range from -0.005 to ~0.01 s⁻¹ (Figures 4.2 to 4.4), and similar values were also obtained from numerous observations (Chapter 6). To ascertain whether such residual geometrical spreading should be significant in attenuation measurements, let us define a reference level γ_Q so that its effect equals that of the effective attenuation, that is $\gamma_Q = \pi f_{obs}/Q_e$. With typical values of $Q_e \approx 1000$ and $f_{obs} \approx 1$ Hz, $\gamma_Q \approx 0.003$ s⁻¹, which is below most of the modeled as well as observed γ values. Therefore, in most practical cases, the residual geometrical spreading should affect the measurements of Q. For a rule of thumb, the larger the value of η, the more significant is the effect of γ.

Interestingly, the level of $\gamma \approx 0.008 \, \text{s}^{-1}$ derived for model Quartz-4 (Figure 4.4b) is close to the measurements from real L_g coda data in this PNE and also to independent numerical coda simulations, from which $\gamma \approx 0.0075 \, \text{s}^{-1}$ (see Section 5.6.4). This suggests that scattered body waves at 0–100-km distances make key contributions to the coda. Also, from a worldwide compilation of S- and L_g-wave results, the same level of $\gamma_D \approx 0.008 \, \text{s}^{-1}$ represents the threshold separating tectonically active ($\gamma > \gamma_D$) and stable ($\gamma < \gamma_D$) regions (Chapter 6). Therefore, it appears that parameter γ correctly captures an important common factor differentiating these lithospheric structures.

Upper-crustal reflectors above the hypocenter deflect the upward-traveling waves and cause steeper geometrical spreading from local earthquakes (see Section 4.3, Figures 4.3b and 4.4a on p. 124–123). As shown in Section 4.3, values of γ also increase when sedimentary layers are present above the hypocenter, and particularly when these layers have lower intrinsic Q (compare Figure 4.3 to 4.2). This suggests a potential explanation to the observed trend for γ decreasing with tectonic age (Section 7.3). Tectonically young structures have generally stronger velocity and attenuation contrasts within the upper crust and above the seismogenic zone. Due to cooling, metamorphism, and dehydration, such contrasts erode with age, which should lead to decreasing γ. Note that a uniform pressure-related crustal velocity gradient has no significant geometrical-spreading effect (Figure 4.4a). Therefore, the upper-crustal structure appears to be the most important for causing increased levels of γ.

The upper-crustal reflectivity also appears to be the only suitable mechanism for reducing γ and particularly for obtaining geometrical-spreading–compensated amplitudes that increase away from the hypocenter (*i.e.*, $\gamma < 0$). To produce such increased amplitudes, reflective layers should be present close to and below the source (Figure 4.3b). Such reflectors enhance the illumination of the pre-critical distance interval for Moho reflections.

Waveform simulations in Chapter 4 show that γ is sensitive to the position of the earthquake source within the lithosphere. For sources located below the zone of upper-crustal reflectivity, modeled geometrical spreading tends to be systematically faster than theoretical, with $\gamma \approx 10^{-3} - 10^{-2} \, \text{s}^{-1}$. The heterogeneous upper crust deflects the source energy downward.

7.5 Attenuation-band model

The observed predominance of linear $\chi(f)$ trends within broad frequency bands is strong evidence and cannot be incidental. This predominance also suggests a simple but useful approach to interpreting seismic attenuation data. Similar to refraction seismic travel-time analysis, linear dependence $\chi = \gamma + \kappa f$ suggests that a high-κ_i layer dominates the observation. Indeed, in a layered Earth, the observed apparent κ factor should be dominated by the effects of zones of

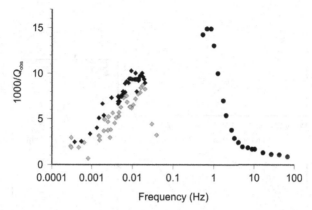

FIGURE 7.6
Observed $1000 \cdot Q^{-1}$ for spherical oscillations, Rayleigh- and P-
waves (gray diamonds), torsional oscillation, Love-, and S-waves
(black diamonds) compiled by Anderson and Given (1980), and
borehole S-wave data from Abercrombie (1998) (black circles).

highest attenuation, κ_i (or lowest Q in conventional language). If such zones exist,
they could be responsible for most of the attenuation, and could partly "mask" the
lower-κ_i layers around them and maintain the near-linear $\chi(f)$ trends.

Figures 7.6 and 7.7 illustrate this idea. Figure 7.6 shows long-period
normal-mode and surface-wave attenuation data compiled by Anderson and Given
(1980), and also borehole S-wave data from Abercrombie (1998). Several other
data compilations exist for the long-period part of this spectrum, and several 1-D
and 3-D mantle Q models were constructed from them. Among the most notable
1-D models are PREM (Dziewonski and Anderson, 1981) and QL6 (Durek and
Ekström, 1996), which will be used below. Significant discrepancies between
these models exist, indicating the still-remaining uncertainty of Q measurement
and inversion[20] (Dahlen and Tromp, 1998, p.355). However, for simplicity of this
example, we do not discuss the differences between data compilations and use the
early one by Anderson and Given (1980) (Figure 7.6). This compilation was also
used as the basis for the mantle absorption band model.

The distribution of data points illustrates the apparent absorption band,
with high Q values ~500 below 0.001-Hz frequency, decreasing to ~60–80
between 0.1–1 Hz, and returning to $Q \approx 1000$ at ~10–30 Hz. In this Q-based
picture, it is difficult to say much more from the data, without performing a
sophisticated inversion for mantle Q distribution. However, if we transform these
Q values into the attenuation coefficient, $\chi = \pi f Q^{-1}$, several further statements can

---

[20] As argued below, the uncertainty of these models is in fact much greater because of
the inversion methods using the Q paradigm.

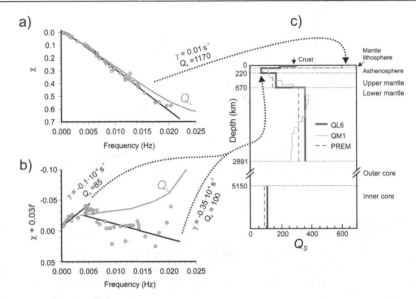

FIGURE 7.7

Interpretation of linear $\chi(f)$ within the mantle. a) and b): S-wave attenuation data from Figure 7.6 in $\chi(f)$ form: a) high-frequency part and b) low-frequency part. In b) quantity $(\chi - 0.03f)$ is plotted to emphasize the difference between two $\chi(f)$ trends which are close to 0.03. Conventional modeled Love-wave Q_L (Anderson *et al.*, 1965) is also shown by gray lines. c) Radial mantle-Q models QL6 (Durek and Ekström, 1996), QM1 (Widmer *et al.*, 1991), and PREM (Dziewonski and Anderson, 1981).

be made without performing a detailed inversion (Figure 7.7a and 7.7b). Toroidal normal modes (Figure 7.7b) form a linear $\chi(f)$ trend with $Q_e \approx 85$ and negative $\gamma \approx -10^{-5}$ s^{-1}, as in Figure 7.2. At frequencies of ~0.004 Hz, this trend breaks into that of traveling long-period Love waves, with $\gamma \approx -3.5 \cdot 10^{-5}$ s^{-1} and $Q_e \approx 100$. The difference between these Q_e values is significant, as emphasized by the "reduction" plotting technique also borrowed from refraction seismology (Figure 7.7b). The high-frequency side of the absorption also shows a linear $\chi(f)$, with $\gamma \approx 1 \cdot 10^{-2}$ s^{-1} and $Q_e \approx 1170$ (Figure 7.7a).

To explain these linear trends, recall that for waves traveling in an approximately uniform medium, χ is given by a ray-path integral:

$$\chi = \frac{1}{T} \int \chi_i dt , \qquad \text{(5.17 repeated)}$$

where T is the travel time. In volumetric form, such as for surface waves,

$$\chi = \frac{1}{\tilde{E}} \int E_D \chi_i dV \, , \qquad\qquad (5.22 \text{ repeated})$$

where E_D is the dissipating energy density, and $\tilde{E} = \int E_D dV$ is the total energy of the field. In both cases, if there exists a layer with high χ_i, integrations in the above expressions (excluding $\tilde{E}$) can be approximated as limited to that layer.

A high-χ_i layer within the Earth can produce constant-κ observations in two ways. By taking χ_i as a linear function in f and differentiating eq. (5.22) in respect to f, we obtain:

$$\kappa = \frac{\partial \chi}{\partial f} = \frac{1}{\tilde{E}} \int \left(\frac{\partial E_D}{\partial f} \chi_i + E_D \kappa_i \right) dV \, . \qquad\qquad (7.2)$$

Therefore, a high and constant κ can be expected in two cases:

1) Dissipating-wave amplitudes (E_D) concentrated within a layer with high κ_i (second term in the integrand of (7. 2)). This type of propagation could be associated with the traveling-Love wave branch in Figure 7.7b.

2) Frequency dependence of the forward kernel E_D/E itself (first term in (7. 2)). In this case, both high γ_i or κ_i contribute to κ. This appears to be the case for the higher-κ, lower-frequency, normal-mode branch in Figure 7.7b. Normal modes exhibit strong and quasi-linear dependences on frequency at any fixed depth (Section 6.1.3), causing the observed variations of χ.

Thus, low-Q_i layers within the mantle should be most important for explaining the data. Tentatively, it appears that the asthenospheric layer with $Q = 70$ at 80–220-km depths in model QL6 (Figure 7.7c) could be responsible for most of the observed apparent $Q(f)$ spectrum (Figure 7.6). Because of its dominant role and likely variations within different tectonic regions, this layer could "mask" the adjoining layers (*i.e.*, trade-off with them), making the rest of the mantle attenuation structure not so well defined. Finally, the crust, despite its having the highest Q in these models, always dominates the shallowest, high-frequency observations (Figure 7.7).

The attenuative zone within the upper mantle is well established by now. From detailed controlled-source studies on continents, this zone also shows reduced seismic velocities and significant variations related to the tectonics (Morozov *et al.*, 1998), may vary in depths within ~60–220-km range (Figure 7.8), exhibit some internal structure and also cause detectable seismic reflections and

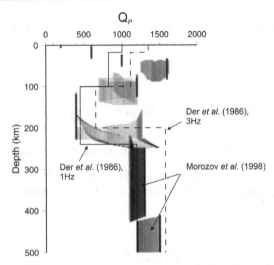

FIGURE 7.8

Short-period P-wave Q in Northern Eurasia from Der *et al.*
(1986) at 1 and 3 Hz and Morozov *et al.* (1998). The 2-D
frequency-independent model by Morozov *et al.* (1998) is
presented by a histogram, with the gray levels proportional to
the numbers of occurrences of the corresponding (*depth*, Q_P)
values.

scattering (Thybo and Perchuc, 1997; Morozova *et al.*, 1998). All these factors
suggest not only high κ_i, but also increased γ_i values within this part of the mantle.

However, some problems still persist in reconciling the surface- and body-
wave models for this attenuative zone. Long-period models show shear-modulus
quality factors of $Q_\mu \approx 70$–100 within this zone (Figure 7.7c), which would
correspond to P-wave $Q_P \approx$ 150–200. At short periods, the corresponding
$Q_P \approx 400$–800 (Figure 7.8). Body P-waves recorded beyond ~25° ranges refract
beneath this low-velocity, high-attenuation zone, and consequently their t^* values
are nearly constant with range. Der *et al.* (1986c) modeled these values and
showed that they were consistent with their short-period Q shown in Figure 7.8.
Further, in Section 6.3, we argued that the difference between the short-period and
long-period t^* could be due to a slight bias in geometrical spreading, and were able
to adjust both of them to values $t^* \approx 0.18$ s, which was close to the short-period t^*.
Therefore, there still remains the problem of making a similar adjustment for
surface-wave and normal-mode attenuation. This problem is significantly more
complicated because of the reliance of the global models on the visco-elastic
theory and Q. A potential solution to this problem will be discussed in the next
chapter.

7.6 *Q* picture revisited

Most current attenuation results are presented in the form of Q. Therefore, let us now review the key observations of Q again in view of the attenuation-coefficient concept developed above. In many cases, Q^{-1} can be viewed just as an alternate form of the temporal attenuation coefficient, $Q^{-1} = \chi/(\pi f)$, and in this sense, these two quantities are equivalent. However, in terms of their physics, relation to scattering and geometrical spreading, and phenomenological behaviors, these quantities should be clearly separated.

Overall, it appears that the general reason for pervasive use of Q in attenuation studies is in an exaggerated reliance on the theory. Theoretical models loom through the entire process of analysis, starting from initial data reduction to Q to its inversion and interpretation. At the same time, despite their mathematical refinement, some models are still rooted not in the fundamental principles of theoretical mechanics, but in heuristic generalizations and simplifying assumptions. This affects the Q picture dramatically, as summarized below.

7.6.1 Bias in Earth *Q* models?

The simple 1-D model of Love-wave Q in Section 6.1.2 shows that the classic 1-D case is still far from being solved. Our solution shows that the Love-wave Q_L in the same mantle model is under-predicted by 10–20% by the conventional method (Figure 6.8). Consequently, we expect that with the use of the new kernels for inversion, the Q values within the mantle would reduce by similar amounts. In addition, the new kernels emphasize the zones of high particle velocities within the waves, as opposed to zones of highest elastic energy in the existing solutions (Figure 6.11). This means that the depth levels of high-attenuation bands may become shallower, and the effects of the crust should increase. Both of these parts of the Earth are subject to strong heterogeneity, which would place additional limitations on inversion for radially symmetric models.

The bias in the predicted Q_L^{-1} in the 1-D solution, which is currently considered as a benchmark (Aki and Richards, 2002), suggests that similar biases should be present in the more sophisticated 3-D modeling and inversion schemes inheriting the same conceptual background. The noted violation of energy balance originates from the fundamental assumptions of the method discussed in Chapters 2 and 3. Unfortunately, this problem is almost certain to affect many recent studies.

7.6.2 Global apparent Q(f)

Let us now consider whether, and in what sense, the observed pattern of attenuation coefficients (Figure 7.2) leads to a frequency-dependent Q within the Earth. In our model, the only meaningful quality-factor type characteristic is the "intrinsic" $Q_i = \pi/\kappa_i$. The observational counterpart to Q_i is Q_e, and this quantity is frequency independent within each of the individual frequency bands in Figure 7.2. It appears that with the exception of normal modes, field seismic measurements never span a continuous range of frequencies wide enough to allow detection of a frequency-dependent Q_e. Deriving Q_i from the observed $\chi(f)$ may involve complex inversion procedures, for example in the cases of normal modes or lab measurements, and in principle, it cannot be guaranteed that a frequency-dependent Q_i will not be required to fit the $\chi(f)$ data. At the same time, given the linear observed $\chi(f)$ patterns, it still appears unlikely that the underlying Q_i should nevertheless be frequency dependent. In any case, frequency dependence of the intrinsic Q_i is certainly *not directly indicated by any data* considered in this book.

The most reliable experimental indications of frequency-dependent attenuation come from comparing different frequency bands (*e.g.*, Sipkin and Jordan, 1979), and often different wave types (Der *et al.*, 1986; Cong and Mitchell, 1988). To re-examine the frequency dependence of crustal attenuation across about two decades in frequency, let us correlate the 3–70 s Rayleigh wave results for South America from Hwang and Mitchell (1987) and from 0.4–1.4 Hz L_g Q_0 and η measurements by Raoof and Nuttli (1984) (Figure 7.9). These data were interpreted by Cong and Mitchell (1988), who concluded that the frequency dependence of crustal Q_S in its tectonic (western) part is weak (with exponent $\zeta < 0.3$ in the *in situ* $Q_S = Q_0 f^\zeta$ law) and within the stable (eastern) part it is much stronger ($\zeta \approx 0.7$). According to Mitchell (1995), strong frequency dependence is typical for stable areas. A strong contrast in attenuation levels was also noted ($Q_0 \approx 900$ in the eastern part changing to $Q_0 \approx 200$ in the western part of the area). However, by looking at the data in Figure 7.9 without tailoring them to the (Q_0, η) model, we arrive at quite different conclusions.

The procedure used by Cong and Mitchell (1988) to derive their ζ values is schematically illustrated by the dotted arrows labeled C&M in Figure 7.9. They first constructed crustal models consistent with the Rayleigh-wave attenuation (left prtions labeled H&M in each plot), then scaled their Q_β values by using trial ζ parameters, and numerically modeled the L_g-phase attenuation. The resulting values of ζ were established by matching Lg attenuation with the measurements by Raoof and Nuttli (1984) at 1-s periods (arrows and dark-gray bars in Figure 7.9a and 7.9b). However, in respect to this procedure, note that: 1) in each case, it used only a single point, namely that at 1 Hz, and ignored the rest of the measured Lg attenuation-coefficient trends and 2) this choice of 1-Hz reference, although well-established by convention, is completely arbitrary, and different choices for this frequency would have changed the values of ζ.

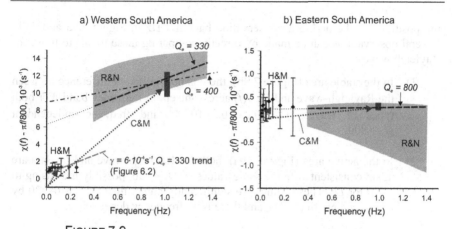

FIGURE 7.9
Comparison of surface-wave and Lg attenuation data converted
to $\chi(f)$ from: a) the western part of South America (tectonically
active) and b) its eastern part (stable). "Frequency-reduction"
was applied to $\chi(f)$, so that linear trends with $Q_e = 800$ appear
horizontal. Error bars show fundamental-mode Rayleigh-wave
data from Hwang and Mitchell (1987) are labeled H&M. Gray-
shaded areas labeled R&N show Lg $\chi(f)$ derived from Q_0 and η
values reported by Raoof and Nuttli (1984). Black dashed lines
are the linear $\chi(f)$ interpretations of Lg waves as in Figures 6.2 (p.
186). Dotted arrows labeled C&M and bars at 1.0 Hz illustrate
the procedure for correlating $Q(f)$ between the Rayleigh and Lg-
waves by Cong and Mitchell (1988).

In $\chi(f)$ diagrams (Figure 7.9a and 7.9b), $Q^{-1}(f)$ values of the Rayleigh and
L_g waves correspond to the slopes of the corresponding radius vectors shown by
the dotted arrows: $Q^{-1} = \chi/(\pi f)$. Consequently, larger values of ζ required to
reconcile these Q's in the stable-area case correspond to the wider angle between
these arrows (Figure 7.9b). Thus, the increased interpreted ζ in the stable area
(Cong and Mitchell, 1988) is actually caused by larger Q_e and lower γ in this area
[$i.e.$, more horizontal and lower-placed grey-shaded (f,χ) distribution for Lg waves
in Figure 7.9b].

Plotting the raw attenuation-coefficient data (for Lg, here reconstructed
from Q_0 and η maps by Raoof and Nuttli, 1984) allows us to see the basic
relationships between them without the use of the assumption-prone frequency-
dependent Q models and numerical modeling. Figure 7.9 shows that Lg $\chi(f)$
distributions have the same patterns as shown in Figure 7.2, and representative
linear trends can be identified for Lg (black dashed lines). As above, we only use
an interpretive approach by drawing "the most likely" linear $\chi(f)$ trends through

the positions of the dark-gray observation bars at 1 Hz in Figure 7.9a and 7.9b. Several observations can be made by directly comparing these trends to those for Rayleigh waves:

1) In the stable area (Figure 7.9b), there is no significant difference between the Rayleigh-wave and Lg $\chi(f)$ across the entire frequency band. For both types of waves, $\gamma \approx (0.2-0.3) \cdot 10^{-3}$ s^{-1} and frequency-independent $Q_e \approx 800$.

2) In the active area (Figure 7.9a), both the Rayleigh-wave and Lg data are again consistent with the same values of $Q_e \approx 330$, possibly increasing to ~400 for Lg. These values are significantly higher than $Q_0 \approx 170-220$ by Raoof and Nuttli (1984), and there is no frequency dependence.

3) However, Lg γ in the active area is much higher, $\sim 6 \cdot 10^{-3}$ s^{-1}, than the Rayleigh-wave γ, which is $\sim 6 \cdot 10^{-4}$ s^{-1} (Figure 7.9a, see also Figure 6.2 on p. 186). This is just below the lower threshold for body- and coda-wave $\gamma_D \approx 8 \cdot 10^{-3}$ s^{-1} proposed for tectonic areas by Morozov (2008). At the same time, note that values above γ_D are still within the uncertainty of the reconstructed Lg $\chi(f)$ data (dash-dotted line with $\gamma \approx 9 \cdot 10^{-3}$ s^{-1}), in which case Q_e would likely increase to ~400.

The similarity of Rayleigh-wave Q_e values in the two areas and the difference of their γ's suggest that the principal difference between them consists in the structure of the upper crust. At 3–70-s periods, Rayleigh-wave Q_e should be principally controlled by the middle and lower crust, whereas the lower-velocity, lower-Q upper crust modifies the distribution of wave amplitudes, which is described by the geometric factor. By contrast, the high-frequency Lg-wave Q_e likely closely corresponds to the Q_i of the upper crust. The corresponding high γ could be explained by the upper-crustal velocity gradients and reflectivity, which for surface waves lead to reduced amplitudes recorded at the surface. Detailed theoretical treatment and numerical modeling of these effects will be given elsewhere; at this time, it is important to ascertain that a phenomenological physical model can correctly account for all observations (see Figures 6.2 (p. 186), 6.4b (p. 190), 7.2 (p. 272), and 7.9) without postulating a frequency-dependent material Q.

If a quality-factor picture is still desired, Figure 7.10 shows the distribution of $\chi(f)$ trends (Figure 7.2) transformed into $Q = \pi f / \chi(f)$ and plotted in logarithmic-frequency scale. As one can see, there exists no "frequency dependence of Q"; instead, both frequencies and Q's depend on the types of seismic observations. Across a very broad frequency band, the apparent Q increases with frequency, but this increase mostly occurs within two bands (0.01–0.1 Hz and 1–100 Hz), below and between which Q is nearly constant. A log-log (f, Q) plot (Figure 7.10b) shows that within the different sub-bands, $Q(f)$ dependences can also be approximately described by the $Q_0 f^{\eta}$ power law, with exponent η varying from 0 to 1 and with

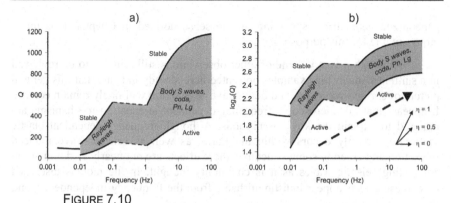

FIGURE 7.10

Apparent $Q(f)$ corresponding to characteristic (γ, Q_e) ranges in Figure 7.2: a) in linear scale, and b) in $\log_{10}Q(f)$ scale. Areas of active and stable tectonic types are indicated. Slopes corresponding to $Q(f) \propto f^{\eta}$, with $\eta = 0$, 0.5, and 1 are indicated by arrows. Bold dashed arrow shows the average $Q(f)$ trend with $\eta = 1/3$. From Morozov (2010b); with permission from Springer.

about ~0.3 scatter in the values of $\log_{10}Q$. However, this appearance could mostly be due to the notorious universality of such plots. If taken across all frequency bands, the average trend of $Q(f)$ is close to $f^{1/3}$, as suggested by Anderson and Minster (1978) and often assumed in mantle studies (dashed arrow in Figure 7.10b). However, this trend still describes not a frequency-dependent property of the Earth's material, but a succession of changes in scale lengths and wave types.

The above reworking of several published datasets shows that changes in the frequency dependence of attenuation indeed occur at ~ 300-s, ~100-s, and ~10–1-s periods (Figure 7.2), but they correspond to the transitions between different wave types dominating these frequency bands. Within each of these wave types, the attenuation quality Q_e and geometrical-spreading γ are regionally variable and correlate with tectonics (for periods shorter than ~100 s) yet are frequency independent within the available observational uncertainties. Moreover, the values of Q_e are generally close for all wave types below ~100-s periods (Figure 7.2). The widespread notion of Q pervasively increasing with frequency may thus be due to the fact that structural effects (γ) are positive and consistently increase during the transitions to higher-frequency wave modes. Note that as mentioned above, such transitions can be formally attributed to "scattering Q." In a broad sense, a significant portion of these increases in γ should indeed be caused by "scattering," in the form of refracting rays, reflections, and conversions within the Earth. However, the association with a "scattering Q" works only within the limited tasks of attenuation measurements and may incorrectly distract the attention from the effects of the first-order Earth's structure. The concept of the generalized geometrical spreading, in the sense defined in Chapter 4, is much more suitable for this purpose.

Although a vast volume of other observations still remains to be reviewed in a similar manner, the examples presented here already indicate that this general picture of surface-wave, Lg, and body-wave attenuation will likely remain correct. It appears that no microscopic frequency-dependent attenuation mechanisms are required to explain the key observations. Although frequency-dependent elastic scattering certainly occurs within the Earth, as well as seismic-wave induced relaxation and creep in some structures, these effects are not nearly as dominant in attenuating seismic waves as it is commonly thought. In the models discussed here, these effects appear indistinguishable from the frequency-independent γ_i and Q_e.

7.6.3 Tidal oscillations and Chandler wobble

Correlation of seismological Q values with those measured from Earth's tide and Chandler wobble observations (Anderson and Minster, 1979) makes another popular argument in favor of the frequency dependence of Q (Figure 7.11). Estimates for Q of the Chandler wobble are 30–35 (Mandelbrot and McCamy, 1970), 40–600 (Okubo, 1982), 36.7 (Liao and Zhou, 2004), and 20–60 (Spiridonov and Tsurkis, 2008). For Earth's tides, $Q \approx 160$–480 (Lambeck, 1977), or 370 (Ray $et\ al.$, 1996), and for the 54-min spheroidal mode $_0S_2$, the estimated Q was 589 ±10% (Sailor and Dziewonski, 1978). When plotted in log-log scales, the estimated ranges of wobble, tidal, and fundamental-mode (f,Q) values fall on an

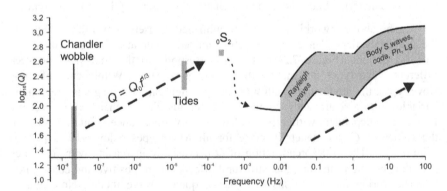

FIGURE 7.11
Simplified Figure 7.10b with Q values for Chandler wobble,
Earth tides, and normal mode $_0S_2$ added (gray bars). For wobble,
the black bar also shows the raw Q values not corrected for non-
dissipative part of rotational energy. Dashed arrows indicate the
main trends.

approximately linear trend corresponding to $Q \approx 8710 f^{1/3}$ (Anderson and Minster, 1979; Figure 7.11).

The interpretation of the above $Q(f)$ trend is based on an extremely bold assumption that the relaxation mechanism inferred for quasi-static creep observations in the lab applies not only to seismological frequencies, but also to whole-Earth oscillation periods over one year. The frequency range spans at least five orders of magnitude (Figure 7.11), and likely even three to six orders more considering the very high characteristic frequencies assumed in Lomnitz's model of creep (Section 2.5.6). The derivation of the $Q(f)$ trend is heavily rooted in the notion of Q^{-1} representing the argument of the complex shear modulus, based on which the connection between the normal-mode Chandler wobble Q's was estimated. In addition, in order to be comparable to the other two Q factors, wobble Q was decreased by a factor of about four as a correction for the non-dispersive part of the rotational energy of the Earth (Anderson and Minster, 1979). Thus, the model is quite elaborate and may allow some scrutiny. In particular, the expectation that Q is determined solely by the frequency (and not, for example, by the oscillation mode) appears to be the most questionable.

By contrast, from the viewpoint of this book, note that: 1) specific physical processes should control different types of attenuation, 2) the shear modulus is real-valued and not directly related to Q, and 3) the Q's of the normal modes and the Chandler wobble correspond to different types of oscillations and may be not directly comparable. Therefore, the above comparison loses its support as an evidence for a frequency-dependent seismological Q.

Observations of Q values increasing with the frequency of oscillations still do not mean that the cause of this increase is the "rhelogical" dependence of Q specifically on the frequency. On the contrary, what we see in the entire frequency range of Figure 7.11 is that both frequencies and Q values change for different types of Earth's oscillations. Within two groups of such oscillations, Q increases together with frequency: 1) for whole-Earth oscillations at frequencies $\sim 10^{-8}$-10^{-3} Hz and 2) for traveling waves above $\sim 10^{-2}$ Hz. Between these groups, a band of lower Q's is located (Figure 7.11), although characterizing this band only in terms of frequency (*i.e.*, as an "absorption band" in $Q(f)$) would be misleading. In reality, this range contains a complex spectrum of high-order normal modes transforming into the traveling surface and body waves.

Interestingly, both bands of correlated increases in Q and f are close to the above $Q \propto f^{1/3}$ trend (Figure 7.11). Within the second of these bands (for traveling waves), the increase in Q is apparently related to γ_i values progressively increasing for higher-frequency waves, as discussed above. Within the first band containing the whole-Earth standing waves, this (Q,f) trend should be caused by different reasons.

Unlike the traveling-wave Q, the whole-Earth quality factor is a well-defined quantity consistent with that of a mechanical vibrating system. As a Q of a mechanical system, this factor is measured near resonance, by methods close to those described in Section 3.9. Similar to the mechanical Q, this factor has no "frequency dependence." Instead, together with the corresponding natural frequency, ω_0, the whole-Earth Q-factor depends on the type of oscillation. As shown in Section 2.4.1, this factor can be safely associated with the argument of the complex-valued natural frequency.

To see how both ω_0 and Q could decrease because of a changing scale-length of the system, let us use a simple mechanical analog. Consider a solid cube of size L oscillating along one of its dimensions (Figure 7.12). To derive the scaling relations of its parameters, we need to express how the coefficients of the kinetic and potential energies of the cube, and also its dissipation function (see Section 2.4.1), scale with L. The kinetic energy is proportional to L^3:

$$E_k\left(\dot{x}\right) = L^3 \int_0^1 du \frac{\rho\left(\dot{x}u\right)^2}{2} = \frac{1}{2}m\dot{x}^2 , \qquad (7.3)$$

where mass-like parameter m equals $L^3\rho/2$, and ρ is the density. The elastic energy is then:

$$E_{el}\left(x\right) = \frac{1}{2}m\omega_0^2 x^2 , \qquad (7.4)$$

where ω_0 is the natural oscillation frequency. For simplicity, let us ignore the effect of Poisson contraction along coordinates Y and Z , i.e., consider only one non-zero component in the strain tensor: $\varepsilon_{xx} = x/L$. Our scaling-law calculation is

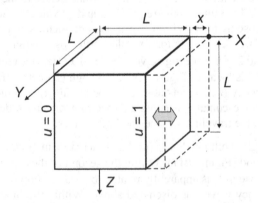

FIGURE 7.12

Mechanical model illustrating scaling of ω_0 and Q with body size L. The cube oscillates along axis X. The face of the cube labeled $u = 0$ is fixed and face $u = 1$ moves; dimensions along axes Y and Z do not change.

insensitive to this approximation. E_{el} is then also given by volume average:

$$E_{el}(x) = L^3 K(L) \int_0^1 du \frac{(\lambda + 2\mu)}{2} \left(\frac{x}{L}\right)^2 = LK(L)\frac{(\lambda + 2\mu)}{2} x^2, \quad (7.5)$$

where λ and μ are Lamé constants. In this expression, we introduced an additional scaling factor for the average strain energy, $K(L)$, to reflect a possible change in the distribution of E_{el} while varying L. Physically, this factor means that not the entire volume equally participates in the elastic-energy accumulation. Taking, for example, $K(L) = aL^{2-2\nu}$ with some a and ν, we have:

$$E_{el}(x) = L^{3-2\nu} \frac{a(\lambda + 2\mu)}{2} x^2, \quad (7.6)$$

and therefore $\omega_0 \propto L^{-\nu}$. This is the first fundamental observation we needed to confirm: larger systems oscillate at lower frequencies, and consequently, $\nu > 0$.

Next, to obtain the corresponding scaling law for Q, consider some hypothetical force of viscosity proportional to the velocities within the cube:

$$f_D = L^3 W(L) \int_0^1 du(-\zeta \dot{x}u) = L^3 W(L)\frac{(-\zeta)}{2}\dot{x}, \quad (7.7)$$

in which a scaling factor $W(L)$ is added similarly to $K(L)$ above. This force corresponds to the following dissipation function:

$$D(\dot{x}) = L^3 W(L)\frac{\zeta}{4}\dot{x}^2 = bL^{3-\psi}\frac{\zeta}{4}\dot{x}^2, \quad (7.8)$$

where form $W(L) = bL^{-\psi}$ is taken for a trial.

With all three components of the Lagrangian model thus defined, the quality factor becomes the ratio of the coefficients in E_k and D:

$$Q = \omega_0 \frac{m/2}{bL^{3-\psi}\zeta/4} \propto L^{\psi-\nu}. \quad (7.9)$$

Note that the specific form of E_{el} in (7.6) is insignificant for determining this scaling law; it is only essential how ω_0, E_k, and D scale with L. Scaling (7.9) expresses the variations of Q with increasing "size" of the system. By inverting the concurrent $\omega_0 \propto L^{-\nu}$ relation, it can also be expressed in terms of the natural frequency:

$$Q \propto \omega_0^{1-\psi/\nu}. \tag{7.10}$$

However, this relation should not be interpreted literally as frequency variations "causing" changes in Q. Both of these quantities simply change together in a varying structure.

Thus, expression (7.10) suggests that the empirical $Q \propto \omega_0^{1/3}$ relation can be explained by $\psi > 0$ and $\psi/\nu \approx 3/2$. This means that energy dissipation should decrease faster than elastic energy when switching to larger-scale (longer-wavelength and lower-frequency) oscillations. Within this crude estimate, such a scenario appears reasonable, because compared to the elastic energy, dissipation should be concentrated closer to the surface of the Earth, which scales slower than its volume. Energy kernels shown in Section 6.1.3 illustrate this point.

7.6.4 Frequency dependence of mantle Q

Apparently the most significant implication of the attenuation-coefficient view relates to the problem of the frequency-dependent Q within the crust and mantle. This problem can obviously never be solved in favor of the frequency-independent model, merely because this model is far more restrictive, and new data conflicting with it may arise. By contrast, the frequency-dependent Q model is extremely permissive, and its inherent trade-offs allow easy reconciling of different datasets. Due to its rich theoretical implications, this model has been favored by most seismologists since the early 1960s. Modern visco-elastic models routinely start by postulating rheological relaxation mechanisms and complex-valued elastic moduli within the Earth (e.g., Dahlen and Tromp, 1998; Borcherdt, 2009), which automatically lead to a frequency-dependent Q. However, because the frequency-independent model is more restrictive, ascertaining its validity would have advanced us much further in understanding the Earth's structure, properties of its materials, and the mechanics of seismic wave propagation. Therefore, this avenue should be explored to the end, and the frequency-independent model is not ruled out until conclusive and unbiased experimental evidence against it is found.

Assuming that the attenuation-coefficient data can be summarized by a collection of piecewise-linear $\chi(f)$ branches (Figure 7.2), such empirical χ trends suggest only a frequency-independent Q within the Earth's crust and mantle. Originally, the dependence of the attenuation coefficient on period (e.g., Figure 6.1 on p. 186) was viewed as the primary indication of the frequency-dependent Q. However, in the present interpretation, this argument is reversed, and the attenuation-coefficient observations only indicate spatially variable but frequency-independent γ and Q_e (Figure 6.2, p. 186). Although we do not discuss inversion here, constant-Q_e data could likely be satisfied with similarly frequency-independent Q_i in the models.

7.6.5 10-Hz transition

Deviations of the geometrical attenuation law $G(\mathbf{r})$ from the assumed background could also explain the "10-Hz transition" problem posed by Abercrombie (1998), as well as strong apparent $Q(f)$ dependencies observed at these frequencies by many authors. In this section, we examine whether such apparent $Q(f)$ could be due to the effects of γ combined with the ambiguity of the power-law form $Q = Q_0(f/f_0)^{\eta}$.

Mapping of parameters (γ, Q_e) onto (Q_0, η) discussed in Section 5.4 show that for a constant cross-over frequency f_c, the apparent frequency-dependence exponent η increases with decreasing observation frequency f. At the same time, with increasing $\eta < 1$ and decreasing $f < f_c$, Q_0 should also decrease as (see eq. (5.25)):

$$Q_0 \approx \frac{\pi f_0^{\eta}}{\gamma} \frac{f^{1-\eta}}{1 + \dfrac{f}{f_c}}. \tag{7.11}$$

Therefore, when using lower-frequency waves, we can expect to observe increased values of apparent η and lower Q_0. This trend in (Q_0, η) is typically observed in borehole body-wave Q measurements and constitutes the 10-Hz transition problem (Abercombie, 1998). In addition, the effects of attenuation (which are proportional to ft/Q_e) become smaller for lower f, leading to increased measurement errors.

By crudely picking the effective $\eta(f)$ dependence from these data (Figure 7.13b) and transforming Q_0 into Q_e by using eq. (7.11), we see that the Q_0 variation with f can be predicted by $Q_e = 1300$ and $f_c = 10$ Hz (gray dashed line in Figure 7.13a), which makes $\gamma \approx 0.02$ s^{-1}. This fit can be improved by making γ or Q_e frequency dependent, or by taking the regional variations of these quantities into account. Here, we only consider an end-member case by taking $Q_e = 1300$ and allowing a frequency-dependent γ (Figure 7.13b). This makes the $Q_0(f)$ fit nearly perfectly (black line in Figure 7.13a). Interestingly, the values of η are practically equal to 20γ in this case (Figure 7.13b).

It is not surprising that the observed frequency-dependent Q_0 and η can be replaced by a frequency-dependent γ. The residual geometrical-spreading factor, γ, is physically more meaningful than both Q_0 and η, and it allows correlating the change in Q_0 and η with the changes in the wavefield. The choice of frequency-dependent $\gamma(f)$ appears reasonable, because γ is the principal characteristic related to the mode content of the wavefield. At low frequencies corresponding to the vertical wavenumbers of lithospheric layering, the deviation of the true spreading

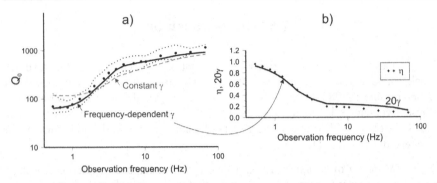

FIGURE 7.13
Interpretation of variations in Q_0 and η as caused by γ: a) Q_0 data by Abercrombie (1998) showing increased frequency-dependent attenuation within the 1–10-Hz frequency range (black circles). Dotted lines are the estimated error bounds. Black and gray dashed lines show Q_0 predicted by eq. (7.11) with frequency-dependent and constant γ, respectively. In both cases, $Q_e = 1300$ is used at all frequencies. b) η values used in transformation (7.11) (dots) and the frequency-dependent γ values multiplied by 20 (line). Modified from Morozov (2008); with permission from John Wiley and Sons.

from the selected geometrical-spreading law ($G_0(t) = t^{-1}$ in this case) should indeed be frequency dependent, and $\gamma(f)$ accommodates this deviation (Figure 7.13b).

Note that by contrast to our $\gamma(f)$ classification (Figure 7.2), Figure 7.13b shows γ values decreasing with the central observation frequency. As explained in the preceding paragraph, this may reflect the "fine structure" of $\gamma(f)$ and relate to the changing structure of the wavefield. However, an alternate, constant-γ, but variable-Q_e model is also able to describe the attenuation data (Figure 7.13a). With this model, variable Q_e would correspond, for example, to layered attenuation, as in the model in Section 7.5. This type of model was used in Figure 7.2. This uncertainty reflects the general trade-off between the background model (layered Q_i in this case) and the empirical model (γ) used to compensate its inaccuracy.

7.6.6 Mantle absorption band

A frequency band of increased absorption of elastic waves within the Earth's mantle was proposed by Jackson and Anderson (1970) and Anderson et al. (1977) and elaborated upon in many subsequent papers, particularly in Liu et al. (1976) and Anderson and Given (1982), where this band was found to be constructed by superposition of multiple microscopic relaxation mechanisms and

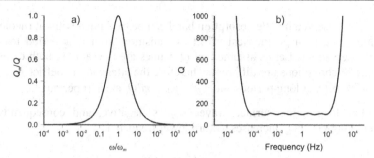

FIGURE 7.14

Absorption band model for the mantle. a) Single standard visco-elastic absorption peak and b) $Q(f)$ modeled by superposing seven such peaks with different ω_m and close to the general pattern inferred for the mantle.

variable with depth (Figure 7.14). By now, the existence of this band is considered well established by many seismologists and is supported by several types of observations (see Doornbos, 1983; Butler, 1987).

The absorption band is based on the visco-elastic relaxation theory, which was discussed in detail in Sections 2.5 and 3.4. As shown there, the essence of this theory consists of representing the observed absorption spectrum $Q^{-1}(f)$ by a superposition of standard peaks (such as in Figure 3.4 on p. 82), each of which corresponding to a single relaxation mechanism. Each of these peaks has the characteristic $Q^{-1} \propto f^1$ and $Q^{-1} \propto f^{-1}$ roll-offs on its flanks, and consequently the entire absorption band is also expected to have such properties (Anderson and Given, 1982).

Although borrowed from atmospheric and plasma physics, the seismic absorption-band model lacks an important component of these theories, which is the mechanical resonance causing the absorption of wave energy. Frequency-selective absorption occurs in gases and plasma when there exist internal resonances of their molecules to which the external field become tuned. A well-known example is the resonance with the 2450-MHz natural oscillations of water molecules used in microwave ovens. However, in seismology, there seem to exist no resonances in the commonly used 10^{-4}–10^2 Hz frequency band, apart from several rare and specific cases of periodic layering found within sedimentary sequences and sometimes near the Moho. The corresponding range of seismic wavelengths is from about 30 m to 100 km, which is the scale-length range for the crustal- to whole-Earth structures, and not for microscopic relaxation mechanisms. Thus, although phenomeonologically, the absorption-band model (Figure 7.14) is valid, it appears that we need to look at the structural factors for its physical explanation. In our approximation, such structural factors are represented by the geometrical term, γ, in the attenuation coefficient.

The observed mantle absorption band can be explained without rheological relaxation, and as a geometrical effect of transition from long-period to short-period observations. Let us assume that Q_e values are the same in both cases, but geometrical corrections are different. Therefore the attenuation coefficient equals, $\chi_{LP} = \gamma_{LP} + \pi f/Q_e$ at long periods and $\chi_{SP} = \gamma_{SP} + \pi f/Q_e$ at short periods.

For long-period surface waves, γ_{LP} is negative, and consequently the apparent long-period Q increases toward lower frequencies:

$$Q_{LP}(f) = \frac{\pi f}{\chi_{LP}} = \frac{Q_e}{1 - f_{c,LP}/f},$$ (7.12)

where the cross-over frequency $f_{c,LP}$ equals $(-\gamma_{LP}Q_e/\pi)$. For Rayleigh waves, $f_{c,LP} \approx 1.5 \cdot 10^{-3}$ Hz, which is significantly lower than the typical surface-wave measurement frequencies. Therefore, $Q(f)$ steeply increases when frequency drops to $f_{c,LP}$ (gray lines in Figure 7.15). For short-period body waves, γ is positive, and the corresponding $f_{c,SP} = \gamma Q_e/\pi$. Consequently, the short-period $Q(f)$ is:

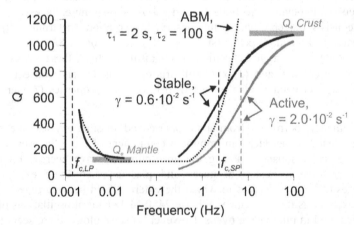

FIGURE 7.15
Apparent absorption band formed by geometrically over-corrected long-period surface waves (cross-over frequency $f_{c,LP}$) and under-corrected short-period body waves ($f_{c,SP}$). For $f_{c,SP}$, two versions of are given, corresponding to the areas of stable (black lines) and active tectonics (gray). Horizontal gray bars indicate the levels of Q_e defining the band (labeled). Dotted line is the mantle absorption band model (ABM) (Anderson and Given, 1982), with relaxation times τ_1 and τ_2 indicated.

$$Q_{SP}(f) = \frac{\pi f}{\chi_{SP}} = \frac{Q_e}{1 + f_{c,SP}/f}. \tag{7.13}$$

With the typical crustal values of $Q_e \approx 1100$ and $\gamma_{SP} \approx 0.006 \text{ s}^{-1}$ for tectonically stable areas and $\gamma_{SP} \approx 0.02 \text{ s}^{-1}$ for tectonically active ones (Section 7.3), we obtain $f_{c,SP} \approx 2.8 \text{ Hz}$ and 7.8 Hz, respectively. Expression (7.13) then shows that $Q_{SP}(f)$ increases with frequency at short periods, and consequently, the two characteristic frequencies $f_{c,LP}$ and $f_{c,SP}$ delineate a band of reduced apparent $Q(f)$ (Figure 7.15). Note that the character of this band is purely geometrical, and its attenuation levels are controlled by only two frequency-independent values of $Q_e = 1100$ (within the crust) and $Q_e = 121$ (within the uppermost mantle) shown by gray bars in Figure 7.15.

Let us also compare the predictions from expressions (7.12) and (7.13) to the mantle absorption band model (ABM) (Anderson and Given, 1982). In the ABM, the high-attenuation Q level is controlled by a fixed minimum of $Q_m = 80$, which is analogous to Q_e in our interpretation. However, in the ABM, Q_m is assumed to be constant within the entire mantle, whereas Figure 7.15 shows that Q_e should mostly be formed by the increased attenuation within the relatively thin and low-Q subcrustal mantle. Such distribution of Q_e would also explain the near-constant values of t^* observed for body waves (Section 6.3).

For $f \gg f_{c,LP}$, eq. (7.12) yields Q values which are inversely proportional to f: $Q_{LP}(f) \approx Q_e f_{c,LP}/f$. These values correspond to the low-frequency flank of the ABM with cut-off parameter $\tau_2 = Q_e/f_{c,LP} = -\pi/\gamma_{LP}$. For body waves at $f \ll f_{c,SP}$, $Q(f)$ is nearly proportional to the frequency (eq. 7.13): $Q_{SP}(f) \approx Q_e f/f_{c,SP}$. Such $Q \propto f$ dependence forms the upper flank of the absorption band, in which parameter τ_1 becomes $\tau_1 = Q_e/f_{c,SP} = \pi/\gamma_{SP}$. Therefore, quantifies τ_1 and τ_2 defined by Anderson and Given (1982) as the cut-off levels in the Debye distributions of relaxation times can instead be explained as reciprocals of the residual geometrical-spreading parameters for body and surface waves, respectively. Note that as shown in Section 2.5.3, these relaxation times can be viewed as fictitious parameters only needed to specify the positions of the flanks of the absorption band. These flanks are controlled by parameters γ in our model.

Note that our model also shows that the short-period flank of the apparent absorption band becomes higher (Q_0 increases) and flatter (η decreases) with tectonic age because of the decreasing $f_{c,SP}$ and γ (Figure 7.15). For comparison, the ABM accounts for the crust by setting the value τ_1 equal zero. The present model also readily incorporates the differences between the oceanic and continental lithospheres by including the corresponding γ and Q_e values in its short-period flank (Figure 7.15).

Thus, the attenuation-coefficient formulation allows constructing a simple model of mantle attenuation in which the absorption band is presented as apparent and related to different geometrical-spreading patterns of the long-period and short-period waves. Further comparison of this model to the ABM is difficult, principally because of the over-parameterized nature of the latter. The ABM uses multiple wave modes (all of them with likely variable geometrical-spreading) and depth-dependent parameters (τ_1, τ_2) in combination with a constant Q_m, and various other assumptions. Potentially, some of these features could be incorporated in a more refined geometrical-band model, which is outlined in the next chapter.

7.7 Frequency dependence of attenuation

Throughout this book, we returned to the problem of the frequency dependence of attenuation parameters several times. The issue of frequency dependence is indeed the one which makes discussions interesting, and also the one providing the link to the mechanisms and fundamental properties of attenuation. This issue is also central to the ongoing discussion (Morozov, 2009a, 2009b, 2010b, 2010c; Xie and Fehler, 2009; Xie, 2010). The key difficulty in solving this problem is in deciding whether Q is an adequate parameter to describe the attenuation, and whether it truly depends on the frequency or on other experimental factors.

Based on the results of this study, this issue appears to be resolved, but at the cost of derailing the concept of Q. If we use the proposed parameters γ_i and κ_i, both of them are frequency independent within the accuracy of the available datasets. Theoretical studies also indicate several end-member cases in which these parameters are frequency independent (Section 5.6). However, such "negative" results certainly cannot be absolutely conclusive. Biot's (1962) theory, and also another hypothetical example in Section 2.4.2 show that κ_i can generally be frequency dependent. Further, in practical measurements involving variable penetration depths and changing wave-mode content, γ_i may also vary with observation frequency (Section 7.6.5).

While discussing the dependence of γ_i, κ_i, Q_0, η, or other attenuation parameters on the frequency of observation, it is important to differentiate their apparent variations concurrently with varying frequency, f, from their physical dependence on f. The former, apparent correlation occurs quite often, commonly because of changing wave types, such as normal-mode orders, each of which has their own frequencies and Q's. The latter, physical dependence is extremely difficult to isolate from the apparent trend above, and it could only be likely found in $\kappa_i(f)$ dependences. However, up to this point, no reliable indications of such dependences have been found. Given the complexity of other aspects of the attenuation problem, such dependences may hardly ever be required to explain seismological data. Following the famous Laplace's quote, "I had no need of that hypothesis[21]."

On the other hand, if the question of frequency dependence is understood in the narrow context of Q, then it is certainly frequency dependent. Considering that by virtue of its definition, Q "naturally" scales with frequency from f^0 to f^1 (Section 3.8.2), it is not surprising that an average trend of about $f^{1/3}$ can be found in most data. However, this general trend ignores much of the fine structure seen on the attenuation-coefficient distributions (see Figures 7.2 and 7.11).

In concluding the subject of frequency dependence, it appears that the problem clarifies greatly when looked at without the prejudice of the model of Q.

---

[21] "Je n'avais pas besoin de cette hypothèse-là," in reply to Napoleon's question why God had not been mentioned in Laplace's book on astronomy.

Chapter 8

Epilogue:

Challenges Ahead

Where is the beginning of the end that comes at the
end of the beginning?

Kozma Prutkov, Fruits of Reflection (1853-54)

To summarize the preceding argument, it appears that the general reason
for the pervasive use of Q in attenuation studies is in an exaggerated reliance on
the theory. Theoretical models loom through the entire process of analysis, starting
from the initial data reduction to inversion and interpretation. At the same time,
despite their mathematical refinement and application to hundreds of datasets,
many models are still not rooted in the fundamental principles of theoretical
mechanics, but rather represent analogies and heuristic generalizations of a limited
number of simple solutions. Frequency- and depth-dependent attenuation models
are also often over-parameterized, which makes them very permissive with respect
to "practical assumptions" that are often poorly verifiable by the data.

Attenuation measurements and the resulting models should be described in
terms of adequate physical quantities which are independent of invalid theoretical
assumptions. In particular, one needs to clearly differentiate between the apparent
(wave) Q (sometimes expressed by t^* or attenuation coefficient) and *in situ*
("medium") Q. Detailed analysis of the conventional measurement procedures

shows that such differentiation in many cases may be complicated by inaccurate knowledge of the geometrical spreading, *i.e.*, again by the reliance on models.

As a solution to this problem, Morozov (2009a) suggested relying on conservative analysis of the data without constructing *a priori* models. The attenuation-coefficient approach offered well-proven and reliable tools for such an analysis. By looking at a variety of datasets without preconceived models, linear patterns of frequency-dependent attenuation coefficients were recognized within several frequency bands from ~500 s to 100 Hz.

Noting the linear patterns in $\chi(f)$ causes major changes in interpretations of attenuation effects in a variety of seismological applications. From independent point-by-point data fitting of seismic amplitude data, we now pass to interpreting the causes of systematic trends in $\chi(f)$ within certain frequency bands. Realization of such trends allows us to reduce the data errors and to isolate various side effects, such as oscillatory behaviors of amplitude spectra due to tuning, multiples, and other resonant effects within the crust. Linearity of the trends (within data errors) observed so far strongly suggests that in reality, frequency-dependent attenuation effects are not nearly as pronounced as most reported $Q(f)$ studies have suggested so far. Frequency-independent κ_i (or Q_i, for those preferring the Q terminology) models were sufficient for describing all observations reviewed in this book. Thus, although frequency-dependent κ_i is possible and even likely for several attenuation mechanisms, the observations still appear to be insufficiently complete and accurate to resolve such frequency dependence.

Identification of the model-independent parameters of attenuation (γ and κ) allowed making a series of observations that were not apparent in the old $Q(f)$ form. The geometrical-attenuation parameter γ was found to be positive for lithospheric body and surface waves, and correlated with tectonic types and ages. It was also predictable by waveform modeling. "Cross-over" frequency $f_c = \gamma/\kappa$ became a convenient parameter showing whether the structural (geometrical) or attenuative effects are dominant for the specific observations in a particular area. Thus, the $\chi(f)$ picture is very quantitative and precise compared to ambiguous and model- and convention-dependent $Q(f)$ descriptions.

The key theoretical message of this book is that the visco-elastic theory is too simple and too preoccupied with its own paradigm for describing the broad variety of attenuation observations. We still need to look for a more complete theory which would specifically address the physical processes taking place during wave propagation within the deep Earth. Judging by the common acceptance of the visco-elastic theory at all scales and, at the same time, our sharp critique of it here, even the basic ideas of such new theory are in question. Although most of the current observations and visco-elastic ideas should of course be incorporated in the new view, much work is nevertheless requited for putting it together. This book only suggests some basic principles for guiding this work and offers some tentative solutions.

The views with respect to the Q and the visco-elastic theory expressed in this book are rather drastic, and unfortunately, they suggest critical revisiting many other attenuation results published in the past 30 to 40 years. This would be a formidable task to perform in a single study, and we have to cut it off at some point. This concluding Chapter therefore presents no finished results; instead, it contains outlines of the most important unsolved problems and sketches of their potential solutions.

8.1 Physics of attenuation and model parameterization

The most serious concern arising from the above discussions is the lack of an adequate theory of attenuation. Generally, two types of theories can be used to describe a physical process: 1) phenomenological or 2) from first principles, or *ab initio*. Phenomenological theories grasp the entirety and complexity of the process, identify its essential constraints, and present ways for explaining and predicting the observations without knowing the details of all interactions. *Ab initio* formulations provide the foundation of the fundamental principles of mechanics, although they may allow practical solutions in only a limited number of end-member cases. A combination of these two approaches, which is strong phenomenology based on clearly understood physics, represents the correct physical picture.

Unfortunately, in the existing theory of seismic attenuation, we observe a reversed picture. The formal theory of visco-elasticity is based not on the *ab initio* principles of mechanics, but on a phenomenology extrapolated to the elementary level (see Section 3.5). Phenomenological descriptions rely not only on physical observations but also on subjective "practical assumptions" and "*a priori* models." As a result, both descriptions originate mostly from theoretical postulates, such as the local material memory, correspondence principle, or scattering, which they are unable to verify.

As illustrated in the preceding Chapters, several fundamental points need to be established before a consistent physical model of mantle attenuation is formulated:

1) The source of energy dissipation (*e.g.*, the kinetic or elastic energy, or separate mechanisms) needs to be clarified. Harmonic oscillatory wave-equation solutions commonly used in visco-elasticity do not help in answering this question. Parameters describing this source need to be specified and related to the observable characteristics of the wavefield. The quality factor, Q, should hardly be among such parameters. The only physically viable constitutive theory known today (Biot, 1962) neither needs an *in situ* rock Q nor leads to such a property.

2) A formulation of equations of motion needs to be found. Again, the visco-elastic differential equations only describe the oscillatory and attenuation solutions *post factum*, but we would need a dynamic principle from which such equations can be derived. The Lagrange-Hamilton variational principle represents a well-recognized way for defining such a principle, and therefore a Lagrangian function needs to be found. Alternately, a dynamic principle could be derived by using homogenization techniques, but again based on microscopic Lagrangian or similar mechanical description of the propagating medium.

3) The attenuation needs to be treated separately and no longer viewed as a complex-plane rotation of the medium velocity (as in the correspondence principle; see Section 2.5.4) The elastic constants should remain real-valued, and local creep functions should disappear from stress-strain relations. The traditional "axiomatic" visco-elastodynamics would become limited in its scope, which would likely reduce to uniform-space or quasi-static creep problems.

Thus, we have to conclude that at least in our opinion, there currently exists no satisfactory theory of attenuation in the "seismological" range of distances, physical conditions, and frequencies. The problem of an *ab initio* physical theory is particularly difficult, yet it is also critical to all other aspects of this research. Biot's (1956) theory gives an example of a well-constructed physical theory of wave attenuation in saturated porous rock which could serve as a prototype, yet this model of free fluids within pores in a solid rock matrix is hardly applicable to the lower-crustal and mantle conditions. The scale-lengths of mantle-wave paths are also vastly different from those used in seismic exploration and ultrasonic lab experiments.

8.2 Reconciling seismological and lab measurements

Reconciling the seismological field and lab measurements is another major and very difficult task (Bourbié *et al.*, 1987). Until now, this task is performed intuitively, assuming that the "material Q" inverted from lab-data analysis corresponds to the "*in situ* rock Q" interpreted from the observations of seismic waves. However, as argued above, neither of these Q values exist. In both cases, the quality factor represents a property of the measurement environment, which may be a lab apparatus or some type of a seismic wave. The resulting value of Q is largely controlled by the structure of the environment, such as the design of the apparatus, type of measurement, or the dimensions and heterogeneity of the structure in which the "*in situ*" wave-Q measurement is performed. As we have seen, many of these factors are poorly understood but can be described by the concept of phenomenological "geometrical spreading".

Seeking a true relation between the attenuative properties of materials used in various lab and *in situ* measurements, a vast range of pressures and scale lengths has to be spanned, which can again be only done based on some solid theory. It is important to use the correct theory, or at least the type of theory which is not critically dependent on the "simplifying" assumptions. As shown in Chapter 2, visco-elasticity appears to be inadequate as such a theory. Visco-elastic models can be constructed for virtually any process exhibiting energy dissipation; however, they are not fundamental enough for correlating the lab and field wave measurements.

It appears that the best way to link the broad variety of attenuation measurements together is to use some form of the Lagrangian mechanics discussed in Chapter 2. An instructive example of such theory was given by Biot's (1962) theory of porous saturated medium. One general lesson from this theory is that the correct mechanical description of attenuation should be specific to the structure and composition of the propagating medium, and it should involve multiple physical properties, some of which may be difficult to assess.

If not based on the concept of Q and the theory of visco-elasticity, reconciling the lab and field observations of seismic attenuation becomes a wide-open question. To answer this question, detailed models need to be constructed from first physical principles for each of the specific observations. Many models and data examples in this book show that, at least in some cases, such models can be based on the local intrinsic attenuation coefficients, α_i or χ_i. As we have seen, using these coefficients is not the same as using Q, and their use often leads to dramatic changes in the interpretations. However, the question of whether even these properties derived, for example, from a lab specimen can be related to those measured from seismic waves propagating through the Earth's mantle remains unsolved.

Generally, we can expect that the "geometrical," or "structural" parts of the intrinsic attenuation coefficient, γ_i, should be practically unrelated in different field and lab measurements. At the same time, under some *favorable conditions*, their frequency-dependent parts, κ_i or Q_i, may correlate in these measurements. Such favorable conditions would include the frequency-independent geometrical spreading and κ_i. For observations with seismic waves within different frequency bands (Chapter 6), these conditions seem to be satisfied at least for the data considered above. However, again, correlating the results to the different oscillation types, scale-lengths, pressures, temperatures, and frequencies used in lab measurements represents spanning a much broader gap. Extensive theoretical research is required in order to evaluate the quantitative effects of this gap.

8.3 Forward kernels

Before looking into the effects of the new theory on the forward and inverse attenuation problems, let us briefly review the existing approach. For a specific example, we will use the whole-Earth normal-mode decomposition described in Chapter 9 of Dahlen and Tromp (1998)[22]. Similar to the rest of the current theory, this approach is based on the visco-elastic Q paradigm. Among several mathematical formulations of this problem offered by Dahlen and Tromp (1998), the Rayleigh variational method is the closest to the Lagrangian approach taken in the present book.

For simplicity of notation, consider a radially symmetric Earth described by its density, ρ, shear modulus, μ, bulk modulus, K, and depths to several important discontinuities, d. The attenuation is described by the inverse shear and bulk attenuation factors, denoted Q_μ^{-1} and Q_K^{-1}, respectively. All these quantities except d are functions of radius. For any given distribution of parameters, there exists a spectrum of normal-oscillation modes, with each mode having a characteristic spatial distribution of amplitudes and the corresponding "eigenfrequency," ω. The modes do not interact with each other, but change when the parameters of the medium are modified. The basic forward problem consists in deriving the variations of ω and the attenuation coefficient, χ, as functions of model-parameter variations, $\delta\rho$, $\delta\mu$, δK, and δd.

To evaluate the sensitivity of ω to the variations in model parameters, we start from the equipartitioning equation:

$$\omega^2 T = U , \qquad \text{(2.113 repeated)}$$

and perturb both its sides,

$$2\omega T \delta\omega + \omega^2 \delta T = \delta U , \qquad (8.1)$$

yielding the desired perturbation of the natural frequency of the mode,

$$\delta\omega = \frac{1}{2\omega}\left(\delta V - \omega^2 \delta T\right). \qquad (8.2)$$

The energy perturbations on the right-hand side of this equation represent integrals over the volume of the model and are considered as linear in terms of model parameters. Therefore, this equation can be written as:

---

[22] Unfortunately, because of collision in notation, we have to rename some of the variables used in Dahlen and Tromp (1998): α to V_P, β to V_S (wave velocities), V to U (potential-energy kernel), γ to χ (apparent attenuation coefficient), κ to K (bulk modulus), and K to F (Fréchet kernels).

$$\delta\omega = \int_0^R \left(F_K \delta K + F_\mu \delta\mu + F_\rho \delta\rho \right) dr + \sum_d \left[F_d \right]_-^+ \delta d \,, \qquad (8.3)$$

where F_K, F_μ, F_ρ, and F_d are the Fréchet kernels corresponding to the respective parameters, R is the Earth's radius, and notation $[\phi]_-^+$ means taking the difference of ϕ values across the discontinuity. The Fréchet kernels therefore represent functional derivatives of ω in respect to the three continuous variables,

$$\frac{\delta\omega}{\delta K} = F_K \,, \quad \frac{\delta\omega}{\delta\mu} = F_\mu \,, \quad \frac{\delta\omega}{\delta\rho} = F_\rho \,, \qquad (8.4a)$$

and an ordinary partial derivative with respect to d,

$$\frac{\partial\omega}{\partial d} = \left[F_d \right]_-^+ . \qquad (8.4b)$$

From eq. (8.2), for each of parameters $p = K$, μ, or ρ, these functional derivatives are related to the derivatives of the normalized elasto-gravitational and kinetic energies,

$$F_p = \frac{1}{2\omega} \left(\frac{\delta V}{\delta p} - \omega^2 \frac{\delta T}{\delta p} \right), \qquad (8.5)$$

with a similar relation with respect to d.

Equations (8.3) and (8.4) give the linearized forward problem for the response of ω to perturbations of mechanical parameters within the Earth. Now, how can one proceed to perturbations involving attenuation? As we saw in Section 2.6, the Lagrangian of the normal mode, $L(u,\dot{u}) = \omega^2 T \dot{u}^2 - V u^2$, does not include attenuation. The current solution is to use the visco-elastic assumptions (Section 3.3) and to extend eqs. (8.3) and (8.4) according to the following procedure (Dahlen and Tromp, 1998; p.347–350).

First, a reference, or fiducial frequency ω_0 is introduced, to serve as a calibration parameter in phase-velocity dispersion equations below. For the PREM model, this frequency is selected as equal to 1 Hz, which was the natural frequency of the instruments broadly used by the World-Wide Standardized Seismographic Network (WWSSN, precursor of today's Global Seismographic Network). Therefore, by convention, $\omega_0 = 2\pi$ Hz; however, it is important to keep in mind that this is still a totally arbitrary number in relation to the Earth's oscillations. Thus, in the end, we should carefully verify that the "observable" predictions of the model do not depend on ω_0^{23}.

Further, the attenuation is interpreted as caused by infinitesimal perturbations in the bulk and shear moduli:

$$K_0 \to K_0 + \delta K_0 + \delta q_K \left(\frac{2K_0}{\pi} \ln \frac{\omega}{\omega_0} + iK_0 \right), \tag{8.6a}$$

$$\mu_0 \to \mu_0 + \delta \mu_0 + \delta q_\mu \left(\frac{2\mu_0}{\pi} \ln \frac{\omega}{\omega_0} + i\mu_0 \right), \text{ and} \tag{8.6b}$$

$$\rho_0 \to \rho_0 + \delta \rho_0. \tag{8.6c}$$

In these expressions: $\delta q_{K,\mu} = Q_{K,\mu}^{-1}$ are the *in situ* attenuation factors associated with the bulk and shear moduli (assuming the unperturbed $Q_{K,\mu}^{-1} = 0$); quantities K_0 and μ_0 correspond to frequency ω_0, and δK_0; and $\delta \mu_0$, and $\delta \rho_0$ denote the attenuation-free perturbations of elastic parameters. The first terms in parentheses in eqs. (8.6a and 8.6b) describe the real-valued shifts in K and μ caused by dispersion[24], which in itself is required by causality (Section 2.2.2)[25].

Finally, the normal-mode frequency also becomes complex and decomposed as:

$$\omega \to \omega + \delta \omega_0 + \delta \omega_d + i\chi, \tag{8.7}$$

where $\delta \omega_0$ is the attenuation-free frequency shift, $\delta \omega_d$ is the shift caused by dispersion, and χ is the logarithmic amplitude decrement, *i.e.*, the temporal attenuation coefficient of the selected normal mode. Assuming that the forward

---

[23] However, to the author's knowledge, this is never verified in the present theory.

[24] Clarification needed here. As shown in Section **2.2.2**, this particular form of dispersion law represents only one possible model. The causality conditions do not constrain $k(\omega)$ to any specific functional form.

[25] Note that this dispersion correction was derived for a continuous spectrum of ω, $Q(\omega)$ = const, and the corresponding wavenumbers, $k(\omega)$. Because we are working with a single normal mode here, this correction does not really apply. Causality only requires that oscillations vanish at negative times, which means that the normal-mode ω (and not K) should be shifted in the presence of attenuation. The eqs. of (8.6) (8.6) represent an attempt for achieving these shifts by means of eq. (8.38.3), but it still does not guarantee causality.

model (8.3) is valid for complex-valued fields $K(r)$ and $\mu(r)$, the predicted attenuation coefficient is obtained:

$$\chi = \int_0^R \left(K_0 F_K \delta q_K + \mu_0 F_\mu \delta q_\mu \right) dr, \qquad (8.8)$$

where the last term related to the discontinuities in eq. (8.3) was dropped as relatively insignificant. Thus, the Fréchet kernels for q_K and q_μ simply equal $K_0 F_K$ and $\mu_0 F_\mu$, respectively. Solving for normal-mode attenuations is practically equivalent to solving for their frequencies.

However, as shown throughout this book, we can be satisfied neither with this derivation nor with its result. Extrapolation of eq. (8.3) to the complex-elastic field domain assumes its holomorphism (analyticity), which one can hardly assume for functions given by arbitrary integral transforms. The use of complex-valued elastic energies removes the foundation of the Lagrange-Hamilton variational mechanics on which eq. (8.3) was originally based. Most importantly, this model uses only two attenuation parameters, q_K and q_μ, heuristically associated with the two elastic moduli, and many potential physical factors causing attenuation become ignored (see Section 3.5). In particular, gradients of particle velocities are among the primary causes of friction, yet they produce no direct effect in eq. (8.8).

Fundamentally, the problem of the visco-elastic model is in its treatment of boundary conditions. Visco-elasticity achieves a compact description of attenuation, which reduces to only one Q-factor per wave, by assuming that boundary conditions include both the elastic and inertial forces across the boundary (Section 2.7.1). Unfortunately, this assumption fails in the problem of the anelastic acoustic impedance (Section 2.7), which appears to be the only case where this assumption can be tested analytically. If these implicit boundary conditions would be changed, terms proportional to $d\delta q_{K,\mu}/dr$ or altogether different Fréchet kernels should appear in the integrand of eq. (8.8).

The difficulty of arguing either for or against solution (8.8) is that like most visco-elastic solutions, it nearly always predicts an oscillatory behavior of the field decaying with time. Based on this simple property, the solution may appear acceptable. Solution (8.8) is also broadly used and has become a part of the standard seismology curriculum. However, exponentially decaying functions of time represent practically the only type of fields considered in linear visco-elasticity, and the solutions constructed in such ways are still not guaranteed to be physically correct. To ensure correctness, we need to rely on the first physical principles rather than on mathematical extrapolations.

A solution alternate to eq. (8.8) and based on the physical principle of conservation of energy was given in Section 6.1.2. As shown there, there still remains a good deal of uncertainty about the actual mechanics of attenuation within the mantle and the shapes of the kernels. The proposed solution was based on the constraint of the total dissipated energy from within the field being equal to the sum of energies dissipated from its parts. The visco-elastic solution (8.8) was shown to violate this constraint, and the predictions for Love-wave Q resulting from these solutions differed by 10–20% for the same model of the mantle (Figure 6.8). This example shows that the forward problem of parameterizing mantle attenuation and predicting its levels on the surface still presents a major theoretical challenge.

8.4 Inversion

Once the forward problem is formulated, the corresponding inverse problem consists in deriving perturbations of the Earth-model parameters from the values of χ observed at several frequencies. In terms of the forward problem (8.8), an appropriate discretization, for example, such as described in Section 6.1.2, needs to be constructed for parameters δq_μ and δq_K, and the problem rendered in matrix form,

$$\chi = \mathbf{F} \delta \mathbf{q},\tag{8.9}$$

where vector $\delta \mathbf{q}$ comprises all model parameters, vector χ includes all observations of the temporal attenuation coefficients, and $\mathbf{F}$ is the matrix of all discretized Fréchet kernels. Even in the simplest, frequency-independent attenuation cases, this problem is often mixed determined, and needs to be solved by some kind of "generalized linear inverse" (Menke, 1989):

$$\delta \mathbf{q} = \mathbf{F}_g^{-1} \chi.\tag{8.10}$$

Not surprisingly, our disagreement with the forward attenuation problem (8.8) should also affect the inversion. Both the parameterization of vector $\delta \mathbf{q}$ and kernels $\mathbf{F}$ should change seriously. To recapitulate the perceived problems of the existing parameterization with respect to the inversion, they appear inadequate because of:

1) Using Q as an *in situ* measure of the seismic attenuation of the medium;

2) Reducing a variety of attenuation mechanisms to only two parameters Q_K and Q_μ;

3) Heuristic association of these Q_K and Q_μ with time-retardation properties of the elastic parameters of the medium;

4) Dependence on further attributes of the visco-elastic model, such as the fiducial frequency ω_0 or the selected forms of dispersion relations; and

5)   Rather too freely allowed frequency dependence of most parameters.

The first four of these issues lead to parameterizations that are both unstable (e.g., trade-off with the assumed background models or reference parameters) and too restrictive in following the visco-elastic theory. At the same time, issue 5) leads to underconstrained inversions, which in turn, need to be regularized using further assumptions and *ad hoc* parameters, such as the "global" $Q \propto f^{1/3}$ trend (Figure 7.11).

A solution proposed in this book (Chapter 5 and Section 6.1.2) is to use a parameterization reflecting the observed linear $\chi(f)$ patterns (Figure 7.2 on p. 272):

$$\chi_i(f) = \gamma_i + \kappa_i f, \tag{8.11}$$

where γ_i and κ_i are the internal parameters of the medium. From the results of Chapters 6 and 7, this frequency-independent γ_i and κ_i should be sufficient for explaining the available data. This model is further explained in Section 8.6 below.

The inverse kernels $\mathbf{F}_g^{-1}$ expected from the new theory should differ from those used in the current visco-elastic kernels in two ways. First, as illustrated in Section 6.1.2, the new predicted Q for surface can be expected to be ~10–20% higher than the one resulting from the visco-elastic model. This suggests that an inversion based on these kernels would lead to similar decreases in the Q values within he mantle, and predominantly within the low-Q layer in the upper mantle.

The second, principal difference of the new inverse kernels should be in their shapes. The sensitivity kernels for χ_i differ from those for frequency- or phase-velocity sensitivity (Figure 8.1). For example, note that the observed 60-s Love-wave Q_L^{-1} is most sensitive to the near surface, where particle velocities (and therefore friction and pore/fault fluid flows) are the fastest (solid black line in Figure 8.1). This is different from the phase velocity, which is most sensitive to the depth near the base of the crust, where the elastic strain is the strongest (dashed black line in Figure 8.1). Although in the present view, this difference appears natural, it contradicts the traditional assumption of Q^{-1} responding to the same structures as the velocity (see Section 6.1.2). Consequently, inversions for the attenuation structure should no longer be similar to velocity tomography.

Unfortunately, it appears that the full inverse problem is going to be much more complex than based on the existing visco-elastic models. Most importantly, it should focus on the specific attenuation mechanisms within the different parts of the Earth, which are poorly understood at present. The example in Section 6.1.2 suggests a simplified way to proceed; however, it still needs to be substantiated and extended to other wave types. Therefore, apart from the general inverse approach of eqs. (8.9) and (8.10), we still have a long way to go before solving the inverse problem of the seismic attenuation within the Earth.

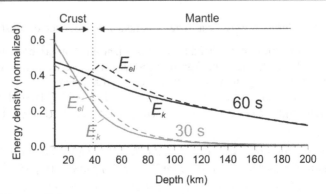

FIGURE 8.1
Part of Figure 6.11, showing distributions of the kinetic (E_k, solid lines) and elastic (E_{el}, dashed lines) energy density for the fundamental Love-wave modes at 60-s (black) and 30-s periods (gray).

8.5 Numerical waveform modeling

Numerical waveform simulations are used in practically all types of seismic interpretation, and it is therefore important that numerical equations accurately reflect the physics included in the theory. All existing numerical schemes for modeling seismic wavefields in attenuative media are based on the visco-elastic approach which is being criticized in this book. Therefore, we cannot use numerical modeling to verify the theoretical statements of Chapters 2 to 7. This is important to keep in mind, given today's tendency for using numerical simulations in theoretical arguments. Instead, we face a major problem of developing an alternate numerical simulator.

Visco-elastic equations are readily amenable to numerical implementations, especially in the frequency domain, in which only shifting of the elastic moduli into the complex plane is required to produce attenuation. Compared to the reflectivity and other effects of velocity-density structure, the effects of attenuation are usually comparatively small and only judged by the presence of a frequency-dependent amplitude decay. Such decay is always reproduced by visco-elastic modeling. However, problems with this modeling become noticeable in the phases of reflections from attenuation contrasts (Figure 8.2). As discussed in Section 2.7, compared to the exact analytical solution (Lines et al., 2008), visco-elasticity predicts an opposite phase shift for a reflection from a contrast in Q. This effect is confirmed by numerical modeling using the popular 1-D numerical propagator (reflectivity) method by Fuchs and Müller (1971) (Figure 8.2). Note that, because of the absence of an accurate numerical solver, the solution shown as "simulated exact" in Figure 8.2 was obtained from the visco-

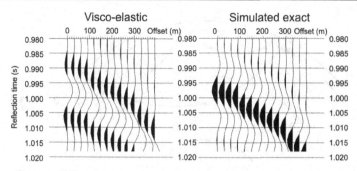

FIGURE 8.2
Numeric modeling of a reflection from a pure attenuation contrast by a visco-elastic approach (Fuchs and Müller, 1971; left) and simulated exact (right). Note the opposite polarities of reflections.

elastic solution by an additional 180° phase rotation, only in order to illustrate the expected difference.

Unfortunately, an alternate numerical method similar to the solution for Love-wave Q in Section 6.1.2 would be difficult to build, because that solution utilized averaging and equipartitioning of energy, which is insufficient for detailed modeling of the entire wavefield. Clearly, a complete variational formulation of the type described in Section 2.4.2 is required; however, we only offered two conceptual examples there, and an exact solution would need to utilize more specific attenuation mechanisms. Thus, in principle, once an acceptable Lagrangian model of the medium is formulated, there should be no difficulty in implementing it in numerical algorithms. As above, the principal, serious difficulty is in developing an acceptable theoretical model of attenuation within the mantle.

8.6 Attenuation model for the Earth

Revising all of the existing models would be a drastic undertaking. At this stage, this task appears to be of insurmountable difficulty, because even the physical mechanisms governing the seismic attenuation within the mantle are unclear and their mathematical descriptions still need to be established. Therefore, at present, we cannot go beyond outlining only relatively schematic, first-order features of this model. Without any detailed inversion, the data appear to suggest the existence of pronounced layering of attenuation within the upper mantle. Such layering is present in all existing mantle Q models and, as shown in Section 7.5, the low-Q layer in the upper mantle may be responsible for most of the attenuation observed on the surface. A more detailed inversion would be difficult, because it would require resolving discrepancies the available data compilations (Section

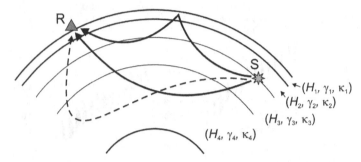

FIGURE 8.3
Schematic phenomenological model of layered mantle
attenuation. S – source and R – receiver. Attenuation properties
within the layers are described by parameters H, γ, and κ. Thick
solid lines show the "specular" (true) wave paths, dashed line is a
"virtual" (off mass shell) ray.

6.2.3) and take into account multiple modes, for example, similar to Durek and
Ekström (1996). However, doing this at present appears difficult.

A phenomenological approximation capable of explaining the available
seismic-attenuation data with reasonable physical sense was given in Chapter 5
and applied to layered Earth in Section 7.5. In a condensed form, this
approximation can be further generalized as follows.

The Earth is considered to have a predominantly layered structure. Within
the n-th layer, the wave attenuation is characterized by parameters H_n, γ_n, and κ_n,
which may also depend on the coordinates and be correlated, for example, to the
tectonic types or other parameters of the regions to which they belong (Figure
8.3). For body-type waves, parameter H_n describes the background geometrical
spreading derived, for example, by solving the dynamic ray-theoretical equations
(see Section 5.6.1). The geometrical spreading can be represented by a path
integral, and H then has the meaning of the modeled wavefront curvature:

$$G_0\left(R\,|\,S\right) = \exp\left(-\int_S^R H ds\right). \qquad (8.12)$$

Note that in principle, H can be frequency dependent in order to incorporate, for
example, critically refracted waves, although this would involve a step beyond the
asymptotic ray theory.

Further, for a given ray, the path factor should be different from the model
one, which is reflected in a scattering-theory correction:

$$P(R,S \mid \mathfrak{R}) = \exp\left\{ -\int_{\mathfrak{R}} \left[H + \frac{1}{V}(\gamma_i + \kappa_i f) \right] ds \right\},$$ (8.13)

where $\mathfrak{R}$ is the ray connecting the source S and receiver R (Figure 8.3). In this expression, the frequency-independent and dependent parts of the correction are explicitly separated as the intrinsic γ_i and κ_i parameters discussed in Chapter 5. The attribution of these parameters to specific physical effects (such as scattering, absorption, diffraction or additional ray-bending) requires careful investigation. Up to this point, these quantities are only treated as phenomenological path corrections. The question of whether these parameters may depend, for example, on the angles of ray propagation or wave types is also not obvious and deserves much attention in the future. From the preceding chapters, it appears that κ_i, when interpreted as caused by anelastic energy dissipation, can be regarded as common, for example, to body and surface S-waves passing through the same area. At the same time, the values of γ_i critically depend on the wave geometry and should be different for all wave types.

Finally, the most general form for representing the total recorded amplitude is by using the functional path integral:

$$A^2(R,S) \propto \int_{\text{all paths}} W[\mathfrak{R}] P^2(R,S \mid \mathfrak{R}),$$ (8.14)

where the integration takes place over all paths connecting R and S. This integration may include "non-specular" rays, i.e., those not obeying the Snell's laws of refraction and reflection, such as the one shown by a dashed line in Figure 8.3. Factor $W[\mathfrak{R}]$ in this expression gives some weight function in the functional space of ray paths selecting the paths. Determination of this weight may also be difficult. For example, one approach could be to make it similar to the "banana" kernels used in finite-frequency tomography (Dahlen et al., 2000). The integration of squares of path factors corresponds to incoherent summation of wave-energy fluxes.

Equations (8.12) to (8.14) give a sketch of a phenomenological ray theory that would naturally incorporate most of the examples in this book. A similar theory can be constructed for non-ray based waves, such as the model for surface waves in Section 6.1.2. This theory, however, is far from being complete and requires further detail and application. In particular, it would be important to combine the ray- and wave-field formulations and see whether the hypothesized single Earth model with common parameters κ_i (Figure 8.3) could indeed quantitatively describe the available surface-, body-, ScS-, PKIKP-, and other types of long- and short-period attenuation data.

8.7 Correlation with geology and tectonics

The ultimate goal of any theory and inversion approach in Earth science is the development of a geologically meaningful model. In this regard, the proposed attenuation concept might become a breakthrough, with its objectivity, simple interpretation, frequency independence, and direct and predictable relation to the structural parameters. If successful in making parameters κ_i common to different wave types, the phenomenological model in Figure 8.3 would get us closer to describing the subsurface in terms of true local properties — unlike the current (Q_0, η) models reflecting the types of waves used for sampling and interpreter's theoretical views. With such an objective model, geological interpretation should finally become possible.

As above, the critical requirement for quantitative correlation of the (γ_i, κ_i) model with geological properties would again be developing a physical theory of wave attenuation within the deep crust and mantle. As suggested by our findings in Chapters 4 and 6, γ_i for the different types of experiments could be predictable by waveform modeling in detailed, realistic structures. Performing such modeling for a set of structures representative of the geological and tectonic environments of various datasets would involve great amounts of effort and time. At this time, it is of course unclear whether our conjectures about γ_i decreasing with increasing tectonic age and homogenizing lithospheric structure (Section 7.3) or being dominated by the upper crust (Section 7.4) would hold in the majority of cases. Nevertheless, the recognition of the existence of γ_i is undoubtedly a step in the right direction, which transformed the enigmatic frequency dependences of Q into measurable and interpretable effects of the structure.

By comparison to γ_i, explaining the inverted values of κ_i and correlating them to geological and tectonic properties should require significantly more effort, which cannot even be envisaged at this time. In particular, if proven by further studies, the suggested commonality of κ_i values derived for different wave types (Section 8.6) could be a major constraint in support of the unified attenuation model. Frequency dependence of κ_i, if reliably demonstrated, could be another major constraint on the nature of seismic attenuation and its relation to the properties of geologic structures.

Bibliography

Abercrombie, R. E. (1998). A summary of attenuation measurements from borehole recordings of earthquakes: the 10 Hz transition problem, *Pure Appl. Geoph.* **153**, 475–487.

Adams, D. A., and R. E. Abercrombie (1998). Seismic attenuation above 10 Hz in southern California from coda waves recorded in the Cajon pass borehole, *J. Geophys. Res.* **103**, 24,257–24,270.

Aki, K. (1969). Analysis of the seismic coda of local earthquakes as scattered waves, *J. Geophys. Res.* **74**, 615–631.

———— (1980). Scattering and attenuation of shear waves in the lithosphere. *J. Geophys. Res.* **85**, 6496–6504

———— (1981). Scattering and attenuation of high-frequency body-waves (1–25 Hz) in the lithosphere, *Phys. Earth Planet. Inter.* **26**, 241–243.

Aki, K. and B. Chouet (1975). Origin of coda waves: source, attenuation, and scattering effects, *J. Geophys. Res.* **80**, 3322–3342.

Aki, K., and P. G. Richards (2002). *Quantitative Seismology*, Second Edition, University Science Books, Sausalito, CA.

Anderson, D. L, and J. W. Given (1982). Absorption band Q model for the Earth, *J. Geophys. Res.* **87**, 3893–3904.

Anderson, D. L., A. Ben-Menahem, and C. B. Archambeau (1965). Attenuation of seismic energy in the upper mantle, *J. Geophys. Res.* **70**, 1441–1448.

Anderson, D. L., and C. B. Archambeau (1964). The anelasticity of the Earth, *J. Geophys. Res.* **69**, 2071–2084.

Anderson, D. L., and J. B. Minster (1979). The frequency dependence of Q in the Earth and implications for mantle rheology and Chandler wobble, *Geophys. J. R. Astr. Soc.* **58**, 431–40.

Anderson, D. L., H. Kanamori, R. S. Hart, and H.-P. Liu (1977). The Earth as a seismic absorption band, *Science* **196** (4294), 1104 – 1106, doi: 10.1126/science.196.4294.1104.

Atkinson, G. M. (2004). Empirical attenuation of ground-motion spectral amplitudes in southeastern Canada and the northeastern United States, *Bull. Seism. Soc. Am.* **94**, 1079–1095.

Azimi, Sh. A., A. V. Kalinin, V. V. Kalinin, and B. L. Pivovarov (1968). Impulse and transient characteristics of media with linear and quadratic absorption laws, *Izvestiya, Phys. Solid Earth* **2**, 88–93.

Bannister, S. G., E. S.Husebye, and B. O. Ruud (1990). Teleseismic *P* coda analyzed by three-component and array techniques: deterministic location of topographic *P*-to-*Rg* scattering near the NORESS array, *Bull Seism. Soc. Am.* **80**, 1969–1986.

Bensoussan, A., J. I. Lions, and S. Papanicolaou (1978). *Asymptotic analysis for periodic structures*, North Holland, Amsterdam.

Benz, H., A. Frankel, and D. Boore (1997). Regional *Lg* attenuation in the continental United States, *Bull. Seism. Soc. Am.* **87**, 600–619.

Biot, M. A. (1956). Theory of propagation of elastic waves in a fluid-saturated porous solid, II. Higher-frequency range, *J. Acoust. Soc. Am.* **28**, 168–178.

——— (1962). Mechanics of deformation and acoustic propagation in porous media, *J. Appl. Phys.* **23**, 1482–1498.

Bland, D. (1960). *The theory of linear viscoelasticity*, Pergamon Press Inc.

Boltzmann, L., (1874). Zur Theorie der elastischen Nachwirkung, *Sitzungsber. Math. Naturwiss. Kl. Kaiserl. Wiss.* **70**, 275.

Borcherdt, R. D. (2009). *Viscoelastic waves in layered media*, Cambridge Univ. Press, ISBN 978-0-521-89853-9.

Bourbié, T., O. Coussy, and B. Zinsiger (1987). *Acoustics of porous media*, Editions TECHNIP, France, ISBN 2710805168.

Brekhovskikh, L. M. (1980). *Waves in layered media* (transl. by R. T. Beyer), Academic Press, New York.

Campillo, M. (1990) Propagation and attenuation characteristics of the crustal phase *Lg*, *Pure Appl. Geophys.* **132**, 1–17.

Canas, J. A., and B. J. Mitchell (1978). Lateral variations of surface wave anelastic attenuation across the Pacific, *Bull. Seism. Soc. Am.* **68**, 1637–1657.

Carcione, J.M. (2007). *Wave fields in real media: Wave propagation in anisotropic anelastic, porous, and electromagnetic media*, Second Edition, Elsevier, Amsterdam.

Carcolé, E., and A. Ugalde (2008). Formulation of multiple anisotropic scattering process in two dimensions for anisotropic source radiation, *Geophys. J. Int.* **174**, 1037–1051, doi: 10.1111/j.1365-246X.2008.03896.x

Castro, R., R. C. Condori, O. Romero, C. Jacques, and M. Suter (2008). Seismic Attenuation in Northeastern Sonora, Mexico, *Bull. Seism. Soc. Am.* **98**, 722–432.

Červený, V. (2001). *Seismic ray theory*, Cambridge University Press, New York, NY.

Chernov, L.A. (1960). *Wave Propagation in a Random Medium*, McGraw-Hill.

Chouet, B. (1976). *Source, scattering, and attenuation effects on high frequency seismic waves*, Ph.D. dissertation, Mass. Inst. of Technol., Cambridge, MA.

———— (1979). Temporal variation of earthquake coda near Stone Canyon, California, *Geophys. Res. Lett.* **6**, 143–146.

———— (2003). Volcano Seismology, *Pure Appl. Geophys.* **160**, 739–788.

Christensen N. I., and W. D. Mooney (1995). Seismic velocity structure and composition of the continental crust: A global view, *J. Geophys. Res.* **100**, 9761–9788

Cong, L., and B. Mitchell (1988). Frequency-dependent crustal Q in stable and tectonically active regions, *Pure Appl. Geophys.* **127**, 581–605.

Connolly, P. (1999). Elastic impedance. *The Leading Edge* **18**, 438-452.

Cottrell, A. H. (1952). The laws of creep, *J. Mech. Phys. Solids* **1**, 53–63.

Dahlen, F. A., S.-H. Hung, and G. Nolet (2000). Fréchet kernels for finite-frequency tomography—I. Theory, *Geophys. J. Int.* **141**, 157–174.

Dahlen, F. A., and J. Tromp (1998). *Theoretical global seismology*. Princeton Univ. Press.

Dainty A. M. (1981) A scattering model to explain seismic Q observations in the lithosphere between 1 and 30 Hz, *Geophys. Res. Lett.* **8**, 1126–1128.

———— (1985). *Air Force Geophysical Laboratory Report*, AFGL-TF-86-0218.

———— (1990). Studies of coda using array and three-component processing, *Pure Appl. Geophys.* **132**, 221–244.

Dainty, A. M. and C. A. Schultz (1995). Crustal reflections and the nature of regional *P* coda, *Bull. Seism. Soc. Am.* **85**, 851–858.

Dalton, C. A., and G. Ekström (2006) Global models of surface wave attenuation. *J. Geophys. Res.* **111**, B05317, doi:10.1029/2005JB003997.

Del Pezzo, E., F. Bianco, S. Petrosino, and G. Saccorotti (2004). Changes in the coda decay rate and shear-wave splitting parameters associated with seismic swarms at Mt. Vesuvius, Italy, *Bull. Seism. Soc. Am.* **94**, 439–452.

Der, Z. A., A. C. Lees, and V. F. Cormier (1986). Frequency dependence of Q in the mantle underlying the shield areas of Eurasia, Part III: the Q model. *Geophys. J. R. Astr. Soc.* **87**, 1103–1112.

Der, Z. A., A. C. Lees, V. F. Cormier, and L. M. Anderson (1986). Frequency dependence of Q in the mantle underlying the shield areas of Eurasia, Part I: analyses of short and intermediate period data, *Geophys. J. R. Astr. Soc.* **87**, 1057–1084.

Der, Z. A., and A. C. Lees (1985). Methodologies for estimating $t^*(f)$ from short-period body waves and regional variations of $t^*(f)$ in the United States. *Geophys. J. R. Astr. Soc.* **82**, 125–140.

Der, Z. A., and T. W. McElfresh (1976). Short-period P-wave attenuation along various paths in North America as determined from P-wave spectra of the Salmon nuclear explosion, *Bull Seism. Soc. Am.* **66**, 1609–1622.

————— (1980). Time-domain methods, the values of t_P^* and t_S^* in the short-period band and regional variations of the same across the United States, *Bull Seism. Soc. Am.* **70**, 921–924.

Der, Z. A., Rivers, W. D., McElfresh, T. W., O'Donnell, A., Klouda, P. J., and Marshall, M. E. (1982). Worldwide variations in the attenuative properties of the upper mantle as determined from spectral studies of short-period body waves, *Phys. Earth Planet Inter.* **30**, 12–25.

Der, Z. A., T.W. McElfresh, R. Wagner, and J. Burnetti (1985). Spectral characteristics of P waves from nuclear explosions and yield estimation, *Bull Seism. Soc. Am.* **75**, 379–390.

Doornbos, D. J. (1983). Observable effects of the seismic absorption band in the Earth, Geophys. *J. R. Astr. Soc.* **75**, 693–711.

Drouet, S. S. Chevrot, F. Cotton, and A. Souriau (2008). Simultaneous inversion of source spectra, attenuation parameters, and site responses: Application to the data of the French accelerometric network, *Bull. Seism. Soc. Am.* **98**, 198-219.

Durek J, and G. Ekström (1996). A radial model of anelasticity consistent with long-period surface-wave attenuation. *Bull. Seismol. Soc. Am.* **86**, 155–158.

————— (1997). Investigating discrepancies among measurements of traveling and standing wave attenuation. *J. Geophys. Res.* **102**, 24,529–24,544.

Dziewonski, A. M., and D. L. Anderson (1981). Preliminary Reference Earth Model (PREM), *Phys. Earth Planet. Inter.* **25**, 297–356.

Dzurisin, D., D. J. Johnson, and J. A. Westphal (1981). Ground tilts during two recent eruptions of Mount St. Helens, Washington, *Eos, Trans., Am. Geophys. U.* **62** (45), 1089.

Evans, B., and D. L. Kohlstedt (1995). Rheology of Rocks, in T. J. Ahrens (ed.) *Rock Physics and Phase Relations — a handbook of physical constants*, American Geophysical Union, Washington, p. 148–65.

Evans, B., and G. Dresden (1991). Deformation of Earth materials — six easy pieces, rheology of rocks, *Rev. Geophys.* **29**, Part 2, Supp. S, 823–43.

Fan, G.-W. and Lay, T. (2003). Strong *Lg* attenuation in the Tibetan Plateau, *Bull. Seism. Soc. Am.* **93**, 2264–2272.

Faul, U. H., J. D. Fitz Gerald and I. Jackson (2004). Shear wave attenuation and dispersion in melt-bearing olivine polycrystals: 2. Microstructural interpretation and seismological implications, *J. Geophys. Res.* **109**, B06202, doi:10.1029/2003JB002407.

Fehler, M., Roberts, P., and Fairbanks T. (1988). A temporal change in coda wave attenuation observed during an eruption of Mount St. Helens, *J. Geophys. Res.* **93**, 4367–4373.

Frankel, A., and R. W. Clayton (1986). Finite difference simulations of seismic scattering: implications for the propagation of short-period seismic waves in the crust and models of crustal heterogeneity, *J. Geophys. Res.* **91**, 6465–8489.

Ford, S., D. D. Dreger, K. Mayeda, W. R. Walter, L. Malagnini, and W. S. Phillips (2008). Regional attenuation in northern California: a comparison of five 1D *Q* methods, *Bull. Seism. Soc. Am.* **98**, 2233–2046.

Frankel, A., A. McGarr, J. Bicknell, J. Mori, L. Seeber, and E. Cranswick (1990). Attenuation of high-frequency shear waves in the crust: measurements from New York State, South Africa, and southern California, *J. Geophys. Res.* **95**, 17,441–17,457.

Fuchs, K., and G. Müller (1971). Computation of synthetic seismograms with the reflectivity method and comparison with observations, *J. R. Astr. Soc.* **23**, 417–433, 1971.

Futterman, W. I. (1962). Dispersive body waves, *J. Geophys. Res.* **67**, 5,279–5,291.

Goldberg, D., and B. Zinszner (1989). *P*-wave attenuation measurements from laboratory resonance and sonic waveform data, *Geophysics* **54**, 76–81.

Greenfield, R. J. (1971). Short-period P-wave generation by Rayleigh-wave scattering at Novaya Zemlya, *J. Geophys. Res.* **76**, 7988–8002.

Griggs, D. T. (1939). Creep in rocks, *J. Geol.* **47**, 225–51.

———— (1940). Experimental flow of rocks under conditions favoring recrystallization, *Bull. Geol. Soc. Am.* **51**, 1001–22.

Gupta, I. N., T. W. McElfresh, and R. A. Wagner (1991). Near-source scattering of Rayleigh to P in teleseismic arrivals from Pahute Mesa (NTS shots, in:

Taylor, S. R., H. J. Patton, and P. G. Richards (Eds.), Explosion Source Phenomenology, *AGU Geophys. Monograph* **65**, 151–160.

Hoshiba, M. (1991). Simulations of multiply scattered coda wave excitation based on the energy conservation law, *Phys. Earth Planet. Inter.* **67**, 123–136.

Hwang, H. J. , and B. J. Mitchell (1987). Shear velocities, Q_β, and the frequency dependence of Q_β in stable and tectonically active regions from surface wave observations, *Geophys. J. R. Astr. Soc.* **90**, 575–613.

Ishimaru, A. (1978). *Wave Propagation and Scattering in Random Media*, Academic Press.

Jackson, D. D., and D. L. Anderson (1970). Physical mechanisms of seismic-wave attenuation, *Rev. Geophys. Space Phys.* **8** (1), 1–63.

Jackson, I., and M. S. Paterson (1993). A high-pressure, high-temperature apparatus for studies of seismic wave dispersion and attenuation, *Pure Appl. Geoph.* **141** (2/3/4), 445–466.

Jackson, I., U. H. Faul, J. D. Fitz Gerald, and B. H. Tan (2002). Grain-size sensitive seismic wave attenuation in polycrystalline olivine, *J. Geophys. Res.* **107** (B12), 2360, doi:10.1029/2001JB001225.

——— (2004). Shear wave attenuation and dispersion in melt-bearing olivine polycrystals: 1. Specimen fabrication and mechanical testing, *J. Geophys. Res.* **109**, B06201, doi:10.1029/2003JB002406.

Jeffreys, H. (1958). A modification of Lomnitz's law of creep in rocks, *Geophys. J. R. Astr. Soc.* **1**, 321–34.

Jin, A., K. Mayeda, D. Adams, and K. Aki (1994). Separation of intrinsic and scattering attenuation in southern California using TERRAscope data, *J. Geophys. Res.* **99**, 17,835–17,848.

Kennett, B. L. N., and R. Engdahl (1991), Traveltimes for global earthquake location and phase identification, *Geophys. J. Int.* **105**, 429–465.

Kinoshita, S. (1994). Frequency-dependent attenuation of shear waves in the crust of the southern Kanto area, Japan, *Bull. Seism. Soc. Am.* **84**, 1387–1396.

——— (2008). Deep-borehole-measured Q_P and Q_S attenuation for two Kanto sediment layer sites, *Bull. Seism. Soc. Am.* **98**, 463–468.

Kjartansson, E. (1979). Constant Q-wave propagation and attenuation, *J. Geophys. Res.* **84**, 4,737–4,748.

Knopoff, L. (1964). *Q. Rev Geophys.* **2**(4), 625–660.

Knopoff, L., F. Schwab, and E. Kausel (1973). Interpretation of *Lg. Geophys. J. R. Astr. Soc.* **33**, 389–404

Lambeck, K. (1977). Tidal dissipation in the oceans: astronomical geophysical, and oceanographic consequences, *Phil. Trans. R. Soc. Lond.* **287**, 545–594.

Landau, L. D, and E. M. Lifshitz (1976). *Course of theoretical physics, Volume 1: Mechanics*, Third edition, Butterworth-Heinemann, ISBN 0 7506 2896 O

———— (1986). *Course of theoretical physics, Volume 7: Theory of elasticity*, Third edition, Butterworth-Heinemann, ISBN 0 7506 2633 X

Lees, A. C., Z. A. Der, V. F. Cormier, M. E. Marshall, and J. A. Burnetti (1986). Frequency dependence of Q in the mantle underlying the shield areas of Eurasia, Part II: analyses of long period data, *Geophys. J. R. Astr. Soc.* **87**, 1085–1101.

Lekić, V., J. Matas, M. Panning, and B. Romanowicz (2009). Measurement and implications of frequency dependence of attenuation, *Earth Planet. Sci. Lett.* **282**, 285–293.

Li., G., J. Hu, H. Yang, H. Zhao, and L. Cong (2009). *Lg* coda Q variation across the Myanmar Arc and its neighboring regions, *Pure Appl. Geophys.* **166**, 1937–1948, doi: 10.1007/s00024-009-0459-4.

Li, H. (2006). *Seismic calibration of northern Eurasia using regional phases from nuclear explosions and 3-D Moho configuration of accreted terranes in western British Columbia*, Ph.D. Dissertation, University of Wyoming.

Liao, C. and Y.-H. Zhou (2004). Chandler wobble period and Q derived by wavelet transform, *Chin. J. Astron. Astrophys.* **4**, 247, doi: 10.1088/1009-9271/4/3/247.

Lin, C.-H., K. I. Konstantinou, H.-C. Pu, C.-C.Hsu, Y.-M. Lin, S.-H. You, and Y.-P. Huang (2005), Preliminary results from seismic monitoring at the Tatun volcanic area of northern Taiwan, *Terr. Atm. Ocean. Sci. (TAO)* **16**, 563–577.

Lin, W. J. (1989) *Rayleigh wave attenuation in the Basin and Range province*. M. Sc. Thesis, Saint Louis Univ, St Louis, MO.

Lines, L., F. Vasheghani, and S. Treitel, S. (2008). Reflections on Q, *CSEG Recorder* **33** (10), 36–38.

Liu, H. P., D. L. Anderson, and H. Kanamori (1976). Velocity dispersion due to anelasticity: implications for seismology and mantle composition, *Geophys. J. R. Astr. Soc.* **47**, 41–58.

Lomnitz, C. (1956). Creep measurements in igneous rocks, *J. Geology* **64**, 473–479.

———— (1957). Linear dissipation in solids, *J. Appl. Phys.* **28**, 201–205.

Londono, J. M. (1996). Temporal change in coda Q at Nevado Del Ruiz Volcano, Colombia, *J. Volcan. Geotherm. Res.* **73**, 129–139.

Macdonald, J. R. (1959). Rayleigh-wave dissipation functions in low-loss media, *Geophys. J.* **2**, 132–135.

Mandelbrot, B.B., and K. McCamy (1970). On the secular pole rotation and the Chandler wobble, *Geophys. J. R. Aastr. Soc.* **21**, 217–232.

Mashinskii, E. I. (2008). Amplitude-frequency dependencies of wave attenuation in single-crystal quartz: Experimental study, *J. Geophys. Res.* **113**, B11304, doi: 10.129/2008JB005719.

Masters, G., and G. Laske (1997) On bias in surface wave and free oscillation attenuation measurements, *Eos, Trans. Am. Geophys. U.* **78**, 46, 485

Matheney, M. P., and R. L. Nowack (1995). Seismic attenuation values obtained from instantaneous frequency matching and spectral ratios, *Geophys. J. Int.* **123**, 1–15.

Mayeda, K., F. Su, and K. Aki (1991). Seismic albedo from the total seismic energy dependence on hypocentral distance in southern California, *Phys. Earth Planet. Inter.* **67**, 104–114.

Mayeda, K., S. Koyanagi, M. Hoshiba, K. Aki, and Y. Zeng (1992). A comparative study of scattering, intrinsic, and coda Q^{-1} for Hawaii, Long Valley, and central California between 1.5 and 15.0 Hz, *J. Geophys. Res.* **97** (B5), 6643–6659.

McNamara, D. E., T. J. Owens, and W. R. Walter (1996). Propagation characteristics of Lg across the Tibetian Plateau, *Bull. Seism. Soc. Am.* **86**, 457–469.

Menke, W. (1989). *Geophysical data analysis: discrete inverse theory*, Academic Press.

Michelson, A. A. (1917). The laws of elastic-viscous flow, Part. 1, *J. Geology* **25**, 405–410.

——— (1920). The laws of elastic-viscous flow, Part 2: *J. Geology* **28**, 18–24.

Minster, J. B. (1978a). Transient and impulse responses of a one-dimensional attenuating medium. I. Analytical results, *Geophys. J. R. Astr. Soc.* **52**, 479–501.

——— (1978b). Transient and impulse responses of a one-dimensional attenuating medium. II. A parametric study, *Geophys. J. R. Astr. Soc.* **52**, 503–524.

Mitchell, B. J. (1995). Anelastic structure and evolution of the continental crust and upper mantle from seismic surface wave attenuation. *Rev. Geophys.* **33**, 441–462

——— (2010). Prologue and invitation to participate in a forum on the frequency dependence of seismic Q, *Pure Appl. Geophys.* **167**, 1129, doi 10.1007/s00024-010-0180-3

Mitchell, B. J., and J. K. Xie (1994) Attenuation of multiphase surface waves in the Basin and Range province, III, Inversion for crustal anelasticity. *Geophys. J. Int.* **116**, 468–484

Mitchell, B. J., and L. Cong (1998). *Lg* coda *Q* and its relation to the structure and evolution of continents: a global perspective, *Pure Appl. Geophys.* **153**, 655–663.

―――― (2005) A new map of *Lg* coda *Q* variation for Eurasia; implications for the relationship of Q to tectonics and for nuclear test ban monitoring, *Seism. Res. Lett.* **76**, 214.

Mitchell, B. J., Y. Pan, J. Xie, and L. Cong (1997). *Lg* coda *Q* variation across Eurasia and its relation to crustal evolution, *J. Geophys. Res.* **102**, 22,767–22,779.

Moncayo, E., C. Vargas, and J. Durán (2004). Temporal variation of coda-*Q* at Calderas volcano, Colombia, *Earth Sci. Res. J.* **8**, 19–24.

Morasca, P., K. Mayeda, R. Cŏk, W. S. Phillips, and L. Malagnini (2008). 2D coda and direct-wave attenuation tomography in northern Italy, *Bull. Seism. Soc. Am.* **98**, 1936 - 1946.

Morozov, I. B. (2001). Comment on "High-frequency wave propagation in the uppermost mantle" by T. Ryberg and F. Wenzel, *J. Geophys. Res.* **106**, 30,715-30,718.

―――― (2008). Geometrical attenuation, frequency dependence of *Q*, and the absorption band problem, *Geophys. J. Int.* **175**, 239–252.

―――― (2009a). Thirty years of confusion around "scattering *Q*"? *Seism. Res. Lett.* **80**, 5–7.

―――― (2009b). Reply to "Comment on 'Thirty years of confusion around 'scattering *Q*'?" by J. Xie and M. Fehler, *Seism. Res. Lett.* **80**, 648–649.

―――― (2009c). On the use of quality factor in seismology, *Eos Trans. AGU* **90**(52), Fall Meet. Suppl., Abstract S44A-02.

―――― (2009d). Earth's structure as the cause of frequency-dependent t^* and *Q*, *Eos Trans. AGU* **90**(52), Fall Meet. Suppl., Abstract S41B-1914.

―――― (2009e). More reflections on *Q*, *CSEG Recorder* **34** (2), 12–13.

―――― (2010a). On the causes of frequency-dependent apparent seismological *Q*. *Pure Appl. Geophys.* **167**, 1131–1146, doi 10.1007/s00024-010-0100-6.

―――― (2010b) Attenuation coefficients of Rayleigh and *Lg* waves, *J. Seismology,* doi 10.1007/s10950-010-9196-5.

―――― (2010c). Anelastic acoustic impedance and the correspondence principle. *Geophys. Prosp.,* doi 10.1111/j.1365-2478.2010.00890.x.

―――― (2010d) Seismological attenuation coefficient and *Q*, *Seism. Res. Lett.* **81**, 307.

Morozov, I. B., E. A. Morozova, S. B. Smithson, and L. N. Solodilov (1998). 2-D image of seismic attenuation beneath the Deep Seismic Sounding profile "Quartz", Russia, *Pure Appl. Geophys.* **153**, 311–348.

Morozov, I. B., and S. B. Smithson (2000). Coda of long-range arrivals from nuclear explosions, *Bull. Seism. Soc. Am.* **90**, 929–939.

Morozov, I. B., C. Zhang, J. N. Duenow, E. A. Morozova, and S. B. Smithson (2008). Frequency dependence of regional coda Q: Part I. Numerical modeling and an example from Peaceful Nuclear Explosions. *Bull. Seism. Soc. Am.* **98**, 2615–2628, doi: 10.1785/0120080037.

Morozova, E. A., I. B. Morozov, S. B. Smithson, and L. N. Solodilov (1999). Heterogeneity of the uppermost mantle beneath Russian Eurasia from ultra-long range profile QUARTZ, *J. Geophys. Res.* **104**, 20,329–20,348.

Mukhopadhyay, S., J. Sharma, R. Massey, and J. R., Kayal (2008). Lapse-time dependence of coda Q in the source region of the 1999 Chamoli earthquake, *Bull. Seism. Soc. Am.* **98**, 2080–2086, doi:10.1785/0120070258.

Nielsen, L., H. Thybo, I. B. Morozov, S. B. Smithson, and L. Solodilov (2003). Teleseismic *Pn* arrivals: Influence of mantle velocity gradient and crustal scattering, *Geophys. J. Int.* **152**(2), F1–F7.

Novelo-Casanova, D. A., A. Martinez-Bringas, and C. Valdés-González (2006). Temporal variation of Q_c^{-1} and b-values associated to the December 2000 – January 2001 volcanic activity at the Popocapetl volcano, Mexico, *J. Volcanol. Geotherm. Res.* **152**, 347–358.

Nussenzveig, H. M. (1972). *Causality and dispersion relations,* Mathematics in science and engineering, v. 95., ed. R. Bellman, Academic Press, New York.

Nuttli, O. W. (1973). Seismic wave attenuation and magnitude relations for eastern North America, *J. Geophys. Res.* **78**, 876–885.

O'Doherty, K. B., C. J. Bean, and J. McCloskey (1997). Coda wave imaging of the Long Valley caldera using a spatial stacking technique, *Geophys. Res. Lett.* **24**, 1545–1550.

Okubo, S. (1982). Theoretical and observed Q of the Chandler wobble—Love number approach, *Geophys. J. R. Astr. Soc.* **71**, 647–657.

Oth, A., D. Bindi, S. Parolai, and F. Wenzel (2008). S-wave attenuation characteristics beneath the Vrancea region in Romania: New insights from the inversion of ground-motion spectra, *Bull. Seism. Soc. Am.* **98**, 2382–2497, doi:10.1785/0120080106.

Panza, G. F., and G. Calcagnile (1975). *Lg, Li* and *Rg* from Rayleigh modes. *Geophys. J. R. Astr. Soc.* **40**, 475–487.

Pasyanos, M. E., E. M. Matzel, W. R. Walter, and A. J. Rodgers (2009). Broadband *Lg* attenuation modeling of the Middle East, *Geophys. J. Int.*, **177**, 1166–1176, doi: 10.1111/j.1365-246X.2009.04128.x.

Patton H. J., and S. R. Taylor (1984). Q structure of the Basin and Range from surface waves. *J. Geophys. Res.* **89**, 6929 –6940.

Phillips, W. S., H. E. Hartse, K. Mayeda (2000). Regional coda magnitudes in Central Asia, *Seism. Res. Lett.* **71**, 211.

Raoof, M. M., and O. W. Nuttli (1984). Attenuation of high-frequency earthquake waves in South America, *Pure Appl. Geophys.* **122**, 619–644.

Ray, R. D., R. J. Eanes, and B. F. Chao (1996). Detection of tidal dissipation in the solid Earth by satellite tracking and altimetry, *Nature* **381**, 595–597.

Razavy, M., 2005. Classical and quantum dissipative systems, Imperial College Press, London, UK, ISBN 1860945252, 334 pp.

Richards, P. G. (1984). On wave fronts and interfaces in anelastic media, *Bull. Seism. Soc. Am.* **74** (6), 2157–2165.

Richards, P. G., and W. Menke (1983). The apparent attenuation of a scattering medium, *Bull. Seism. Soc. Am.* **75**, 1005–1021.

Richards, P.G., 1984. On wave fronts and interfaces in anelastic media, *Bull. Seism. Soc. Am.* **74** (6), 2157 - 2165.

Roecker, S. W., B. Tucker, J. King, and D. Hatzfeld (1982). Estimates of Q in central Asia as a function of frequency and depth using the coda of locally recorded earthquakes., *Bull. Seism. Soc. Am.* **72**, 129–149.

Romanowicz, B., and B. Mitchell (2007). Deep Earth structure: Q of the Earth from crust to core. In: Schubert, G. (Ed.), *Treatise on Geophysics*, 1. Elsevier, p. 731–774

Roult G, and E. Clévédé, (2000). New refinements in attenuation measurements from free-oscillation and surface-wave observations. *Phys. Earth Planet. Inter.* **121**, 1–37.

Ryberg, T., and F. Wenzel (1999). High-frequency wave propagation in the uppermost mantle, *J. Geophys. Res.* **104**, 10,655-10,666.

Ryberg, T., K. Fuchs, A. V. Egorkin, and L. Solodilov (1995). Observations of high-frequency teleseismic Pn on the long-range Quartz profile across northern Eurasia, *J. Geophys. Res.* **100**, 18,151-18,163.

Sailor, R. V., and A. Dziewonski (1978). Measurements and interpretation of normal mode attenuation, *Geophys. J. R. Astr. Soc.* **53**, 559–582.

Sams, M., and D. Goldberg (1990). The validity of Q estimates from borehole data using spectral ratios, *Geophysics* **55**, 97–101.

Sato H, and M. Fehler (1998). *Seismic wave propagation and scattering in the heterogeneous Earth*, Springer-Verlag, ISBN 0-387-98329-5.

Sato, H. (1977). Energy propagation including scattering effects: single isotropic scattering, *J. Phys. Earth* **25**, 27–41.

Schreiber, E., O. L. Anderson, and N. Soga (1973). *Elastic constants and their measurements*, McGraw Hill.

Sharrock, D.S., I. G. Main, and A. Douglas (1995). Observations of Q from the northwest Pacific Subduction Zone recorded at teleseismic distances, *Bull. Seism. Soc. Am.* **85**, 237–253.

Sipkin, S.A., and T. H. Jordan (1979). Frequency dependence of Q_{ScS}. *Bull Seism. Soc. Am.* **69**, 1055–1079.

Smith, M. L. (1977). Wobble and nutation of the arth, *Geophys. J. R. Astr. Soc.* **50**, 103–140.

Spiridonov, E. A., and I. Ya. Tsurkis (2008). On the period and quality factor of the Chandler wobble, *Izvestiya Phys. Solid Earth* **44**, doi: 10.1134/S1069351308080077

Streeter, V. L., and E. B., Wylie (1981). *Fluid Mechanics*, McGraw-Hill.

Thybo, H. and E. Perchuc (1997). The seismic 8° discontinuity and partial melting in continental mantle, *Science* **275**, 1626-1629.

Tonn, R. (1991). The determination of the seismic quality factor Q from VSP data: a comparison of different computational methods, *Geophys. Prosp.* **39**, 1–27.

Wang, Y. (2008). *Seismic inverse Q filtering*, Blackwell, ISBN 978-1-4051-8540-0.

Warren, N. (1972). Q and structure, *Earth, Moon, and Planets* 4, Springer, Netherlands, pp. 430–441.

Weeraratne D., D. W. Forsyth, Y. Yang, and S. C. Webb (2007). Rayleigh wave tomography beneath intraplate volcanic ridges in the South Pacific. *J. Geophys. Res.* 112, B06303, doi:10.1029/2006JB004403.

Weinberg, S. (1995). *The quantum theory of fields*. Cambridge University Press. ISBN 0-521-67053-5.

Wessel, P., and W. H. F. Smith (1995). New version of the Generic Mapping Tools released, *Eos Trans Am. Geophys. U.* **76**, 329.

White, R. E. (1992). The accuracy of estimating Q from seismic data, *Geophysics* **57**, 1508–1511.

Widmer, R., G. Masters, and F. Gilbert (1991). Spherically-symmetric attenuation within the Earth from normal mode data, *Geophys. J. Int.* **104**, 541–553.

Wiggins, R.A. (1976). A fast, new computational algorithm for free oscillations and surface waves, *Geophys. J. R. Astr. Soc.* **47**, 135–150.

Wu, R.-S. (1985). Multiple scattering and energy transfer of seismic waves, separation of scattering effect from intrinsic attenuation, *Geophys. J. R Astr. Soc.* **82**, 57–80.

Xie, J. (2007). *Pn* attenuation beneath the Tibetan Plateau, *Bull. Seism. Soc. Am.* **97**, 2040–2052, doi:10.1785/0120070016.

——— (2010). Can we improve estimates of seismological *Q* using a new "geometrical spreading" model? *Pure Appl. Geophys.* **167**, 1147–1162, doi: 10.1007/s00024-010-0188-8.

Xie, J., and O. W. Nuttli (1988) Interpretation of high-frequency coda at large distances: stochastic modelling and method of inversion, *Geophys. J. Int.* **95**, 579–595.

Xie, J., and B. J. Mitchell (1990). A back-projection method for imaging large-scale lateral variations of *Lg* coda *Q* with application to continental Africa, *Geophys. J. Int.* **100**, 161–181.

Xie, J., and M. Fehler (2009). Comment on "Thirty years of confusion around scattering *Q*" by Igor. B. Morozov, *Seism. Res. Lett.* **80**, 646–647.

Xie, J., R. Gok, J, Ni, and Y. Aoki (2004). Lateral variations of crustal seismic attenuation along the INDEPTH profiles in Tibet from *Lg Q* inversion, *J. Geophys. Res.* **109**, B10308, doi: 10,1029/2004JB002988.

Yang, X., T. Lay, X. B. Xie, an M. S. Thorne (2007). Geometric spreading of *Pn* and *Sn* in a spherical Earth model. *Bull. Seism Soc. Am.* **97**, 2053–2065, doi: 10.1785/0120070031.

Zener, C. M. (1948). *Elasticity and Anelasticity of Metals*, University of Chicago Press.

Zhu, T., K.-Y. Chun, and G. F. West (1991). Geometrical spreading and *Q* of *Pn* waves: an investigative study in western Canada, *Bull. Seism. Soc. Am.* **81**, 882–896.

Kramers-Krönig Relations

In many physical problems, the response of a variable $r(t)$ to some "force" $f(t)$ is given by the convolution of $f(t)$ with response function $G(t)$,

$$r(t) = \int_{-\infty}^{\infty} f(\tau) G(t-\tau) d\tau. \qquad \text{(A1.1)}$$

Let us assume that $G(t)$ is "causal", meaning that $G(t) = 0$ for all $t < 0$ in the time domain. Mathematically, this means that $G(t)$ satisfies the identity,

$$G(t) = \theta(t) G(t), \qquad \text{(A1.2)}$$

where $\theta(t)$ is the step function.

In the time domain, this identity determines the values of $G(t)$ within half of the time axis, $t < 0$; consequently, a related constraint exists in the frequency domain. Specifically, eq. (A1.2) allows expressing all values of $\text{Re}G(\omega)$ through $\text{Im}G(\omega)$, and *vice versa*. This can be shown as follows.

Transform eq. (A1.2) into the frequency domain:

$$G(\omega) = F\left[\theta(t)G(t)\right] = \int_{-\infty}^{\infty} \theta(t) G(t) e^{i\omega t} dt, \qquad \text{(A1.3)}$$

and substitute into it:

$$G(t) = \frac{1}{2\pi}\int_{-\infty}^{\infty} G(\omega')e^{-i\omega't}d\omega',$$ (A1.4)

which results in:

$$G(\omega) = \frac{1}{2\pi}\int_{-\infty}^{\infty} d\omega'G(\omega')\int_{-\infty}^{\infty} dt\theta(t)e^{i(\omega-\omega')t}$$

$$= \int_{-\infty}^{\infty} d\omega'G(\omega')\theta(\omega-\omega').$$ (A1.5)

The Fourier transform of the step function in the above equation equals:

$$\theta(\omega) = F\big[\theta(t)\big] = \frac{1}{2}\delta(\omega) + \frac{i}{2\pi\omega},$$ (A1.6)

and consequently we have,

$$G(\omega) = \frac{1}{2}G(\omega) + \frac{i}{2\pi}P\int_{-\infty}^{\infty} d\omega'\frac{G(\omega')}{\omega-\omega'},$$ (A1.7)

where P denotes the Cauchy's principal value of the integral. Finally,

$$G(\omega) = \frac{i}{\pi}P\int_{-\infty}^{\infty} d\omega'\frac{G(\omega')}{\omega-\omega'}.$$ (A1.8)

This identity is often called the Kramers-Krönig relation, and it can be expressed in terms of the real and imaginary parts of G:

$$\begin{cases} \mathrm{Re}\,G(\omega) = \frac{-1}{\pi}P\int_{-\infty}^{\infty} d\omega'\frac{\mathrm{Im}\,G(\omega')}{\omega-\omega'}, \\ \mathrm{Im}\,G(\omega) = \frac{1}{\pi}P\int_{-\infty}^{\infty} d\omega'\frac{\mathrm{Re}\,G(\omega')}{\omega-\omega'}. \end{cases}$$ (A1.9)

Further, if $G(t)$ is real-valued, then $\mathrm{Re}G(\omega)$ and $\mathrm{Im}G(\omega)$ are even and odd functions of ω, respectively. In this case, the integrals in eq. (A1.9) can be further limited to positive frequencies:

$$\begin{cases} \operatorname{Re}G(\omega) = \dfrac{2}{\pi}P\displaystyle\int_0^\infty d\omega' \dfrac{\operatorname{Im}G(\omega')}{\omega'^2 - \omega^2}, \\[3mm] \operatorname{Im}G(\omega) = -\dfrac{2\omega}{\pi}P\displaystyle\int_0^\infty d\omega' \dfrac{\operatorname{Re}G(\omega')}{\omega'^2 - \omega^2}. \end{cases} \qquad (A1.10)$$

Appendix 2

Lagrangian Model of Creep

Here we describe a simple Lagrangian model of material creep in a vibrating-system experiment. The purpose of this exercise is to demonstrate how creep can be described by instantaneous dynamic equations which are commonly used in mechanics, instead of the phenomenological, visco-elastic creep functions (Section 2.5.1).

For simplicity, the model is constructed for a 1-D case, using the mechanical-oscillator analog discussed in Sections 2.4.1, 2.5.8, and in other parts of this book. The following derivation focuses on the general description of material creep, and functional forms of the various terms can be altered in numerous ways in order to bring in new physical ideas or match the specific observation conditions.

Consider a specimen of mass m subjected to time-variant force $f(t)$. The sample is mounted in the measurement apparatus. Let us denote its displacement by u (analog for strain in continuum mechanics). From creep experiments, we know that "fast" and "quasi-static" deformations of a solid exhibit different elastic constants, which we denote k_U ("unrelaxed") and k_R ("relaxed"), respectively. By their definitions, and in agreement with the direction of creep, $k_U \geq k_R$. Further, energy dissipation may exist, related to both movement of the body itself and to its "internal deformation," *i.e.*, creep. All these statements can be summarized in the mechanical model shown in Figure A2.1. This arrangement resembles the Burgers model (Carcione, 2007, p.77) both ends of which are "shunted" to an immovable boundary. However, when describing a vibrating mechanical system (and not only

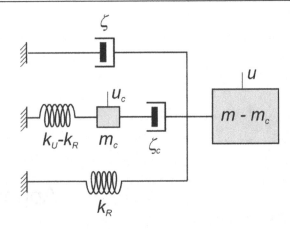

FIGURE A2.1
Equivalent mechanical creep model in eqs. (A2.1 to A2.3).

a constitutive equation), it is also important to specify to which points in the model the masses are attached.

The key difference of this approach compared to the phenomenological visco-elastic constitutive equations consists in using the "internal" (potentially not directly observed in a creep or vibration experiment) "creep" variable u_c. In general, we therefore also add a "creep mass" m_c and viscosity ζ_c (Figure A2.1). For small ζ, large ζ_c, and rapidly varying stresses, the response of the system is equivalent to that of mass m suspended on a spring with modulus k_U, and for very slow movements (creep) is equivalent to that of the mass $\left(m - m_c\right)$ acted on by the relaxed modulus k_R. Mass m_c and viscosity ζ can be set equal to zeros, with the same limiting-cases behaviors unchanged.

Physically, the model in Figure A2.1 describes a medium in which some internal structure is relatively mobile but connected to the main rock matrix by strong friction. Under quick deformation, the structure moves together with the matrix, but under a nearly static strain, it relaxes with time and withdraws its resistance to the external stress (Figure A2.2). Obviously, other mechanical models can also be constructed to achieve a similar behavior.

The diagram in Figure A2.1 corresponds to the following Lagrangian:

$$L\left(u, u_c, \dot{u}, \dot{u}_c\right) = \frac{m - m_c}{2}\dot{u}^2 + \frac{m_c}{2}\dot{u}_c^2 - \frac{k_R}{2}u^2 - \frac{k_U - k_R}{2}u_c^2, \qquad (A2.1)$$

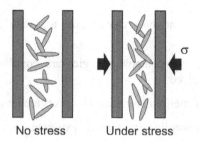

No stress   Under stress

FIGURE A2.2
Conceptual physical creep model (repeated Figure 2.7). Internal parameter u_c in eqs. (A2.1) to (A2.3) may describe the orientations of internal structures, which slowly relax under strain.

combined with dissipation function,

$$D\left(u,u_c,\dot{u},\dot{u}_c\right) = \frac{\zeta}{2}\dot{u}^2 + \frac{\zeta_c}{2}\left(\dot{u}-\dot{u}_c\right)^2.$$ (A2.2)

The external force $f(t)$ acting at point u is described by the corresponding term added to the Lagrangian:

$$L \rightarrow L - f\left(t\right)u.$$ (A2.3)

Because quadratic forms were chosen for both L and D above, the resulting differential equations of motion are linear:

$$\begin{cases} \dfrac{d}{dt}\left(\dfrac{\partial L}{\partial \dot{u}}\right) - \dfrac{\partial L}{\partial u} + \dfrac{\partial D}{\partial \dot{u}} = \left(m-m_c\right)\ddot{u} + k_R u + \zeta\dot{u} + \zeta_c\left(\dot{u}-\dot{u}_c\right) = f\left(t\right), \\[2mm] \dfrac{d}{dt}\left(\dfrac{\partial L}{\partial \dot{u}_c}\right) - \dfrac{\partial L}{\partial u_c} + \dfrac{\partial D}{\partial \dot{u}_c} = m_c\ddot{u}_c + \left(k_U - k_R\right)u_c - \zeta_c\left(\dot{u}-\dot{u}_c\right) = 0. \end{cases}$$ (A2.4)

Because of the linearity of these equations, one can consider a harmonic force $f(t) = f(\omega)\exp(-i\omega t)$ and find the corresponding solutions for $u(t)$ and $u_c(t)$ in the same exponential form. This would give rise to the effective "visco-elastic modulus" $M(\omega) = |f(\omega)/u(\omega)|$ and the creep function. The entire model in Figure A2.1 could probably be approximated by one or two "standard linear solids." However, note once again, that such replacement would simply be a heuristic description of the resulting oscillatory solution. The true physics of the system is contained in the functional forms of the Lagrangian model of (A2.1) to (A2.3). From these functional forms, we see:

- What physical variables characterize the state and dynamics of the system;

- What kinetic and potential energies are contributed by the different processes involved;

- What types of energy-dissipation processes are involved; and

- What types of non-linearities can be expected.

Answers to all of these questions about the process of deformation and wave propagation are lost when adhering to visco-elastic descriptions explaining nearly everything through heuristic equivalent models and relaxation times.

Appendix 3

Acoustic Impedance
with Attenuation

In this Appendix, we develop a general definition of the acoustic impedance which directly applies to media with attenuation. Note that the well-known expression $Z = \rho V$ is not the true definition but only the final formula for an isotropic and attenuation-free medium. An extrapolation of this expression to the visco-elastic case by using the correspondence principle for V (Section 2.5.4) would be incorrect.

Generally, the acoustic impedance and its extensions to oblique incidence are defined so that the reflection coefficient r_{12} from a boundary of two media is solely determined by the ratio of their impedances Z_1 and Z_2:

$$r_{12} = \frac{\dfrac{Z_2}{Z_1} - 1}{\dfrac{Z_2}{Z_1} + 1} \cdot \qquad \text{(2.115 again)}$$

Fundamentally, this relation arises from Z describing the general linear property of the boundary conditions involved in the determination of r_{12}. In acoustics, the impedance is therefore defined as the ratio of pressure to particle velocity component across the boundary (Brekhovskikh, 1980):

$$Z = \frac{p}{\dot{u}_z}, \qquad (A3.1)$$

which allows extending formula $Z = \rho V$ to oblique incidence at angle θ.

$$Z_A = \frac{\rho V}{\cos \theta}. \qquad (A3.2)$$

In more complex cases, such as elastic wavefield with attenuation, a yet more general definition for Z from first principles is required. Such a definition can be derived from the role the impedance plays in elasto- and electro-dynamics.

Note that the well-known identity for the transmission coefficient $r_{12} = 1 - t_{12}$ also follows from the displacement-continuity boundary condition and does not depend on parameters of the media. Equation (2.115) also indicates the fundamental scale ambiguity of Z, which can be multiplied by an arbitrary complex factor c,

$$Z \rightarrow Z' = cZ, \qquad (A3.3)$$

without altering the observed reflectivity.

Based purely on expression (2.115), several extensions of the normal-incidence acoustic impedance (eq. (2.116)) to oblique incidence were proposed and called Elastic Impedances (*e.g.*, Connolly, 1999). However, all of these definitions were based on the scaling ambiguity of impedance (eq. (A3.3)) and heuristic integrations of reflectivity time series, and did not rigorously represent properties of the medium. Such definitions will not be considered here.

Seeking a rigorous definition of the acoustic impedance which would incorporate both oblique incidence and attenuation, we need to look into the wave boundary conditions. Note that eq. (2.115) follows from solving the boundary conditions on the reflecting interface. In the acoustic case, these conditions require the continuity of normal displacement and stress across the interface:

$$\begin{cases} u_z|_1 = u_z|_2, \\ \sigma_{zz}|_1 = \sigma_{zz}|_2, \end{cases} \qquad (A3.4)$$

where σ_{zz} is the normal stress component, and subscripts "1" and "2" indicate the propagating media. From Hooke's law, stress is always proportional to spatial derivatives of the displacement:

$$\begin{pmatrix} \sigma_{zz} \\ \sigma_{zx} \end{pmatrix} = -\widetilde{\mathbf{Z}} \begin{pmatrix} u_z \\ u_x \end{pmatrix}, \qquad (A3.5)$$

where $\tilde{\mathbf{Z}}$ is a linear matrix differential operator. For example, for an elastic field, it equals:

$$\tilde{\mathbf{Z}} = -\begin{pmatrix} (\lambda + 2\mu)\partial_z & \lambda\partial_x \\ \mu\partial_x & \mu\partial_z \end{pmatrix}, \tag{A3.6}$$

where λ and μ are the Lamé constants. Equations (A3.5) and (A3.6) give the most general definition of impedance (Morozov, 2010c).

In the acoustic case and for a planar harmonic wave (in which $u_x/u_z = \tan\theta$), an additional, scalar impedance of each medium can be defined as:

$$\tilde{Z}_A = -\frac{\sigma_{zz}}{u_z} = \left[-\tilde{\mathbf{Z}}\begin{pmatrix} 1 \\ \tan\theta \end{pmatrix} \right]_z. \tag{A3.7}$$

Since $\sigma_{zz} = -\tilde{Z}_A u_z$ in each of the two media, the reflection amplitude becomes determined entirely by the values of $\tilde{Z}_A\big|_1$ and $\tilde{Z}_A\big|_2$, and it can be easily seen that equation 2.115 is satisfied with:

$$r_{12} = \frac{u_z^{reflected}}{u_z^{incident}}\Bigg|_1. \tag{A3.8}$$

Thus, the impedance can be uniquely defined as a factor relating the stress and displacement boundary conditions in the incident, reflected, or transmitted waves (eqs. (A3.5) to (A3.7)). Note that in an acoustic case, definition (A3.1) has practically the same meaning, except that the velocity is used instead of the displacement, and consequently factor $c = -i\omega$ is added, where ω is the angular frequency. Therefore, as it can be verified from eq. (A3.7), that the following quantity is identical to the standard acoustic impedance (eq. (A3.2)) at all angles:

$$Z_A = \frac{\tilde{Z}_A}{-i\omega}. \tag{A3.9}$$

Factor $-i\omega$ in eq. (A3.9) is insignificant for describing the reflectivity (see expression (A3.3)), but it may become important when considering non-stationary waves. However, here we consider harmonic fields and mostly use the conventional Z_A defined in eqs. (A3.2) and (A3.9).

Subject and Author Index

Lists of Tables and Figures

Tables

Figures